Vladislav A. Sadykov, Serguei F. Tikhov, Lyubov A. Isupova
Heterogeneous Catalytic Redox Reactions

Also of interest

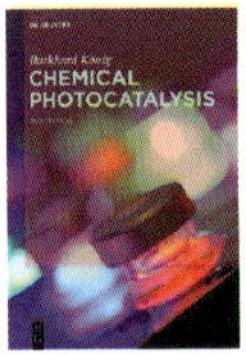

Chemical Photocatalysis.
Koenig, 2019
ISBN 978-3-11-057654-2,
e-ISBN 978-3-11-057676-4

Grignard Reagents and Transition Metal Catalysts.
Formation of C-C Bonds by Cross-Coupling
Cossy (Ed.), 2016
ISBN 978-3-11-035266-5,
e-ISBN 978-3-11-035272-6

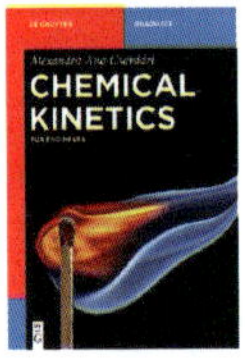

Chemical Kinetics.
For Engineers
Csavdari, 2021
ISBN 978-3-11-047515-9,
e-ISBN 978-3-11-047517-3

Catalytic Reactors.
Saha (Ed.), 2015
ISBN 978-3-11-033296-4,
e-ISBN 978-3-11-033298-8

Chemical Reaction Technology.
Murzin, 2015
ISBN 978-3-11-033643-6,
e-ISBN 978-3-11-033644-3

Vladislav A. Sadykov, Serguei F. Tikhov,
Lyubov A. Isupova

Heterogeneous Catalytic Redox Reactions

Fundamentals and Applications

DE GRUYTER

Authors
Prof. Vladislav A. Sadykov
Russian Academy of Science
Boreskov Institute of Catalysis
pr. Akademika Lavrentieva 5
Novosibirsk State University
Pirogova str., 2
Novosibirsk
630090
Russia
Sadykov@catalysis.ru

Dr.Sc. Serguei F. Tikhov
Russian Academy of Science
Boreskov Institute of Catalysis
pr. Akademika Lavrentieva 5
Novosibirsk
630090
Russia
tikhov@catalysis.ru

Dr.Sc. Lyubov A. Isupova
Russian Academy of Science
Boreskov Institute of Catalysis
pr. Akademika Lavrentieva 5
Novosibirsk
630090
Russia
isupova@catalysis.ru

ISBN 978-3-11-058586-5
e-ISBN (PDF) 978-3-11-058777-7
e-ISBN (EPUB) 978-3-11-058592-6

Library of Congress Control Number: 2019948031

Bibliographic information published by the Deutsche Nationalbibliothek
The Deutsche Nationalbibliothek lists this publication in the Deutsche Nationalbibliografie; detailed bibliographic data are available on the Internet at http://dnb.dnb.de.

Cover image: Serguei F. Tikhov
Typesetting: Integra Software Services Pvt. Ltd.
Printing and binding: CPI books GmbH, Leck

www.degruyter.com

Preface

Heterogeneous catalytic redox reactions play a tremendous role in providing required stability of humanity existence on the Earth by producing huge amounts of vital chemicals in industrial chemical plants, ensuring life quality by preventing environment pollution and decreasing the level of greenhouse gases (carbon dioxide, methane, nitrous oxide, etc.) emissions into atmosphere by exhausts clean-up, more efficient and pure transformation of fuels into energy as well as a broader usage of biofuels. In the first group of reactions, nitric acid production by selective oxidation of ammonia is among the leaders by the production scale. In the second group, catalytic combustion of fuels is a very attractive approach to solve environmental problems. Moreover, dry reforming of natural gas/biogas into syngas with its subsequent transformation into pure sinfuels allows to deal with the third group of problems.

In all these reactions as dependent upon their actions oxide catalysts, including those promoted by precious metals, play a leading role as industrial catalysts or efficient competitors with such expensive systems as Pt-Rh gauzes in selective ammonia oxidation into NOx at high pressures. Their optimization and further improvement is to be based upon detailed studies of atomic-scale fundamental factors, determining their catalytic properties in redox reactions. This requires application of all modern structural, spectroscopic, kinetic and theoretical methods to characterize the density of surface sites as dependent upon their real/defect structure controlled by the structural type, chemical composition, method of preparation and reaction media effect. For redox reactions such decisive factors as oxygen bonding strength and surface coverage by reactive oxygen species are determined by the particles morphology, their real/defect structure and the chemical composition of the surface layers. For catalytic reactions proceeding on the oxide surface, their efficiency and selectivity depend upon the reaction mechanism comprising the sequence of elementary steps required to be determined. Hence, basic concepts and approaches of the solid-state chemistry, surface science, quantum chemistry and chemical kinetics are to be systematically applied to deal with the problem of the design of efficient oxide catalysts of redox reactions. For catalysts operating in large-scale industrial reactors, their thermal conductivity and spatial design (honeycombs, microchannel plates, etc.) controlling heat and mass transfer along the reactor length are to be optimized as well.

These problems are considered in this book with a special accent on the authors' experience in this field at Boreskov Institute of catalysis and Novosibirsk State University (Novosibirsk, Russia) supported by collaboration with leading scientific centers in Russia and throughout the world. This book is based on materials presented in hundreds of original papers, three monographs, six chapters in collective monographs and a lot of review articles. The most specific feature of our research reflected in this book is that a great attention is paid to the effect of the real/

https://doi.org/10.1515/9783110587777-201

defect structure of oxide catalysts on their surface sites and catalytic properties as well as to its variation due to interaction with the reaction media, thus applying solid-state chemistry concepts to enrich heterogeneous redox catalysis theory and practice. This allowed to reliably identify the nature of the active surface sites, which for some oxide types and structures are related to clusters of cations (characterized by infrared spectroscopy of adsorbed CO) located at surface outlets of extended defects. Mechanism of such catalytic reactions as N_2O decomposition and CH_4 dry reforming, studied with the help of a combination of transient methods, was formulated with a due regard for specificity of surface sites nature. Application of new methods of synthesis based on polymerized polyester precursors, mechanochemistry, plasmochemistry as well as hydrothermal treatment of aluminum-based alloys obtained by mechanochemical treatment allowed to control the real/defect structure of oxide catalysts. This allowed to design new efficient catalysts of such redox reactions as catalytic combustion, ammonia oxidation into nitrogen oxides, N_2O decomposition, fuels transformation into syngas, water-gas shift reaction and Fisher–Tropsch synthesis. Structural monolithic catalysts (honeycombs, etc.) were successfully tested in these reactions in real conditions and demonstrated high activity and performance stability.

Research related to the design of monolithic oxide catalysts for replacing a part of Pt-Rh gauzes used in industrial reactors for the oxidation of ammonia to No_x under pressure for the production of nitric acid in nitric acid plants has received award from the Russian government in the field of technology in 1998.

Contents

1 Specific activity of simple oxides of transition metals and factors that can determine it

1.1 Differences in specific activity of transition metal oxides toward complete oxidation reactions

Specific activity of simple oxides is the most important quantity in heterogeneous catalysis both theoretically – for determining the nature of active sites and mechanisms of catalytic reactions – and practically – for choosing the methods to synthesize the most active catalysts. The difference in specific activity of transition metal oxides obtained by different methods toward a relatively simple reaction of CO oxidation is a well-known phenomenon. Of particular interest are publications where specific activities are compared under similar experimental conditions. For copper oxides, this phenomenon has been discovered in the 1980s [1, 2] and verified in later studies [3]. The indicated difference is determined to a great extent by nature of the precursor of oxide catalyst. Thus, activity of the copper oxide obtained by the oxidation of metallic copper was higher by an order of magnitude as compared to the oxide obtained by decomposition of hydroxide [3]. According to [4], activity of the copper oxide obtained by precipitation from a copper nitrate solution in the presence of hydrogen peroxide also exceeded by more than an order of magnitude the activity of the least active commercial copper oxide produced at the Ural chemical plant (Verkhnyaya Pyshma, Russia). In [5], the activity of copper oxides with the specific surface area 38.6–22 m^2/g, which varied from 1 to 6.4 mol CO/m^2 s, was studied at 100 °C. Turn-over frequency (TOF) of different copper oxides at 100 °C changed from 0.29×10^3 to 3.41×10^3 s^{-1} [4]. A comparative analysis of the data reported by different authors [6] showed that specific activity of copper oxides differs by more than two orders of magnitude. This is much greater than a possible measurement error.

A comprehensive study on specific activity of the copper oxides obtained by precipitation of copper nitrate and sulfate salts with subsequent thermal decomposition at 275–700 °C was reported in [7]. In the process, specific surface area of CuO changed from ~ 1 to 90 m^2/g, and coherent scattering regions (CSR) from > 200 to ~10 nm. Kinetic experiments were carried out by different methods to investigate the initial oxidation steps and activity in the flow (gradient PFR method) and flow-circulatory (gradientless FCR method) modes. Flow rate and catalyst weight were varied to obtain similar surface areas of the tested catalysts [7].

It was shown that at minimum conversions (a mode close to the differential one), the activity of highly dispersed samples differs no more than by a factor of 4. For CuO (1) annealed at 750 °C, the activity dropped by more than an order of magnitude (Table 1.1).

For cobalt oxides obtained by different methods, the activity in CO oxidation at 300 °C differed in the range from 35 to 120 $mlCO_2/(m^2$ min), whereas the activity in

https://doi.org/10.1515/9783110587777-001

Table 1.1: Catalytic activity of CuO powders measured upon analysis of light-off curves at the same contact time (*t*) or total surface area (SA) [7].

Sample (SSA, m^2/g)	PFR, Rs · ×10^{-12} (molecules CO/cm^2 s)	FCR, Rs · ×10^{-12} (molecules CO/cm^2 s)	
	100 °C, t	100 °C,t	130 °C, SA
CuO (90)	1.8	0.5	–
CuO (49)	2.5	0.6	3.9
CuO (13)	6.0	1.7	6.5
CuO (8)	7.6	1.6	9.0
CuO (1)	–	–	0.26

C_2H_4 oxidation differed from 3.5 to 10.5 ml $CO_2/(m^2$ min) upon variation of specific surface area from 0.44 to 14.3 m^2/g [8]. The difference in specific activity at 140 °C by approximately an order of magnitude was noted in [9] for samples obtained by decomposition of cobalt nitrates, carbonates and hydroxides upon variation of calcination temperature and specific surface area in the range from 0.4 to 41 m^2/g. The gradientless mode in a microreactor with vibro-fluidized bed ensured the absence of diffusion resistance. In [10], as specific surface area of cobalt oxide samples was increased from 10 to 85 m^2/g, their specific activity at –60 °C changed from 3.46×10^{-8} to 1.76×10^{-7} mol CO/m^2 s.

For iron oxides with the corundum structure that were obtained from different precursors, specific activity measured at 140 °C by gradientless method changed nearly by two orders of magnitude when specific surface area of the oxides was varied from 14 to 90 m^2/g [11]. Specific activity of iron oxides obtained by thermal decomposition of goethite at different temperatures changed by an order of magnitude [12]. The activity of chromium oxides produced by calcination of nitrate and bichromate at different temperatures changed approximately by two orders of magnitude [12].

For titania, the ignition temperature varied from 10 to 27 °C [13]. A more detailed comparison of specific activity with recalculation of the reaction rate to the unified pressure and temperature in the reactions of complete oxidation of hydrogen and CO revealed a significant difference in specific activity of titania samples with the rutile structure obtained by different methods [14]. This difference substantially exceeds a possible discrepancy of experimental conditions (Table 1.2).

The synthesis conditions affected also the hysteresis of complete oxidation of hydrogen [14]. Titania sample with a higher activity had a significantly more pronounced hysteresis loop when the temperature of catalytic experiment was raised and lowered (Fig. 1.1).

Table 1.2: Specific activity of different TiO_2 samples in the deep oxidation of H_2 and CO reported in the literature [14–17].

No.	Preparation method	SSA, m^2/g	Conditions of experiment	Reaction rate, molecules $H_2(CO)/m^2$ s		Ref.
				P_{H2} = 1 vol.%, T = 350 °C	P_{CO} = 1 vol.%, T = 300 °C	
1	Coprecipitation from solution of titanium chloride and ammonia	84	P_{H2} = 1 vol.%, P_{O2} = 99 vol.% Without water freezing	7.8×10^{15}	–	15
2	Thermal decomposition of titanium propylate in oxygen	50	$P_{H2(CO)}$ = 1 vol.%, P_{O2} = 99 vol.% Without water freezing	1.7×10^{16}	1.3×10^{18}	15–16
3	Precipitation with ammonia from titanium tetrachloride solution	15	P_{H2} = 3.5 vol.%, P_{O2} = 58 vol.% Without water freezing	3.0×10^{13}	–	17
4	Precipitation with ammonia from titanium tetrachloride solution	21	P_{CO} = 3.5 vol.%, P_{O2} = 50 vol.%	–	2.3×10^{13}	18

Thus, for various transition metal oxides having close phase composition, specific catalytic activity in simple reactions of CO or hydrogen oxidation can vary in a very broad range. In this connection, the analysis of relative activity series of different transition metal oxides [18–20] should take into account that they can significantly shift in both directions depending on the synthesis conditions of oxides.

1.2 Analysis of factors affecting the specific activity

a) Conditions of experiment

Unfortunately, in many cases it is very difficult to compare data on specific activity reported in different publications. Prevailing are the flow methods that create

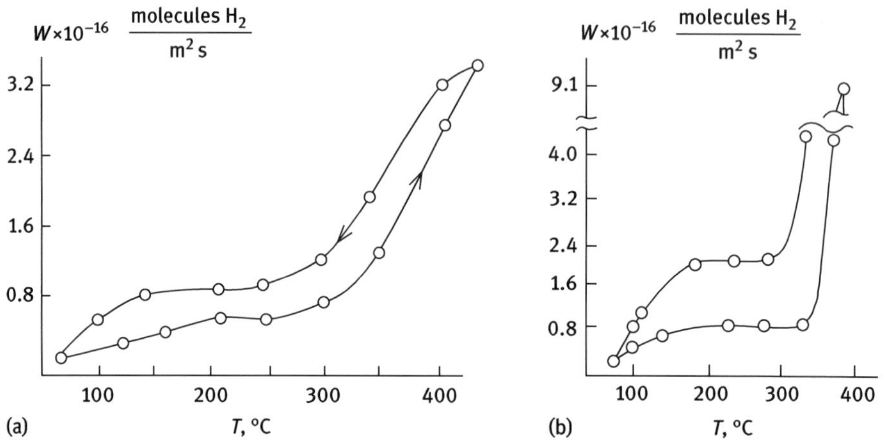

Fig. 1.1: Dependence of hydrogen oxidation rate on temperature upon its raising and lowering for TiO_2 samples: (a) – 1; (b) – 2 (Table 1.2). Composition of the mixture: 1% H_2 in oxygen.

a gradient of reagent concentrations over the bed. The main characteristics used in the studies are the temperatures at which a certain conversion is reached, or the temperature dependence of conversion at certain weights of the catalysts but with no regard to their specific surface area. Even at small weights of samples, high conversions in the flow mode create considerable gradients in the gas-phase composition, oxidation state of the oxide surface and concentrations of adsorbed compounds along the catalyst bed. The temperature-programmed mode is very convenient for the proximate comparison of activities of one-type catalysts; however, it does not always reflect the actual steady state of the catalyst at a specified temperature. This can be seen, for example, in [21–23], but actually refers to the absolute majority of publications. So, it is difficult to compare the activity values obtained by different research teams. In our opinion, which is consistent with [7], the most objective information on the specific activity of the unit surface, TOF, can be obtained only by gradientless methods, although such methods are more laborious. Time on run and closeness to the steady state of the oxide surface are also important. Particularly, prolonged relaxations of activity were observed for several hours in [2, 22–25].

b) Effect of impurities

The effect of impurities in precursors is rarely analyzed; however, they can essentially affect the specific activity of oxides. The effect of "anionic modification" is known: residual anions of carbonates, sulfates, chlorides and hydroxyl groups as impurities can affect the activity of oxide catalysts [26, 27]. The impurities represented by alkali metal cations can significantly deteriorate the activity of oxides.

Thus, the removal of sodium impurity from copper oxide increased approximately 3.5-fold the activity of the oxide toward complete oxidation of butane and CO [28].

Specific activity is often analyzed upon raising the calcination temperature of the same precursor. However, such treatment exerts a considerable effect on the concentration of impurities in the surface layer (Fig. 1.2).

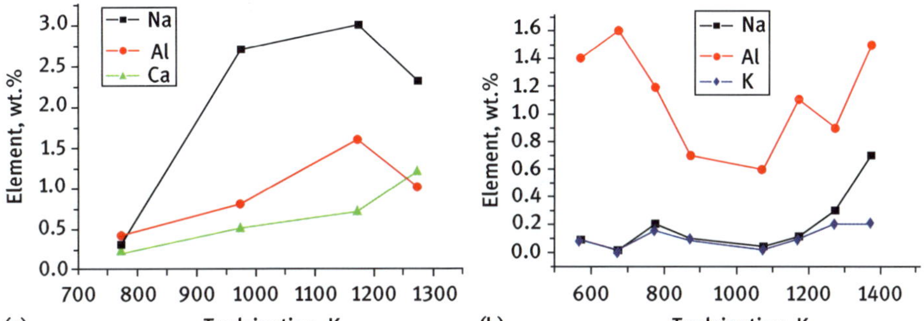

Fig. 1.2: The content of impurities in the 15-Å surface layer versus calcination temperature for hematite samples obtained by decomposition of ammonium oxalatoferriate (a) and goethite (b).

It seems interesting that dependence of the concentration of impurities on calcination temperature changes with the features of precursor and impurity (Fig. 1.2a, Fig. 1.2b).

c) Effect of particles' dispersion and morphology (exposition of different planes)

An increase in dispersion of oxide particles is considered by many authors as the most important factor for enhancing the catalytic activity of oxide [21, 22, 29–31]. This is technologically feasible if the oxide is stable in the reaction medium. From a practical standpoint, there are problems with the granulation of such powders and with the low density of oxides having a high specific surface area [7], which decreases activity of the catalyst unit volume. In terms of fundamental knowledge, specific activity of the oxide unit surface can decrease with its growth [7].

Since nanodispersed oxides are convenient objects for transmission electron microscopy, a detailed study of such oxides interrelated different planes with the catalytic activity. Thus, for copper nanooxides it was found that specific activity increases with increasing the exposition of different planes according to transmission electron microscopy (TEM) data: (111) > (001) > (011) (Fig. 1.3) [5].

For cobalt nanooxide, as revealed by TEM, most active were the crystallites with the more developed (011) plane as compared to the (001) plane [23]. In [10], a comparison of intensities of X-Ray diffraction pattern (XRD) peaks showed that

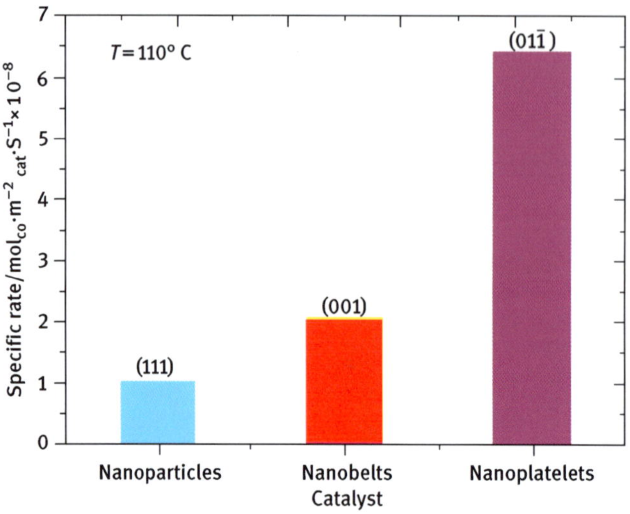

Fig. 1.3: Specific rate of CO conversion over CuO nanoparticles, nanobelts and nanoplatelets at 110 °C. The predominantly exposed crystal planes are indexed in the figure [5].

the [440] planes are more active than the [111] plane, which is consistent with the conclusions made in [30]. In [32], cobalt spinel with the most developed {011} plane shaped as nanobelts was more active than {001} nanocubes.

In [11–34], the activity of α-Fe_2O_3 with different morphology of particles (plates of different shape and needles) with the developed (0001), (01–10) and (11–00) planes was studied. Specific activity differed approximately by two orders of magnitude, but a definite trend was not revealed.

For partial oxidation, of special importance is the relation between the structure of planes and reactivity (catalytic activity) of disperse oxides, particularly the molybdenum [34–36], vanadium [37] and cobalt oxides [38]. Figure 1.4 illustrates the dependence of acrolein yield on the contribution of the (010) plane to the total surface area estimated from electron microscopy data and specific surface area of molybdenum oxides.

In later studies, the use of high-resolution electron microscopy and tunneling scanning microscopy made it possible to detect thin atomic steps at the lateral planes of MoO_3 particles [39, 40], which can at least change the surface area of this plane, not to speak of changes in coordination of atoms on such steps. Note that in table 2 presented in [36] the yield does not exceed 12.3%, whereas in the figure it reaches 15% and even ~17%! It seems interesting also that, according to [41], the oxidative catalysis on MoO_3 shows "structural sensitivity on the type of crystallographic plane in the oxidation of propene to acrolein, with (100) plane being more active."

At present, there are no detailed quantitative estimates of the contribution of individual planes of crystallites to the total specific surface area of oxide. To obtain

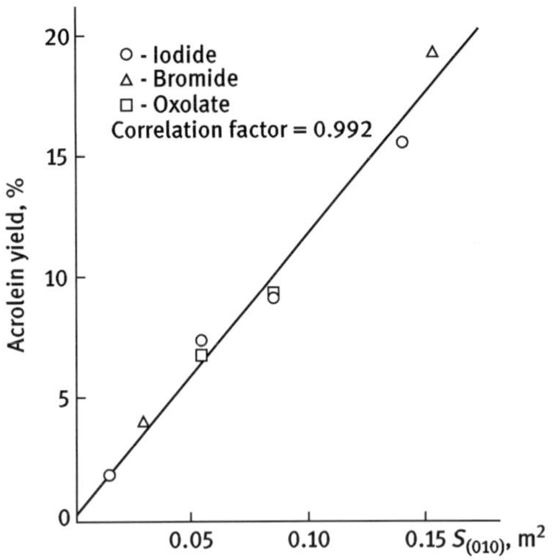

Fig. 1.4: The yield of acrolein in the reaction of allyl compounds at the surface of different MoO_3 samples as a function of the surface area of the basal (010) plane of the catalyst [35].

such estimates, it is necessary to make a detailed analysis of the morphology taking into account three-dimensional (3D) images of particles, and their quantitative averaging over a quite large sample. In particular, in [5] it was noted for the most active sample with the predominant (001) plane that its lateral planes have the developed fringes. There is no quantitative comparison of XRD data on intensity of the peaks from different crystallographic lattices with electron microscopy (EM) data on the particle morphology. One cannot disagree that

> neither particle size nor grain size independently showed a significant correlation with catalytic activity [42].

It seems that the presented results can be considered as a preliminary hypothesis. Hypotheses on the activity of some planes on the surface of oxides can be validated only in the case of complete balance of the data obtained by different methods – a quantitative comparison of the texture, morphology and structural features of single-phase particles.

d) Effect of phase composition and reaction medium

In principle, even the formally single-phase oxides may include other phases as impurities, which can strongly affect the specific activity. From this point of view, it is interesting to compare the activities of copper oxides having different phase composition. In [43], the catalytic oxidation on Cu^0, CuO and Cu_2O thin films has been studied using a static batch reactor. The following activity series has been found: $Cu^0 > Cu_2O >$

CuO. According to [2], the activity of cuprous oxide and metallic copper, measured in the pulse mode in a gradientless reactor, was much lower as compared to copper oxides without sodium impurities. In [3], the activities of CuO and the Cu_2O/CuO double system, which formed during topochemical reduction of tenorite, were compared in the pulse mode (Fig. 1.5). It was shown that in the initial step of cuprite nuclei formation, the activity increases twofold and then remains constant upon growth of the new phase nuclei (an increase in the reduction rate) and their merging (a decrease in the reduction rate) (Fig. 1.5). In [4], the activity of Cu_2O, estimated from the temperature at which conversion occurs, was lower as compared to all CuO samples.

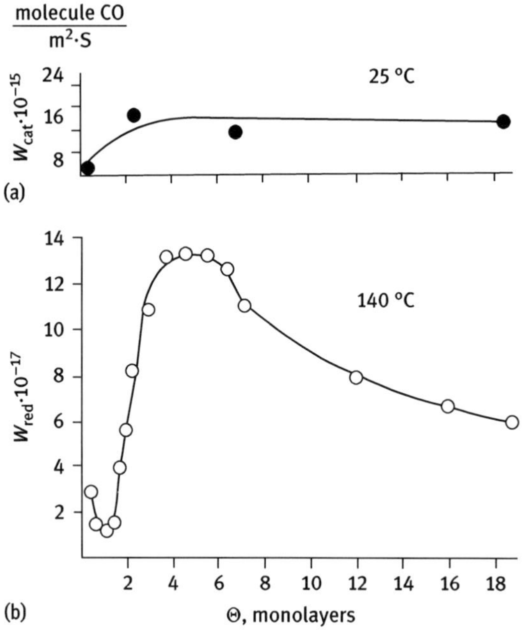

Fig. 1.5: Dependence of the catalytic reaction rate at 25 °C (a) and reduction rate at 140 °C (b) on effective reduction degree of "hydroxide 1" CuO (monolayers) at 140 °C in the pulse mode. Feed compositions: reduction – 3 vol.% CO in He; reaction – 1 vol.% CO + 1% O_2 in He [3].

Thus, according to [3], the cuprite phase itself is not much more active than the tenorite phase, and its impurities cannot affect the activity of copper oxides.

In [44], activities of different cobalt oxides with the spinel structure were compared with the activity of CoO having the NaCl structure at 140 °C in a flowing mixture of 1% CO + 1% O_2 in He. Specific activity of CoO was found to be intermediate between activity values of the most active and inactive spinel samples. For iron oxides, specific activity of γ-Fe_2O_3 was at a level with the most active α-Fe_2O_3 samples [33].

The pulse mode or the static unit with a low concentration of components exerts a relatively weak effect on the state (structure, stoichiometry, etc.) of oxide surface, whereas in the flow mode the interaction between surface and reaction medium is much more intensive. This interaction is determined not only by the phase or chemical composition of oxide but also by the composition of reaction medium, temperature and, certainly, time. Thus, in [2] Cu_2O showed a lower activity after treatment with pulses of a stoichiometric mixture. After switching to a flow with excess oxygen, the activity went through a maximum for some tens of hours. In [2], an increase in activity of metallic copper was observed during an hour in a mixture with oxygen excess at 140 °C. After that, CO conversion slowly decreased for 6 h by some percents. Activity of Cu_2O/CuO did not change for 7 h. At 205 °C the behavior of Cu^0 and Cu_2O was similar to that of Cu^0 at 140 °C, whereas the activity of CuO went through a small maximum at ~30 min time on run (Fig. 1.6). Heating of the catalytic layer with time was observed. Even a more complicated pattern was observed for the reaction mixture with an excess of CO. The authors of [23, 24] concluded that

> non-stoichiometric metastable copper oxide species (clusters) formed during reduction are very active in the course of CO oxidation because of its excellent ability to transport surface lattice oxygen.

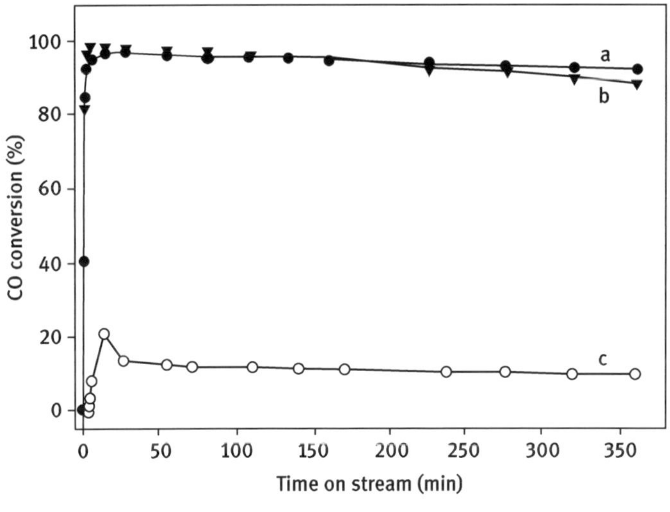

Fig. 1.6: CO oxidation activities of Cu, Cu_2O and CuO under oxygen-rich conditions: (a) Cu, (b) Cu_2O, (c) CuO [23]. Reaction temperature 205 °C.

Metastability of the active sites for CuO is confirmed also by the effect of pretreatment conditions of the oxide [23]. Training in oxygen leads to a lower activity than a milder treatment in He at 310 °C (decreasing the surface oxygen concentration without phase transition into Cu_2O [3]) resulting in a more significant (by more than an order of magnitude) increase in activity in pulse regime (Fig. 1.7). The oxidation treatment

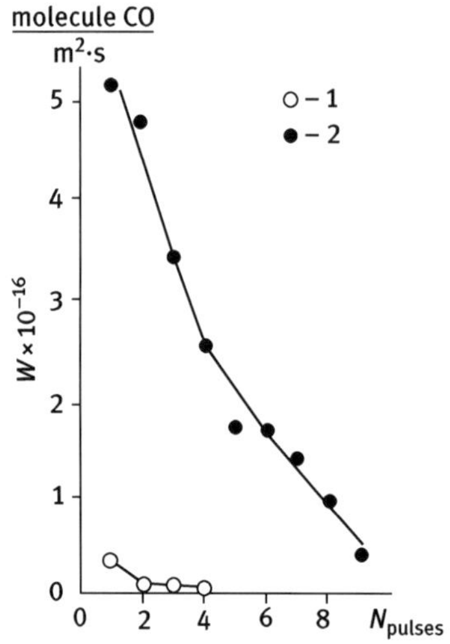

Fig. 1.7: Dependence of the catalytic reaction rate measured at 25 °C on the pulse number for 1% CO + 1% O_2 in He feed after different pretreatments at 310 °C: (1) pretreatment I (oxygen); (2) pretreatment II (O_2, then He). Pulse volume 6.1 cm^3, helium flow rate 60 cm^3/min, sample "hydroxide 1" CuO [3].

of CoO at 350 °C, resulting in the formation of spinel on the surface, decreased the specific activity approximately by a factor of 3, and the subsequent treatment in helium, similar to [3], produced nearly a 40-fold increase in activity in the pulse mode [44].

For titanium oxides, similar activation effects in complete oxidation reactions during reduction treatments were observed [45–47]. For cobalt oxide, a slow deactivation during 70 min in the reaction medium was noted [25]. Thus, it is necessary to distinguish the activity in the initial period of testing, when it is determined exclusively by the state of the oxide surface depending on the synthesis and treatment conditions. A prolonged action of the reaction medium can essentially change the state of the oxide surface.

e) Real structure (defectness) of simple oxides

In most publications devoted to specific activity of oxides, even the possibility of formation of any extended defects is neglected. In some studies, on the contrary, it is emphasized that defect structures, such as grain boundaries, are of key importance for catalytic reactions [48]. Along with this, the works considering the real structure of oxides provide many examples of the existence of extended defects.

For the tenorite structure (CuO was obtained by evaporation of a copper nitrate solution followed by calcination under air), extended defects were detected for the first time and their structures were described in [49, 50] (Fig. 1.8). After repeated redox treatments, similar to [26], the concentration of extended defects decreased, thus lowering their specific activity [49].

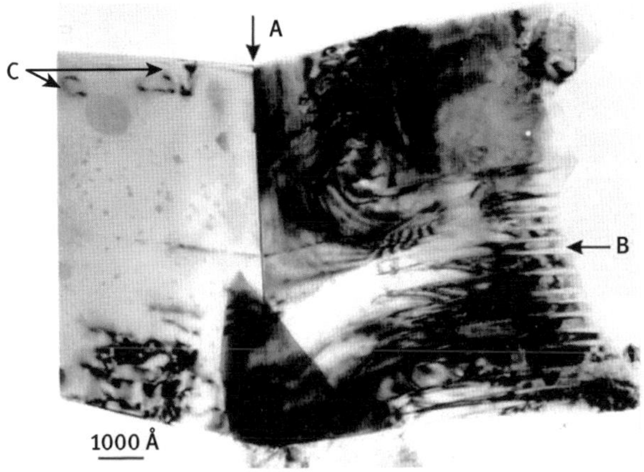

Fig. 1.8: A typical bright-field image of extended defects in disperse copper oxide particles (44): A – (100) twin, B – polysynthetic twins in (001) plane, C – edge dislocations with the Burgers vector in [101] plane [49].

For iron oxides obtained by thermal decomposition of goethite at ca. 400 °C, not only the elongated aggregates of particles are observed but also the layered packing of the primary particles, which form stacking faults (SF) or twins (Fig. 1.9) [50, 51].

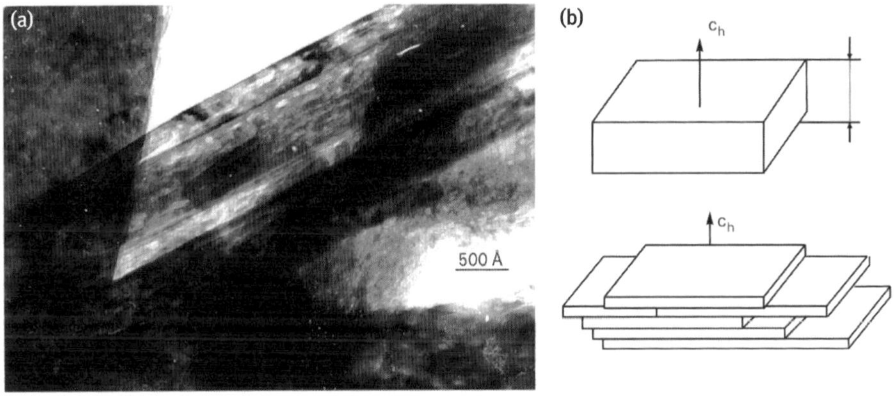

Fig. 1.9: Electron microscopy image of hematite particles obtained by decomposition of goethite (a) and their structural scheme showing the formation of defects by overlapping of layers (stacking faults and twins) (b) in the [0001] direction [51].

At a higher calcination temperature, owing to a loss of water from the bulk of oxides, spherical micropores are generated in the particle volume by the coalescence of point defects, which is followed by collapse of the micropores. Incoherent junction of the lattice in such places leads to the appearance of dislocations and incoherent inclined interblock boundaries (Fig. 1.10) [51].

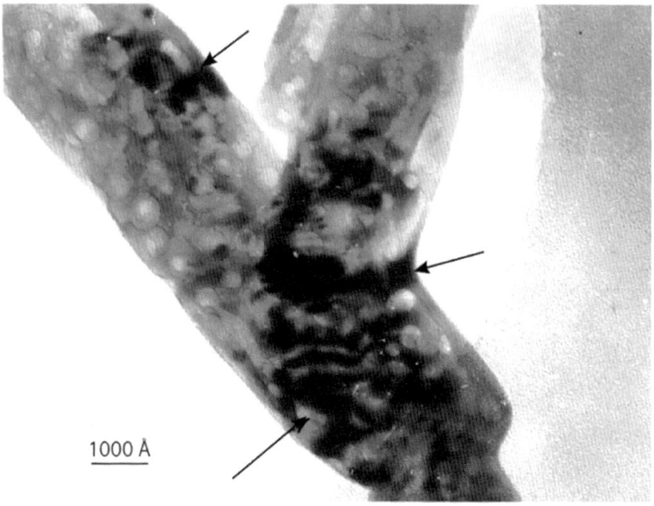

Fig. 1.10: The particle of hematite sample from the goethite series, which was calcined at 600 °C. Arrows indicate the boundaries of incoherent intergrowth of primary particles (sample 7, Table 3.4, Chapter 3) [51].

It should be noted that the presence of bulk extended defects is not a mandatory feature of any low-temperature oxide samples of the types considered above. Our results demonstrate that for the oxide structure of a certain type, the relative frequency of emergence of some or other types of extended defects depends on both the synthesis conditions and the chemical composition of oxide. Thus, for hematite samples obtained by decomposition of iron oxalate and ammonium oxalatoferriate in air at 400–500 °C, only subsurface defects of SF type and/or purely surface defects represented by steps were observed [52] (Fig. 1.11).

Aging of amorphous iron hydroxide in hydrochloric acid solutions (230) can be used to obtain highly dispersed single-crystal particles of hematite that are virtually free of any extended defects (Fig. 1.12) [53]. It seems interesting that exactly this sample has the lowest specific activity toward CO oxidation [33].

For Cr_2O_3 prepared by thermal decomposition of nitrate solution, extended defects generated by excess of oxygen were found [12]. These oxide particles were found to consist of α-Cr_2O_3 (corundum structure) and CrO_2 (rutile structure) slabs. These slabs are stacked in large particles with the $[211]_R$ axes of symmetry having a great number of surface steps of nano (Fig. 1.13) and atomic size (Fig. 1.14a). According to high-resolution electron microscopy (HREM), in the < 111 > direction a superstructure with a period of ca. 13 Å was observed close to α-Cr_2O_3 cell length in the < 0001 > direction. Microdiffraction for such particles is complex due to splitting of reflections but remains of a single crystal type (Fig. 1.14b).

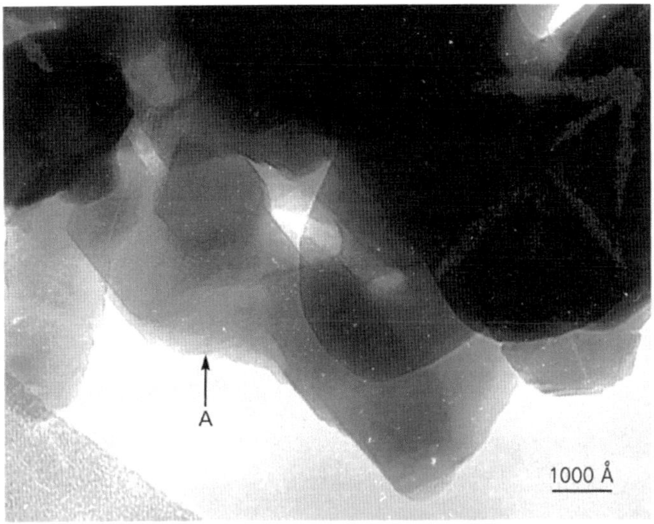

Fig. 1.11: Steps (A) on the lateral planes of hematite particles obtained by decomposition of ammonium oxalatoferriate (sample 3, Table 3.4, Chapter 3).

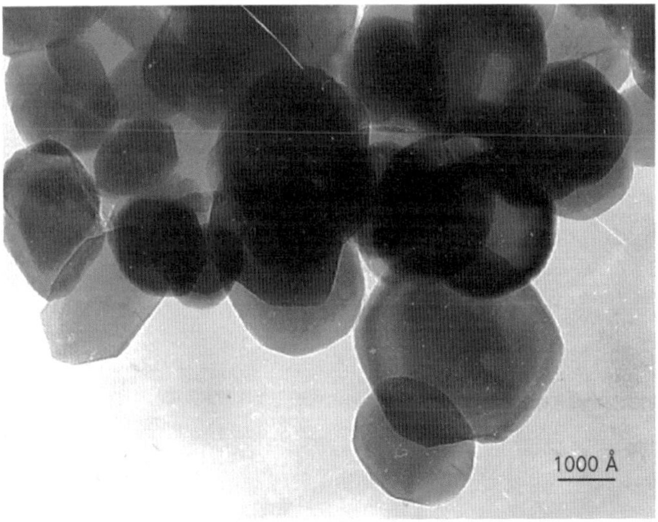

Fig. 1.12: A typical image of thin hematite particles obtained by aging of amorphous iron hydroxide in acid solutions. Predominant orientation of the particles is (0001) (sample 6, Table 3.4, Chapter 3).

Some extended defects can appear due to oxygen deficiency, and crystallographic shear structures are well known. Thus, for titanium oxides it was shown that at a stoichiometry within $TiO_{2.000-0.9994}$, clusters of interstitial Ti^{3+} ions with

Fig. 1.13: Image of a particle of the low-temperature sample of "nitrate" chromium oxide [12].

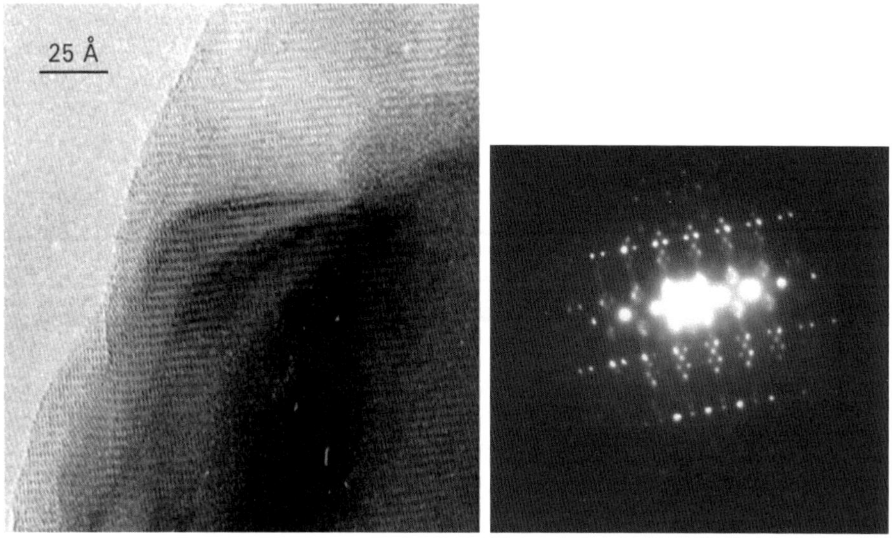

Fig. 1.14: Image of the edges of nitrate chromium oxide particle (400 °C) at a high resolution (×10^6) and the corresponding microdiffraction [12].

the diameter 0.5–3.0 nm are present in the sample cooled from 1,000 °C to room temperature [54, 55]. The reduction of titanium oxides is accompanied by the formation of crystallographic shear structures [56, 57]. Extended defects for cobalt oxides are described in Chapter 3. In [42], for CoO obtained by different methods, the concentration of grain boundaries and the degree of faceting of individual planes were studied quantitatively [58].

High-temperature annealing of oxides is characterized by several opposite trends. On the one hand, the size and crystallinity of primary crystallites increase and extended defects are annealed, but, on the other hand, sintering of crystallites increases grain boundaries in the emerging aggregates. Table 1.3 lists the most abundant types of defects for different types of transition metal oxides. It should be noted also that any, even perfectly crystallized particles of finite sizes will have surface defects represented by edges at the interface of perfect planes. The concentration of edges per unit surface area of oxides increases with the dispersion of oxide particles (Fig. 1.15). Such one-dimensional extended defects are the absolutely indestructible objects on the surface of disperse powders of a solid, which are neglected by the majority of specialists in surface science of oxides.

Table 1.3: Main types of defects in simple transition metal oxides [12].

Oxide(s)	Main type of defects
CuO	Twins (001) and (100); screw dislocations along < 101 >, microstrains, misfit dislocations, grain boundaries.
α-Cr_2O_3	Intergrowth of CrO_2 and α-Cr_2O_3, twins and stacking faults (1120), surface steps.
α-Fe_2O_3	Cation vacancies and interstitials, twins and stacking faults (0001); twins (1120), (1012) and (1123); screw dislocations < 1120 > {0001}, grain boundaries, surface steps, surface spinel precipitates.
Co_3O_4	Cation vacancies and interstitials, twins and stacking faults (111), grain boundaries, microstrains, misfit dislocations on the phase board Co_3O_4/CoO.
MnO_2	Dislocations and (100) stacking faults; intergrowth of ε and β phases.
Fe_3O_4 –γ-Fe_2O_3	Cation vacancies and superstructure; twins and stacking faults (111).
CoO	Clusters of point defects; (110) twins, surface steps, dislocations, spinel microinclusions.
NiO	Microstrains, surface steps.

The presented information on the defect structure of simple oxides is mostly qualitative. It is impossible to estimate the concentration of defects per unit mass (or particle surface) of oxides from electron microscopy data without their statistical processing. With the current level of experimental equipment, it is impossible also to estimate differentially the activity of individual site on the surface. A conventional kinetic

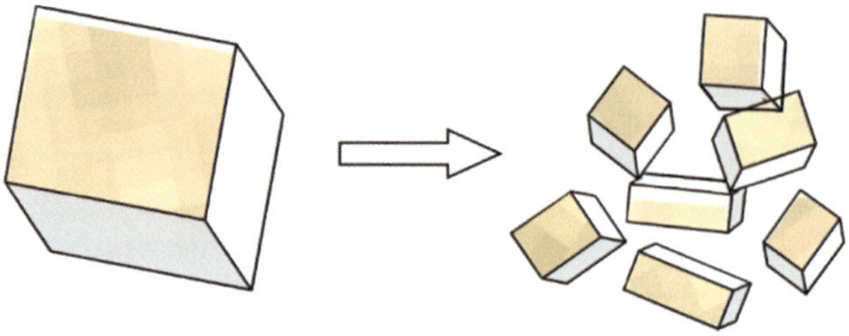

Fig. 1.15: A scheme of dispersion of oxide phases that produces more crystal edges and corners [48].

experiment always reflects some averaged properties of the oxide surface even in the case of several types of sites with different activity.

For example, in [59] it was found for $LaMnO_{3+x}$ samples that plasmochemical synthesis produces a catalyst with the highest activity toward CO oxidation and NO reduction by carbon monoxide. The specificity of this sample consists in a combination of quite high maximum temperatures in a plasmochemical reactor (up to 1,200 °C), high rate of thermolysis of the solution drops, their small residence times in the reaction zone and high rates of cooling. For oxides with the firm lattices (e.g., lanthanum manganite), the particles of which have no time to take an equilibrium form, this leads to the formation of microspheres (not greater than 500–1,000 Å in diameter) with the nanodomain structure, which have a high concentration of planar defects – domain boundaries (Fig. 1.16). Splitting of reflections in the <100> direction indicates that junction of nanodomains in (100) planes is almost coherent and results in the formation of twin boundaries. Modulation of the structure observed in the high-resolution transmission electron microscopy (HRTEM) image, which leads to splitting of X-ray diffraction maxima, is related most likely to spatial inhomogeneity in the distribution of superstoichiometric oxygen, and, hence, Mn^{3+} and Mn^{4+} cations over crystallites. These structural properties can be used to explain high activity of the plasmochemical sample, but only on a qualitative level. No quantitative estimates of nanodomain boundaries terminating on the surface were made to compare with other samples.

f) A comparison of specific activities of oxides and their averaged substructural characteristics

In terms of analysis of the averaged characteristics, quite informative is a quantitative comparison of specific activity of unit surface area with some averaged but quantitative characteristics of the surface or bulk structure. In the latter case, it is possible if bulk and surface characteristics of oxides change symbatically.

The simplest case of indirect evaluation of the density of extended defects on the CuO surface is estimation of the concentration of potential nucleation sites of

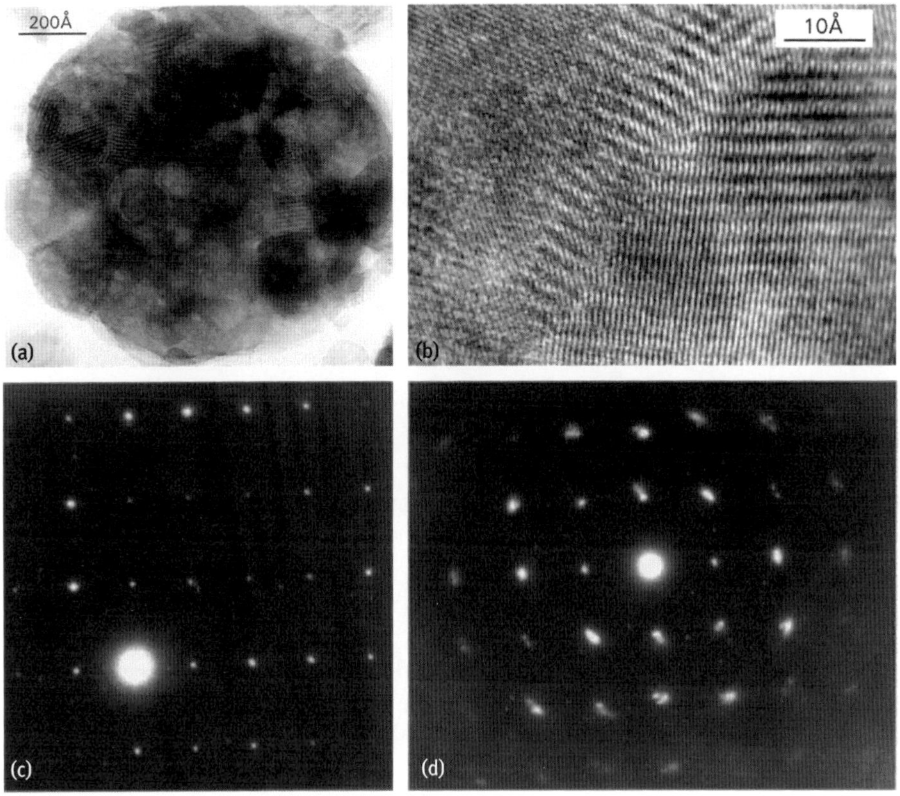

Fig. 1.16: Typical image of a spherical particle of lanthanum manganite synthesized by plasmochemical method (a), high-resolution image of domain boundary (b), and microdiffraction in < 110 > (c) and < 100 > (d) directions.

the cuprous oxide phase upon reduction of the CuO sample by CO based on the analysis of reduction curve in the two-phase region (the rate increases after the point of minimum, Fig. 1.4c) described by the Roginsky–Schultz equation [28, 60]:

$$W_{het} - W_{min} = AW_{spec.}(n_{min} - n_{het})^{2/3},$$

where W_{het} and W_{min} are the reduction rates in heterogeneous region and the point of minimum (see Fig. 1.5b), W_{spec} is the specific rate at CuO/Cu_2O interface, and n_{min} and n_{het} are the amounts of consumed oxygen at the point of minimum and current (heterogeneous) region. Parameter A is proportional to the amount of Cu_2O nuclei formed at the CuO surface. Assuming that Cu_2O nuclei are formed predominantly on the defect surface sites and activity of these sites is approximately equal, it can be concluded that there is quite a good correspondence between concentration of active surface sites and specific initial activity of copper oxides (Fig. 1.17) [28].

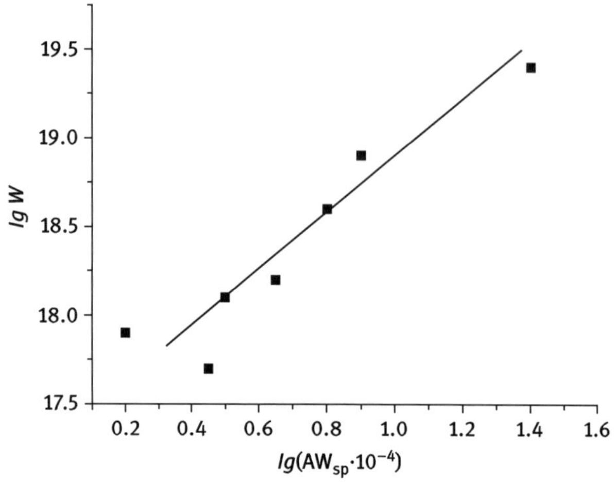

Fig. 1.17: Correlation dependences of initial rates of the catalytic oxidation of CO over different copper oxide samples and values of topochemical parameter AW^*_{sp} [28].

The averaged information on the relative concentration of extended defects can be acquired from small-angle X-ray scattering (SAXS) data [50, 61]. This method provides information on the size and relative number of the regions with nonuniform electron density, which differs from density of the solid matrix determined by the perfect crystallographic lattice of the phase. In the region of sizes below 300 Å, the integrated intensity of larger aggregates is determined by both the crystallite sizes and the related sizes and total length of grain boundaries. In the case of large crystallites, SAXS is determined mostly by extended defects in these crystallites [50].

One can see in Fig. 1.18 that a good correlation between the activity toward CO oxidation and the concentration of extended defects according to SAXS data is observed for quite a wide range of oxides. This dependence is most pronounced for the systems of similar genesis (Fig. 1.18a, 1.18d). The dependence is more distinct for the initial activity than for the steady one (Fig. 1.18b, Fig. 1.18c), which is caused by the leveling effect of the reaction medium.

Another method to estimate the averaged characteristics of defectness is a detailed analysis of diffraction patterns based on the harmonic analysis of the profile of X-ray peaks, which gives the averaged quantitative information about parameters of the crystal structure [62–65]. Here, one can mention also the study of catalysts for the synthesis of ammonia [65], where XRD data were used to estimate density of dislocations that correlated with activity. In Fig. 1.19, the dependences of CO oxidation rate for three types of nickel oxide are shown in logarithmic coordinates. The oxide sample obtained by decomposition of nitrates at 400 °C has a higher concentration of microdistortions and a higher activity.

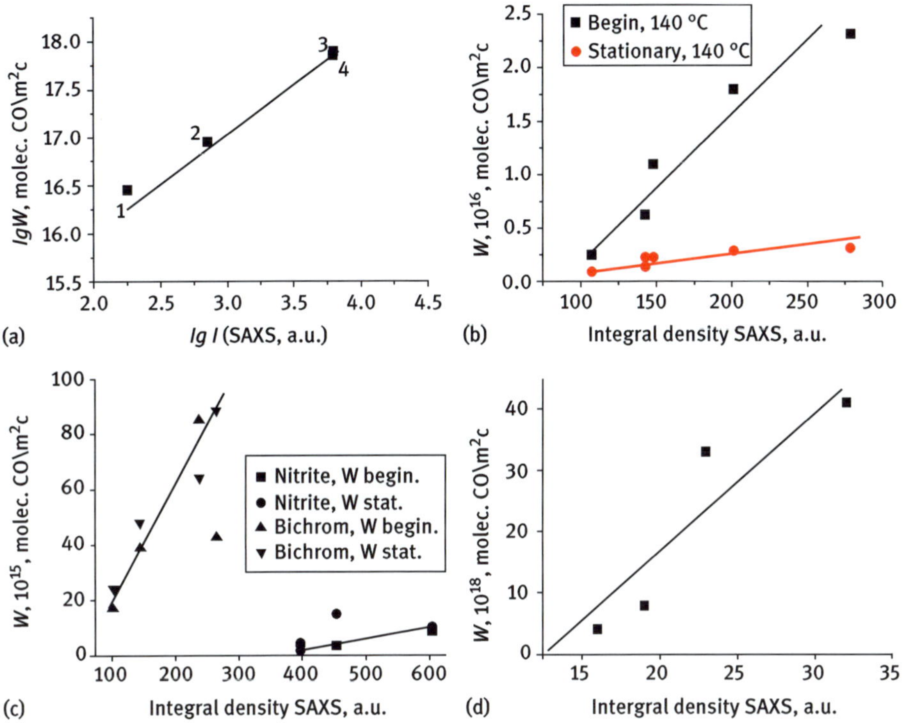

Fig. 1.18: Correlations of specific rate of the catalytic oxidation of CO over Fe_2O_3, Cr_2O_3 and Co_3O_4 samples with the density of bulk defects according to SAXS data:
(a) 1–4 – Fe_2O_3 samples synthesized at 600–1,000 °C, 4 – the sample synthesized by plasmochemical method from iron pentacarbonyl; (b) correlation of initial and steady rates of the catalytic oxidation of CO at 140 °C over the Fe_2O_3 hematite sample subjected to mechanochemical activation (54) (density of defects decreases in the course of activation) [51]; (c) correlation dependences of initial and steady activity of α-Cr_2O_3 samples from nitrate and bichromate series in CO oxidation at 185 °C [11]; (d) dependence of specific rates of the catalytic oxidation of CO over Co_3O_4 (ultra-pure grade series, initial rate at 25 °C).

A more distinct dependence of activity on the density of defects in the basal plane was found for the iron oxide samples obtained by thermal decomposition of goethite (Fig. 1.20).

Thus, in a wide range of activities, quite a good correlation is observed between specific activity toward oxidation of CO and the averaged concentration of extended defects for simple oxides. Such correlations cover a much broader range of quantities than any other dependences for simple oxides; so, they cannot be neglected when analyzing the structure of active sites, reaction mechanism and possible structure of intermediates.

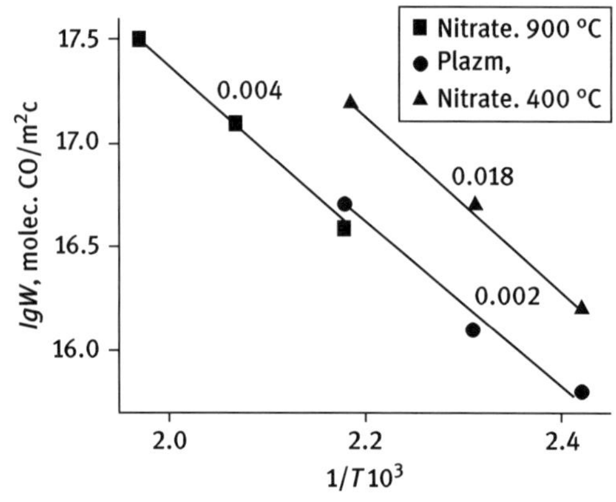

Fig. 1.19: Temperature dependences of the steady rate of CO oxidation over nickel oxide samples. Numerals on the curves indicate the concentration of microstrains (a.u.).

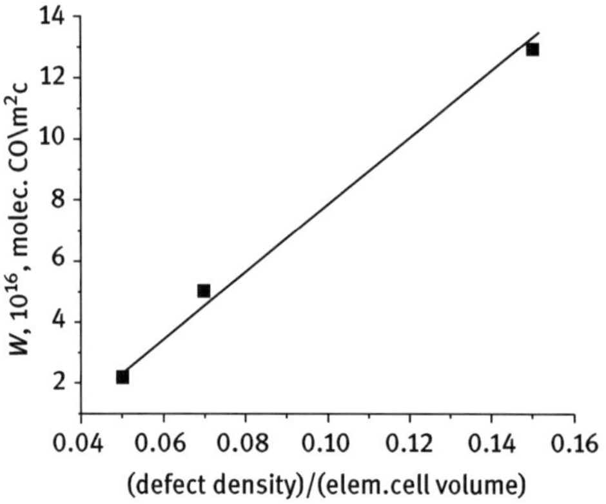

Fig. 1.20: Correlation of the steady activity at 227 °C for the low-temperature Fe_2O_3 samples of goethite series (calcination temperature from 300 to 500 °C) with the total density of plane stacking faults in the basal plane estimated by the method reported in [62, 63, 65].

The quantitative relation with the defectness of oxides and their catalytic properties is revealed not only for simple redox reactions. There is a very interesting but little-known work [66] considering the dependence of rate constants of two parallel reactions – dehydrogenation and dehydration of isopropyl alcohol – for cobalt ferrites ($CoFe_2O_4$) with the spinel structure obtained by thermal decomposition of oxalates ($Co_{0.33}Fe_{0.67}C_2O_4{\cdot}2H_2O$), schoenites ($Co_{0.33}Fe_{0.67}(NH_4SO_2){\cdot}6H_2O$), and coprecipitated cobalt and iron hydroxides calcined at different temperatures. This was supplemented with the quantitative analysis of parameters of the thin crystal structure. According to [66], microstrains estimated by harmonic analysis can be related to the phase inhomogeneity and nonuniform distribution of point defects in crystallites which are characterized by CSR. Stacking faults, which are also estimated within this analysis, correspond to deformation or twin type. Figures. 1.21–1.23 display the corresponding dependences of activity and some parameters of the crystal lattice.

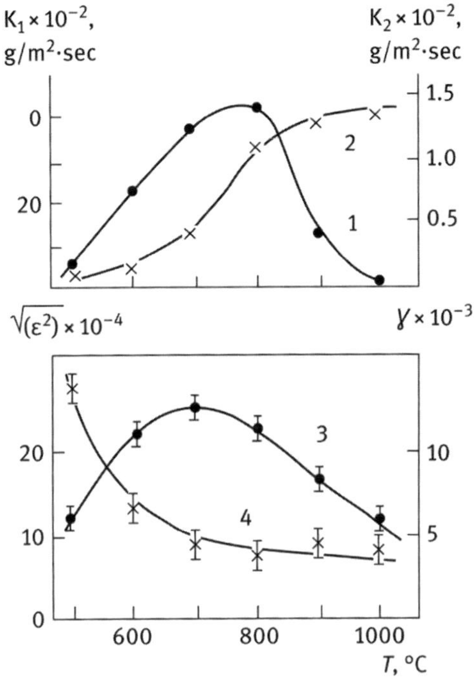

Fig. 1.21: Dependence of specific rate constants in dehydrogenation (K_1) – 1 and dehydration (K_2) – 2 of isopropyl alcohol, microstrains (root of < ε^2 >) – 3, and concentration of SF (gamma x 10^{-3}) – 4 on calcination temperature of spinel samples prepared from hydroxides [66].

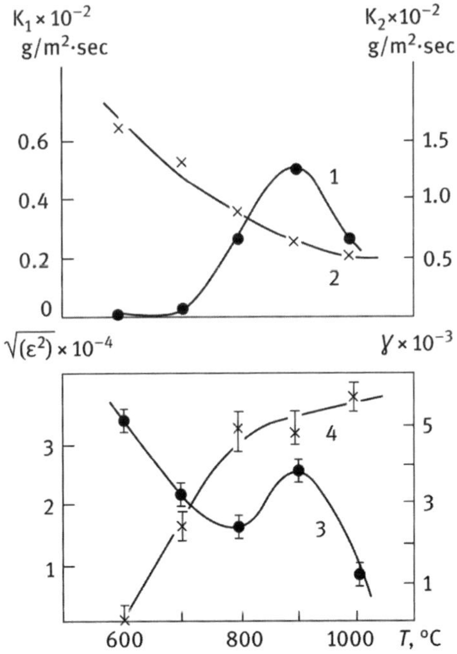

Fig. 1.22: Dependence of specific rate constants in dehydrogenation (K_1) – 1 and dehydration (K_2) – 2 of isopropyl alcohol, microstrains (root of < ε^2 >) – 3, and concentration of SF (gamma x 10^{-3}) – 4 on calcination temperature of spinel samples prepared from oxalates [66].

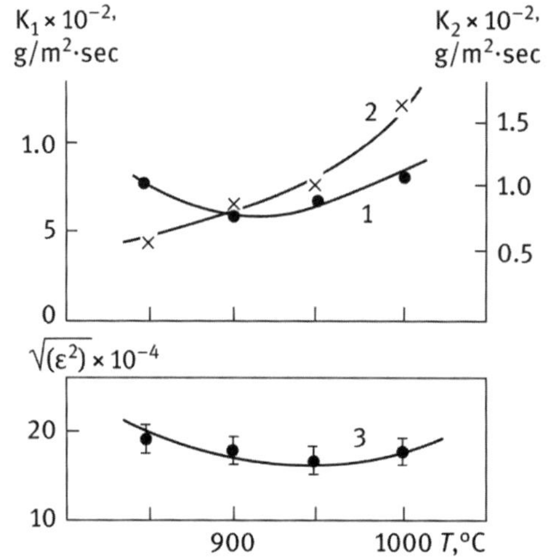

Fig. 1.23: Dependence of specific rate constants in dehydrogenation (K_1) – 1 and dehydration (K_2) – 2 of isopropyl alcohol, microstrains (root of $< \varepsilon^2 >$) – 3 on calcination temperature of spinel samples prepared from schoenites [66].

On the one hand, one can see in Figs. 1.21–1.23 that changes in dehydrogenation rate and residual microstrains are very similar. As the calcination temperature is raised, these characteristics for hydroxides and oxalates (Figs. 1.21 and 1.22) go through a maximum, whereas for schoenites (Fig. 1.23) they go through a small minimum. In addition, for oxalates and hydroxides the dehydration rate decreased with increasing the concentration of SF.

On the other hand, absolute values of activity and concentrations of defects do not give a unified dependence. It should be noted also that the total activity level for all types of samples substantially differs depending on the features of precursor. This may be related both to the effect of impurities because different precursors were used and to the effect of particles morphology, which was not studied. Thus, extended defects certainly affect the complex catalytic reactions, too.

An interesting combination of full-profile X-ray analysis and detailed high-resolution transmission electron microscopy was used in the works where structural characteristics and specific activity of perovskites with the composition $La_{1-x}Ca_xMnO_3$ were compared upon variation of x from 0 to 1 [66–69]. The effect of reaction medium (0.9% CH_4 + 9% O_2 in nitrogen) at 350–600 °C was also examined. It was shown that the initial samples have the rhombohedral structure in the region of x from 0 to 0.4 and the tetragonal structure in the region of x from 0.5 to 0.8. At a higher calcium content the monoclinic structure is formed. Along with this, extended planar defects in the [101] direction were found in the region of phase transitions at $x = 0.5$. The phase in the region of x up to 0.8 was a homogeneous solid solution. However, according to the full-profile analysis, microstrains reached the maximum values just in the region of $x = 0.7$–0.8, whereas CSR sizes in this region of compositions were the least. The maximum activity at 550–500 °C was observed in the region close to the composition with $x = 0.7$–0.8 (Fig. 1.24) [69]). At $x = 0$–0.1,

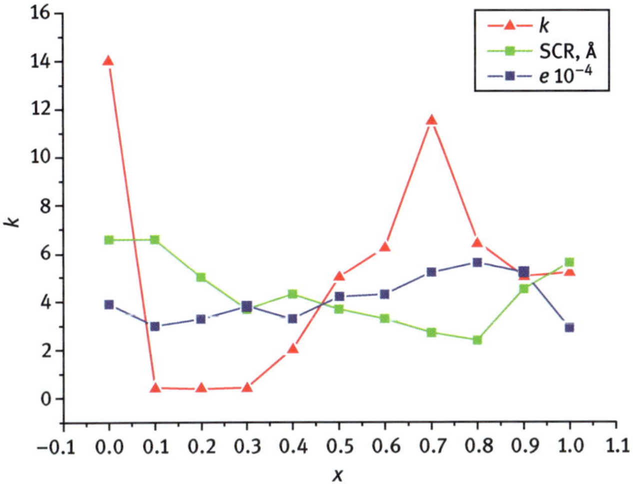

Fig. 1.24: Effect of the stoichiometry of mixed oxides with the perovskite structure and composition $La_{1-x}Ca_xMnO_{3+\delta}$ on the rate constant of methane oxidation at 550 °C(k), microdistortions (*e*) and sizes of coherent scattering regions (SCR) (according to [67–69]).

changes in activity and structural parameters did not correlate. Therewith, the dependence on superstoichiometric oxygen was completely absent.

The fact that the activity maximum did not coincide with extrema of the averaged characteristics of defectness may be related to the effect of reaction medium. It was found that keeping in the reaction medium changes the phase composition of the most active samples [69]. In the region of compositions up to $x = 0.8$, only the rhombohedral phase is preserved. In addition, the Mn_3O_4 spinel phase is released on the surface of aggregates as nanoparticles formed along (001) planes of planar defects. The authors supposed that an increase in the concentration of planar defects and nanosized spinel particles with increasing *x* from 0.3 to 0.7 enhances the specific activity of perovskites. At a higher calcium content, manganese oxide nanoparticles are aggregated and planar defects become ordered, thus lowering the activity.

1.3 Conclusions

The analysis of literature shows that specific activity of unit surface area of oxide catalysts essentially differs for many redox (and other) reactions. A possible effect of the particle morphology (exposition of different planes) and differences in the phase composition on specific activity is noted. However, the most significant relationship was revealed for activity and concentration of extended defects estimated by X-ray methods, which give the averaged information on the defectness of oxides.

Thus, the real microstructure (extended defects) can substantially affect and even determine the activity of oxides in the catalytic reactions. In this connection, it is reasonable to remind the earlier cited [3] conclusion made by G. Ertl:

> One should keep in mind that bond breaking is generally facilitated by defects, so that overall kinetics of a catalytic reaction might be governed by the properties of these active sites rather than by those of the low-index crystal planes [70].

This does not mean that only the defect sites on the surface are active. But it is clear that defect surface sites can substantially affect the activity and selectivity of catalytic reactions.

References

[1] van Dillen, JAJ., Geus, JW., de Jong, KP., & van der Meijden, J. The effect of the extent of oxidation of copper on the activity for CO oxidation. J. Chim. Phys. 1981, 78, 979–986.

[2] Sadykov, VA., Tikhov, SF., & Popovskii, VV. Interaction of copper oxides with the reaction medium in the heterogeneous catalytic oxidation of carbon monoxide by molecular oxygen. 5. Formation of the steady state on the surface of copper oxides. Kinet. Catal. 1986, 27, 133–140.

[3] Sadykov, VA., Tikhov, SF., Bulgakov, NN., & Gerasev, AP. Catalytic oxidation of CO on CuOx revisited: Impact of the surface state on the apparent kinetic parameters. Catal. Today 2009, 144, 324–333.

[4] Svintsitskiy, DA., Chupakhin, AP., Slavinskaya, EM., Stonkus, OA., Stadnichenko, AI., Koscheev, SV., & Boronin, AI. Study of cupric oxide nanopowders as efficient catalysts for low-temperature CO oxidation. J. Molec. Catal. A 2013, 368–369, 95–106.

[5] Zhou, K., Wang, R., Xu, B., & Li, Y. Synthesis, characterization and catalytic properties of CuO nanocrystals with various shape. Nanotechnology 2006, 17, 3939–3943.

[6] Tikhov, SF. Study by electromotive force method of the interaction of manganese and copper oxides with reaction media medium in heterogeneous catalytic oxidation of carbon monoxide with molecular oxygen. PhD Thesis, Novosibirsk, Russia, Boreskov Institute of Catalysis, 1985.

[7] Svintsitskiy, DA., Pakharukov, IYu., Slavinskaya, EM., Kardash, TYu., Parmon, VN., & Boronin, AI. Influence of the copper(II) oxide dispersion on its catalytic properties in carbon monoxide oxidation: A comparative study by using two types of catalytic reactors. ChemCatChem 2016, 8, 3546–3555.

[8] Yu Yao, Y-F. The oxidation of hydrocarbons and CO over metal oxides: III. Co3O4. J. Catal. 1974, 33, 108–122.

[9] Razdobarov, VA., Sadykov, VA., Veniaminov, SA., Bulgakov, NN., Kovalenko, ON., Pankratiev, YuD., Popovskii, VV., Kryukova, GN., & Tikhov, SF. Nature of the active oxygen of Co_3O_4. React. Kinet. Catal. Lett. 1988, 37, 109–114.

[10] Alvarez, A., Ivanova, S., Centeno, MA., & Odriozola, JA. Sub-ambient CO oxidation over mesoporous Co_3O_4: Effect of morphology on its reduction behavior and catalytic performance. Appl. Catal. A: General 2012, 431–432, 9–17.

[11] Sadykov, VA., Tikhov, SF., Tsybulya, SV., Kryukova, GN., Veniaminov, SA., Kolomiichuk, VN., Bulgakov, NN., Paukshtis, EA., Ivanov, VP., Koshcheev, SV., Zaikovskii, VI., Isupova, LA., & Burgina, EB. Role of defect structure in structural sensitivity of the oxidation reactions

catalyzed by dispersed transition metal oxides. J. Molec. Catal.: A: Chemical 2000, 158, 361–365.

[12] Sadykov, VA., Tikhov, SF., Tsybulya, SV., Kryukova, GN., Veniaminov, SA., Kolomiichuk, VN., Bulgakov, NN., Paukshtis, EA., Zaikovskii, VI., Kustova, GN., & Burgina, EB. Structural sensitivity of the oxidation reactions catalyzed by dispersed transition metal oxides: role of defect structure. Stud. Surf. Sci. Catal, Proc.3rd World Congress on Oxidation Catalysis. (R.K. Graselli et al. Eds.), 1997, 110, 1155–1164.

[13] Grigorieva, AV., Goodilin, EA., Dubova, KL., Anufrieva, TA., Derlyukova, LE., Vyacheslavov, AS., & Tretyakov, YD. Titania nanotubes, nanorods and nanopowder in the carbon monoxide oxidation process. Solid State Sci. 2010, 12, 1024–1028.

[14] Kovalenko, ON. Preparation and study of the deep oxidation catalysts over porous metallic supports, PhD Thesis, Novosibirsk, Russia, Boreskov Institute of Catalysis, 1987.

[15] Gass, SL., Jullet, F., & Feichiner, SJ. Prèparation des aèrosols d'oxides mètallique par decomposition thermique des alcoolates en phase vapeur et propertètès de ces solides. II Prèparationdes a Prèparationrosols d'oxides de titane, d'alimine et de silice.// Bull. Soc. Chin. France 1973, 2(1), 429–435.

[16] Svintsova, LG., & Korneychuk, GP. Study of mechanism of hydrogen oxidation over titanium oxides. Kinet. Catal. 1981, 22, 424–430.

[17] Svintsova, LG., Golodets, GI., & Shimanovskaya, VV. Kinetics and mechanism of carbon oxide oxidation over titanium oxides. Kinet. Catal. 1980, 21, 1469–1475.

[18] Heck, RM., & Farrauto, RJ. Catalytic Air Pollution Control, 2nd edn., Wiley, 2002, 69–129.

[19] Kummer, J.T. Advances in Chemistry Series, Mc Evoy J., ed., American Chemical Society, Washington DC, 1975, 178–192.

[20] Popovskii, VV. Oxidation of substances on solid oxide catalysts. Kinet. Catal. 1972, 13, 1190–1203.

[21] Kou, T., Si, C., Gao, Y., Frenzel, J., Wang, H., Yan, X., Bai, Q., Eggelerc, G., & Zhang, Z. Large-scale synthesis and catalytic activity of nanoporous Cu–O system towards CO oxidation. RSC Adv. 2014, 4, 65004–65011.

[22] Guo, MY., Liu, F., Tsui, J., Voskanyan, AA., Ching, AM., Djurišĭ, AB., Chan, WK., Chan, K-Yu., Liao, C., Shihd, K., & Suryae, C. Hydrothermally synthesized CuxO as a catalyst for CO oxidation. J. Mater. Chem. A 2015, 3, 3627–3632.

[23] Huang, T-J., & Tsai, D-H. CO oxidation behavior of copper and copper oxides. Catal. Lett. 2003, 87, 173–178.

[24] Nagase, K., Zheng, Y., Kodama, Y., & Kakuta, J. Dynamic study of the oxidation state of copper in the course of carbon monoxide oxidation over powdered CuO and Cu_2O. J. Catal. 1999, 187, 123–130.

[25] Jansson, J., Palmqvist, AEC., Fridell, E., Skoglundh, M., Ősterlund, L., Thormählen, P., & Langer, V. On the catalytic activity of Co_3O_4 in low-temperature CO oxidation. J. Catal. 2002, 211, 387–397.

[26] Khassin, AA., Minyukova, TP., & Yurieva, TM. Genesis of catalysts for methanol synthesis. Mendeleev Commun. 2014, 24, 67–74.

[27] Khassin, AA., Minyukova, TP., & Yurieva, TM. Role of anionic impurities in the formation of the active state of catalysts based on transition metals. Kinet. Catal. 2014, 55, 502–508.

[28] Sadykov, VA., Tikhov, SF., Popovskii, VV., & Kryukova, GN. Interaction of copper oxides with the reaction medium in heterogeneous catalytic oxidation of carbon monoxide with molecular oxygen. 3. Nature of reactive centers on copper oxide surface. Kinet. Catal. 1985, 26, 144–152.

[29] Svintsitskiy, DA., Kardash, TYu., Stonkus, OA., Slavinskaya, EM., Stadnichenko, AI., Koscheev, SV., Chupakhin, AP., & Boronin, AI. In situ XRD, XPS, TEM, and TPR study of highly active in CO oxidation CuO nanopowders. J. Phys. Chem. C 2013, 117, 14588–14599.
[30] Xie, X., Li, Y., Liu, Z-Q., Haruta, M., & Shen, W. Low-temperature oxidation of CO catalyzed by Co_3O_4 nanorods. Nature 2009, 458, 746–749.
[31] Zhang, Y., Chen, Y., Zhou, J., Wang, T., & Zhao, Y. Synthesis and high catalytic activity of mesoporous Co_3O_4 nanowires for carbon monoxide oxidation. Solid State Commun. 2009, 149, 585–588.
[32] Hu, L., Sun, K., Peng, Q., Xu, B., & Li, Ya. Surface active sites on Co_3O_4 nanobelt and nanocube model catalysts for CO oxidation. Nano. Res. 2010, 3, 363–368.
[33] Tikhov, SF., Sadykov, VA., Kryukova, GN., & Razdobarov, VA. Dipole coupling of CO(NO) adsorbed on iron surface centers: A typical feature of surface active sites. Mendeleev Commun. 1994, 2, 69–71.
[34] Volta, JC., Tatibouet, JM., Dhicktu, C., & Germain, JE. Structural-sensitive catalytic oxidation on MoO_3 catalysts. In: Proc. 8th Int. Congr. Catal. Proceeding, Weinheim. Verlag Chemie 1984, 4, 451–461.
[35] Bruckman, K., Grabowski, R., Haber, J., Mazurkiewicz, A., Sloczynski, J., & Wiltowski, T. The role of different MoO_3 crystal faces in elementary steps of propene oxidation. J. Catal. 1987, 104, 71–79.
[36] Haber, J. The Concept of Structure-Sensitivity in Catalysis by Oxides, Morterra C, Zecchina A, Costa G, eds., Structure and Reactivity of Surfaces. Stud. Surf. Sci. Catal., Amsterdam, Elsevier, 1989, Vol. 48, 447–468.
[37] Mari, K., Miyamoto, A., Hi, T., & Murakami, Y. Active sites for the oxidation of carbon monoxide on V_2O_5 catalysts. J. Chem. Soc. Chem. Commun. 1982, 4, 260–261.
[38] Ziolkowski, J., & Barbaux, Y. Identification of sites active in oxidation of butene-1 to butadiene and CO_2 in Co_3O_4 in terms of the crystallochemical model of solid surfaces. J. Molec. Catal. 1991, 67, 199–215.
[39] Dominguez-Esquivel, JM., Guzman-Mondujano, O., & Garcia-Borquez, A. Evidence for short-range order and steps in MoO_3 crystal surfaces. J. Catal. 1987, 103, 1–19.
[40] Rohrer, GH., Lu, W., Smith, RL., & Hutchinson, A. Imaging surface/crystallographic shear plane intersections on the $Mo_{18}O_{52}$ (100) surface using scanning tunneling microscopy. Surf. Sci. 1993, 292, 261–266.
[41] Volta, JC., & Tatibouet, M. Structure sensitivity of MoO_3 in mild oxidation pf propylene, J. Catal. 1985, 93, 467–470.
[42] Mingle, K., & Lauterbach, J. Synthesis–structure–activity relationships in Co_3O_4 catalyzed CO oxidation. Front. Chem. 2018, 6, article, 185.
[43] Jernigan, GG., & Somorjai, GA. Carbon monoxide oxidation over three different oxidation states of copper: Metallic copper, copper (I) oxide, and copper (II) oxide – A surface science and kinetic study. J. Catal. 1994, 147, 567–572.
[44] Sadykov, VA., Lokhov, YuA., Tikhov, SF., Krjukova, GN., Bredikhin, NM., Popovskii, VV., Bulgakov, NN., Razdobarov, VA., Solovjova, LP., Olenkova, IF., & Golovin, AV. Nature of Surface Active Centers and Mechanism of CO Catalytic Oxidation on Cobalt Oxide Catalysts, Shopov D. et al. eds., Proc. VI Int. Symp. Heterog. Catalysis, Sofia, July 13–18, Publ. house of the Bulgarian Academy of Sciences, 1987, Part I, 359–364.
[45] Bazanov, SS., Boreskov, GK., Gridasova, GV., Keyer, NP., Kefrely, LM., Kudinov, VM., Maly, VI., & Sazonova, IS. Influence of the blow shrinkage on the catalytic properties of oxide catalysts. Kinet. Catal. 1967, 8, 1348–1355.
[46] Long, J., & Teichner, SJ. Activitè catalytique de dioxide de titane (anatase) dans la reaction d'oxidation de l'oxide de carbon. Bull. Soc. Chim. France 1965, 9, 2625–2635.

[47] Vainchbock, NT., Vergnon, P., Juillet, F., & Teichner, SJ. Oxidation catalytique de l'oxide de carbone en presence de dioxide de titane (anatase) active par differents traitements rèducteure. 1.Activation sous pression rèduite. Bull. Soc. Chim. France 1970, 8–9, 2806–2812.
[48] Landau, MV., Vidruk, R., Vingurt, D., Fuks, D., & Herskowitz, M. Grain boundaries in nanocrystalline catalytic materials as a source of surface chemical functionality. Rev. Chem. Eng. 2014, 30, 379–401.
[49] Kryukova, GN., Zaikovskii, VI., Sadykov, VA., Tikhov, SF., Popovskii, VV., & Bulgakov, NN. Study of the nature of extended defects of copper oxide. J. Solid State Chem. 1988, 74, 191–199.
[50] Sadykov, VA., Tikhov, SF., Kryukova, GN., Bulgakov, NN., Popovskii, VV., & Kolomiichuk, VN. The structure, energetics and reactivity of extended defects of copper oxide. J. Solid State Chem. 1988, 74, 200–208.
[51] Kryukova, GN., Tsybulya, SV., Solovyeva, LP., Litvak, GS., & Andrianova, MP. Effect of heat treatment on microstructural evolution of the haematite derived from synthetic goethite. Materials Sci. Eng. A 1991, 149, 121–127.
[52] Sadykov, VA., Isupova, LA., Tsybulya, SV., Cherepanova, SV., Litvak, GS., Burgina, EB., Kustova, GN., Kolomiichuk, VN., Ivanov, VP., Paukshtis, EA., Golovin, AV., & Avvakumov, EG. Effect of mechanical activation on the real structure and reactivity of iron (III) oxide with corundum-type structure. J. Solid State Chem. 1996, 123, 191–202.
[53] Golovin, AV., Kryvoruchko, OP., Buyanov, RA., & Zolotovskii, BP. Study of hydrogels Fe(III), amorphous products of their ageing and thermal dehydrogenation by magnetic methods. Izvestiya SO AN USSR. Seriya khimicheskih nauk 1981, 4, 70–78, in Russian.
[54] Bursill, LA., Blanchin, MG., Mebarek, A., & Smith, DJ. Point linear and extended defect structures in non-stoichiometric rutile. Radiat. Eff. 1984, 74, 253–265.
[55] Yagi, E., Koyana, A., Sakairi, H., & Hasiguti, RR. Observation of defect structures of slightly reduced rutile (TiO_2) by channeling method. IPCR Cyclotron Progr. Rept. 1976, 10, 96–98.
[56] Blanchin, MG., Bursill, LA., & Smith, DJ. Precipitation phenomena in non-stoichiometric oxides. 1. Pairs of crystallographic shear planes in reduced rutiles. Proc. Roy. Soc. 1984, A391, 351–372.
[57] Bursill, LA., Blanchin, MG., & Smith, DJ. Precipitation phenomena in non-stoichiometric oxides. 2. 100 platelet defects in reduced rutiles. Proc. Roy. Soc. 1984, A391, 373–391.
[58] Gai-Boyes, PL. Defects in oxide catalysts: fundamental studies of catalysis action. Catal. Rev. Sci. Eng. 1992, 34, 1–54.
[59] Tikhov, SF., Kimkhai, ON., Sadykov, VA., Moroz, EM., Bogdanov, SV., Kryukova, GN., Pak, EA., Alikina, GM., Kalinkin, AV., Kustova, GN., Ivanov, VP., Tsybulev, PN., Voronin, PN., & Barton, J. The Influence of the Genesis on Catalytic and Physicochemical Properties of Mn Lanthanides. Proc.7th Symp. Heterogeneous Catalysis, Burgas, Bulgaria,1991, 423–428.
[60] Rozovskii, AYa., Stytsenko, VD., & Tretyakov, VF. Kinetics of the reaction of gas with solid. Specificity and models. Kinet. Catal. 1979, 20, 79–93, in Russian.
[61] Kolomiichuk, VN. Determination of metal particle size in supported catalysts by small-angle X-Ray scattering technique. React. Kinet. Catal. Lett. 1982, 2, 123–128.
[62] Kagan, AS., Portnoi, VK., & Fadeeva, VI. Diffraction picture under mistakes of dense packing in spinel structure. Krystallography 1974, 19, 489–497, in Russian.
[63] Warren, BE. X-ray studies of deformed metals. Progr. Metal. Phys. 1959, 8, 147–202.
[64] Tsybulya, SV., & Kryukova, GN. Nanocrystalline transition aluminas: Nanostructure and features of x-ray powder diffraction patterns of low-temperature Al_2O_3 polymorphs. Phys. Rev. B 2008, 77, 024112.
[64] Yatsimirskii, VI., Vyazmitina, OM., & Kozlova, TP. Elucidation of role of dislocation in working iron catalyst of ammonia synthesis. Theor. Exp. Chem. 1971, 7, 645–651.

[65] Cherepanova, SV., & Tsybulya, SV. Simulation of X-ray powder diffraction patterns for one-dimensionally disordered crystals. Mater. Sci. Forum. 2004, 443–444, 87–90.
[66] Fadeeva, VI., Voynov, ID., & Tretyakov, YuD. Influence of the fine parameters of crystal structure on the catalytic activity of cobalt ferrite. Kinet. Catal. 1978, 19, 625–628, in Russian.
[67] Gerasimov, EYu., Zaikovskii, VI., Tsybulya, SV., & Isupova, LA. A study of the microstructure of $La_{1-x}Ca_xMnO_3$ (x=0.5, 0.8) solid solutions. J. Surf. Invest. X-Ray Synchrotron and Neutron Tech 2009, 3, 756–760.
[68] Isupova, LA., Gerasimov, EYu., Zaikovskii, VI., Tsybulya, SV., Kulikovskaya, NA., & Saputina, NF. Synthesis of homogeneous $La_{1-x}Ca_xMnO_3$ solid solutions by the Pechini method and their activity in methane oxidation. Kinet. Catal. 2009, 50, 886–891.
[69] Isupova, LA., Gerasimov, EYu., Zaikovskii, VI., & Tsybulya, SV. Effect of the reaction medium on the structure of the $La_{1-x}Ca_xMnO_3$ (x=0–1) solid solutions prepared by the Pechini method. Kinet. Catal. 2011, 52, 104–110.
[70] Ertl, G. Dynamics of reactions at surfaces. Adv. Catal. 2000, 45, 1–69.

2 Surface oxygen forms on oxide catalysts: nature of adsorption sites, bonding strength, surface coverage, reactivity and correlation with specific catalytic activity

Importance of oxygen bonding strength in catalysis of oxidation reactions on oxide catalysts was stressed in classical works of Boreskov [1] and Haber [2]. Development of this concept up to date is considered in this chapter.

Analysis of the atomic-scale reasons for various oxygen species appearing on the catalyst surface should be based upon the discrete nature of the energy spectrum of oxygen species on the oxide surface. This means that there are always several surface oxygen species differing by the strength of their bonding (heat of adsorption). These species appear as separate maxima in the spectra of the oxygen thermal desorption [3–5] or as steps on the dependencies of the heat of oxygen adsorption versus coverage registered using isosteric methods [6, 7], calorimetry or solid electrolyte potentiometry [5, 8].

For a given oxide, the existence of different surface oxygen species is caused by the atomic structure of its faces, by the appearance of point (vacancies) and extended (steps) surface defects, including those emerging in the vicinity of outlets of bulk extended defects [9–13] (see Chapter 1).

A large amount of data on the atomic structure of the oxide single-crystal faces were accumulated using methods such as low-energy electron diffraction (LEED), scanning tunneling microscopy (STM), atomic force microscopy (AFM) and photoelectron diffraction (see, e.g., reviews [14, 15]). For finely dispersed oxides, the atomic structure of their surfaces was studied with the help of methods such as infrared spectroscopy of adsorbed test molecules [16], Electron Spin Resonance (ESR) of adsorbed nitroxides [17], tunneling electron microscopy (TEM) [18–20], Ion Scattering Spectroscopy (ISS) [21], and neutron diffraction in the argon adlayer [22, 23]. The development of quantum-chemical semi-empirical [10, 11, 24–26] as well as ab initio methods of the surface energy and adsorption heat calculation [27–35] allowed analysis of possible structures of adsorption centers capable to retain oxygen with the strength required for the process considered. Some results are given in Table 2.1.

According to [9], in an oxidative medium, the well-developed densely packed (111) faces of spinel oxides and (0001) faces of corundum-type oxides are mainly covered by strongly bound nonreactive bridged M_2O oxygen species, whose adsorption heats range within 100–120 kcal/mol. Figures 2.1 and 2.2 demonstrate that indeed after removal of ~20% of oxygen monolayer from the surface of hematite, the heat of oxygen adsorption achieves this level. Rather weak dependence of the rate of reduction on the oxygen bonding strength indicates on the reconstruction of the surface layer facilitating CO activation on clusters of reduced iron cations.

https://doi.org/10.1515/9783110587777-002

Table 2.1: Calculated and experimental oxygen desorption heats (kcal/mol) [7].

Oxide	ΔH calculated		ΔH_{exp}
	Regular sites	**Defect sites**	
CuO	MO: 9 M_2O: 60	M–O near (001) twin outlet onto (101) face: 3	10;25; 50
Co_3O_4	MO(Oh): 40–50 MO(Td): 60–70 M_2O: 130	M–O near outlet of (110) stacking fault onto (110) face: 15	15; 40; 120
Fe_2O_3	MO: 35–40 (Oh) 60–70 (Td) M_2O: 130	M–O near outlet of (0001) stacking fault onto (0001) and (112-0) face: 18–30	20; 40; 60; 120
MnO_2	MO: 32–40 M_2O: 60	M–O on (110) face, relaxation of surface Mn–O bonds caused by oxygen atom removal: 14	16; 30; 45

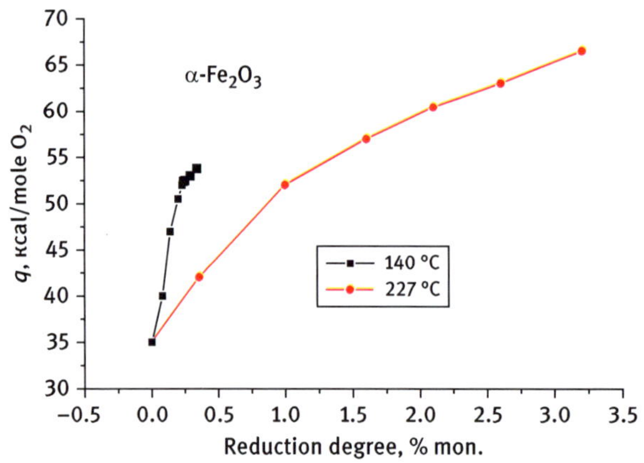

Fig. 2.1: Dependence of the heat of oxygen adsorption on the amount of oxygen removed from the hematite surface by pulses of 0.3% CO in He after pretreatment in O_2. Microcalorimetric experiments.

The structures of such faces annealed in high vacuum and, thus, weakly reduced, most likely correspond to localization of tri-coordinated cations in the upper layer [35]. Theoretical analysis of these structures [27–35] has revealed that they must experience a strong relaxation forcing the cation to sink into the surface oxygen layer. At the same time, STM data for the (0001) face of thin hematite layer grown epitaxially on the Pt (111) suggest the existence of domains with the upper layer represented by both oxygen and iron atoms [35].

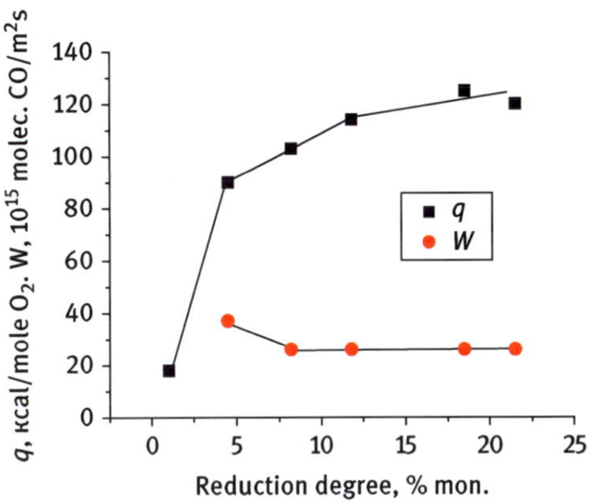

Fig. 2.2: Dependence of the heat of oxygen adsorption on the amount of oxygen removed from the hematite surface by pulses of 1% CO in He at 224 °C. Microcalorimetric experiments.

For the (110) spinel face of Co_3O_4 (Table 2.1 and Fig. 2.3), a variety of regular and defect sites provide a broad variation of the bonding strength of adsorbed surface oxygen species.

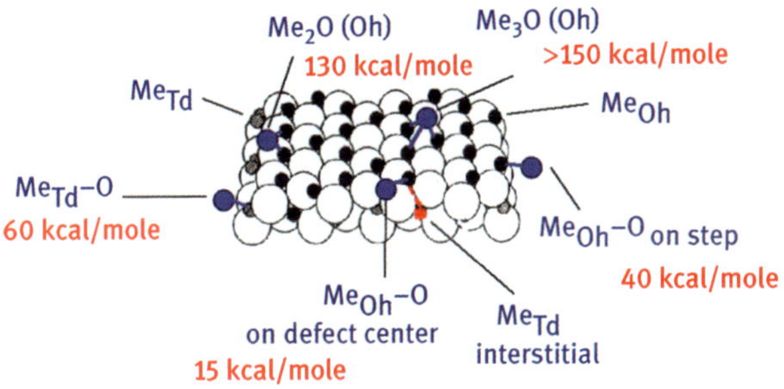

Fig. 2.3: Surface sites at Co_3O_4 spinel (110) face [5].

In CuO (tenorite) and MnO_2 (pyrolusite), M_2O bridged oxygen species bound to the regular sites of the most developed faces have the bonding strength ~50–60 kcal/mol) (Table 2.1 and Fig. 2.4) [9, 36]. Weakly bound surface oxygen forms appear at surface steps, including those at outlets of extended defects (Fig. 2.4 and Table 2.2).

Under oxidative conditions, for corundum-like chromium oxide, the densely packed basal (0001) face is usually covered by chromate groups (tetrahedrally coordinated Cr^{6+} cations), where the terminal oxygen has the adsorption heat close to 60 kcal/mol [9]. Note that in this case, the basal face cations are tetra-coordinated that agrees well with predictions based on the data obtained with high-vacuum methods (vide supra). Apparently, under oxidative conditions, the chromium cation changes its oxidation state easily, thus favoring the thermodynamic stability of such configuration, which is not possible in the case of iron or aluminum oxides.

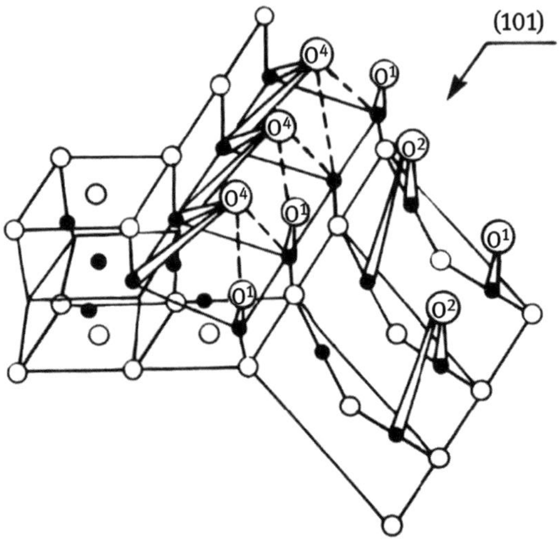

Fig. 2.4: Outlet of (001) twin on (101) face CuO: surface oxygen species on regular and defect sites [36].

Table 2.2: Heat of oxygen adsorption on different types of sites for CuO.

Face	Q MO, kcal/mole	Q M_2O, kcal/mole	Type of site
(101)	9.2	60.4	Cu at step bound with 3 O and 4 subsurface Cu
(331)	7.8	61.1	
(101)	3		Outlet of (001) twin on (101) face
(010), (001)	~10		Outlet of screw dislocation
(110)	10		Outlet of (001) twin

For oxides with the close-packed oxygen sublattice, the terminal (M–O) oxygen species with the adsorption heat ~40–60 kcal/mol are mostly located on the open faces: (100) and (110) types in spinel structures; (112-3) and (101-2) in corundum ones. In the case of oxides obtained via the thermal decomposition of precursors (salts, hydroxides) synthesized from water solutions, these faces are not well represented, though their exposition increases with an increase of the calcination temperature. Thus, according to [37], for a sample of cobalt oxide prepared from the nitrate salt, as the calcination temperature increases from 400 to 800 °C, the shape of particles changes from platelets with the mostly developed (111) faces to pyramids with the mostly developed (110) faces.

Rhombohedral (101-2) and pyramidal (112-3) faces of hematite are the natural growth faces [32, 38], and in dispersed samples, their share also increases with increasing calcination temperature, since their surface energies are lower than that of the basal face [11]. On rhombohedral and pyramid faces, the terminal oxygen species are linked to cations in octahedral or tetrahedral coordination [11, 12]. Moreover, the terminal oxygen species with a similar bonding strength may appear near steps formed due to the high-temperature reconstruction of prismatic and basal faces [3, 39].

In stoichiometric MeO oxides of the rock salt structure, the (100) faces are not able to adsorb oxygen without changing the cation charge, though they do contain penta-coordinated cations required for such an adsorption. As a result, the stoichiometric perfect face of NiO is practically inert to reduction by CO or hydrogen at moderate temperatures [14, 15, 40]. For dispersed nickel oxide, such faces are well developed only in samples obtained via plasma thermolysis of nitrate solution [41]. At the same time, in samples obtained via traditional thermal decomposition of nitrate solution, the particles' morphology is mainly determined by (111) faces covered by strongly bound oxygen species probably stabilized by hydroxyl groups [39].

In oxides with a broad range of the surface/bulk nonstoichiometry (cobalt and manganese spinel oxides, nickel oxide), the terminal oxygen species appear on the densely packed faces when cation vacancies are generated due to the oxygen excess [9, 42–44]. This excess oxygen is removed at 400–500 °C.

In transition metal oxides, the most weakly bound oxygen species (heat of adsorption 10–30 kcal/mol) are mainly located on coordinatively unsaturated cluster sites appearing in the vicinity of outlets of extended defects such as dislocations, twins, stacking faults and intergrain boundaries [9–11]. Usually, the density of extended defects is determined by the oxides' preparation procedure. The dislocation network in the surface layers of oxide particles is also generated during phase transitions, when, for example, magnetite phase nuclei emerge during the hematite reduction [9, 11], or when Co_3O_4 transforms to CoO at high temperatures [39, 43]. Such cluster sites may also appear due to reconstruction of the (100) and (110) faces of spinel [45–48] and (111) face of rock salt-like oxides [40] caused by a loss of oxygen and/or hydroxyls. For Co_3O_4 oxide this feature reflected in O_2 Temperature-Programmed Desorption (TPD) spectra is demonstrated in Fig. 2.5. These weakly bound oxygen species determine the activity of some transition metal oxides [9] and perovskites [49–54] in oxidation of CO and hydrocarbons at temperatures up to 400 °C. This trend is demonstrated in Figs. 2.6 and 2.7 for Co_3O_4 and hematite samples in the case of CO catalytic oxidation.

In a more general case, we could expect correlation of catalytic activity in oxidation reactions with the chemical potential (or thermodynamic activity/fugacity) of the surface oxygen species which depends on both the bonding strength and coverage of the surface oxygen species. Indeed, for different phases of manganese oxides this correlation was demonstrated using solid electrolyte (pyrex glass) potentiometry (SEP) [55] (Fig. 2.8). For initial oxidized or reduced samples, activity was estimated in the first pulse of the reaction feed. True steady state of all manganese oxides in CO oxidation reactions corresponds to partially oxidized disordered spinel phase [56]. Clearly reaction media effect has a strong impact in this case as well as for some other oxide systems (Fig. 2.9).

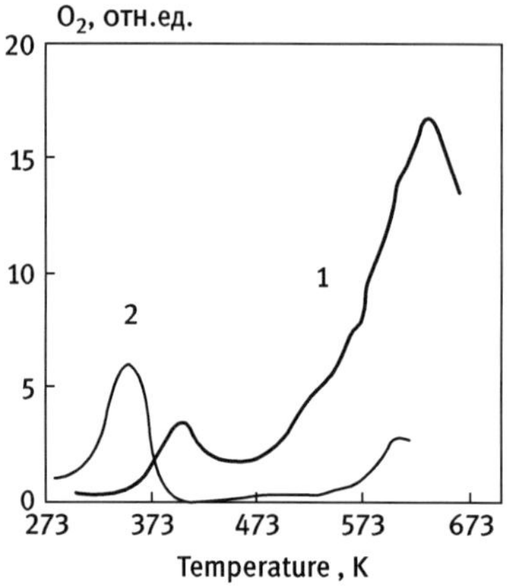

Fig. 2.5: O_2 TPD spectra for Co_3O_4 samples with different pretreatment. 1 – pretreatment in O_2, 2 – pretreatment in He at 400 °C with O_2 adsorption at 25 °C.

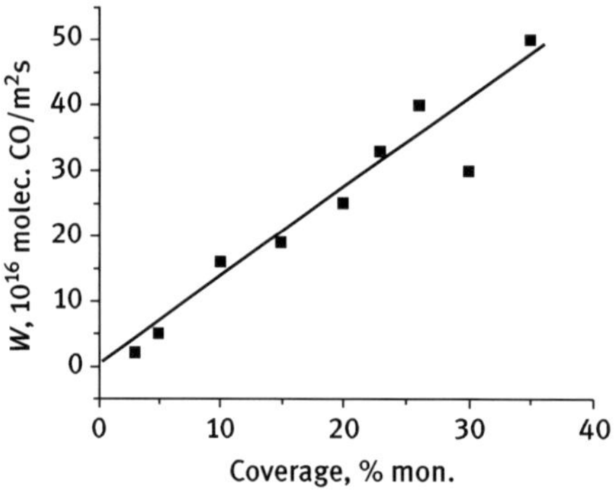

Fig. 2.6: Correlation of Co_3O_4 samples' steady-state activity in CO oxidation with the surface coverage by weakly bound oxygen. Feed 1% CO + 1% O_2 in He, 140 °C.

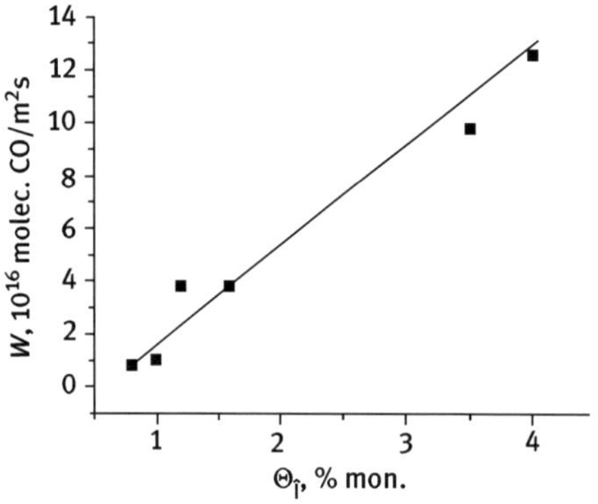

Fig. 2.7: The steady-state rate of CO oxidation 1% CO + 1% O_2 in He, 227 °C vs. the amount of weakly bound oxygen estimated by TPD method for α-Fe_2O_3 samples prepared from goethite.

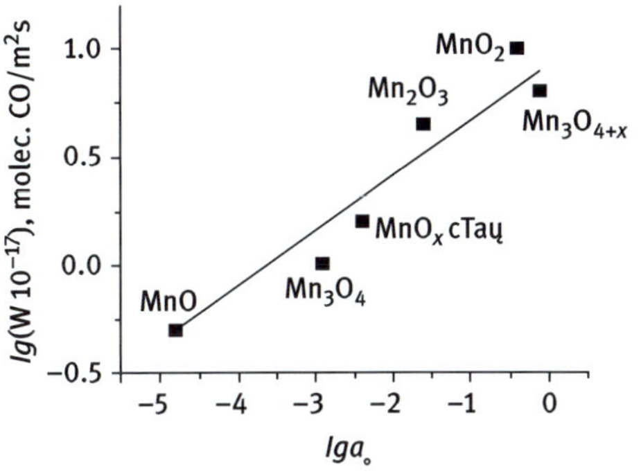

Fig. 2.8: Correlation of manganese oxides' catalytic activity with the surface oxygen activity at 228 °C estimated by SEP method. Feed 1% CO + 1% O_2 in He [55].

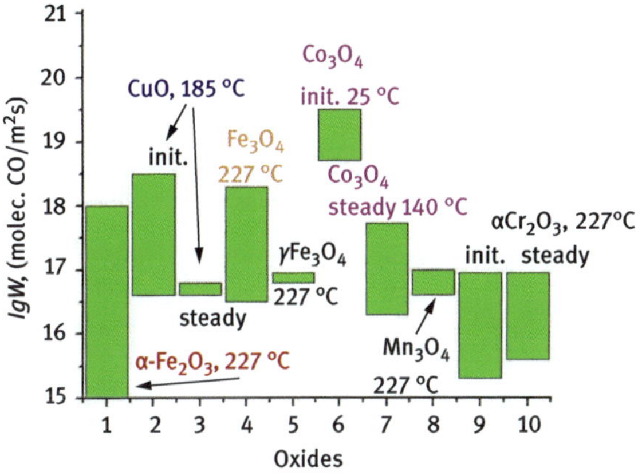

Fig. 2.9: Structural sensitivity in CO catalytic oxidation on simple transition metal oxides.

The highest structural sensitivity is observed for oxides with rigid structures (hematite, etc.). Rearrangement of all Cu–O and Mn–O phases under the reaction media effect into one steady-state phase based on disordered cuprate [57] and spinel [56] structures, respectively, with a moderate catalytic activity was observed.

For perovskite systems, the particles' morphology may be presented by spheres comprised of the cubic micrograins [58–60], platelets including hexagonal ones [58, 59], prisms with sharpened tops, pyramids and octahedrons [61, 62]. These data combined with the results of the selective electron diffraction [58–61, 63] and High-Resolution Electron Microscopy (HREM) [60–62, 64] imply that the surface of particles may be represented by the (100), (110) and (111) faces as indexed in the cubic space group.

Since the perovskite structure is formed by the dense packing of mixed lanthanum–oxygen layers, transition metal cations being located between the layers in oxygen octahedrons, each type face may have a different termination. Thus, along the [100] direction, alternated are layers LaO and MeO_2, along [110]; layers O_4, MeO and $La_4Mn_2O_2$, along [111] and densely packed lanthanum–oxygen layers in the cubic sequence *abc* with the layers of transition metal cations in between (Fig. 2.10).

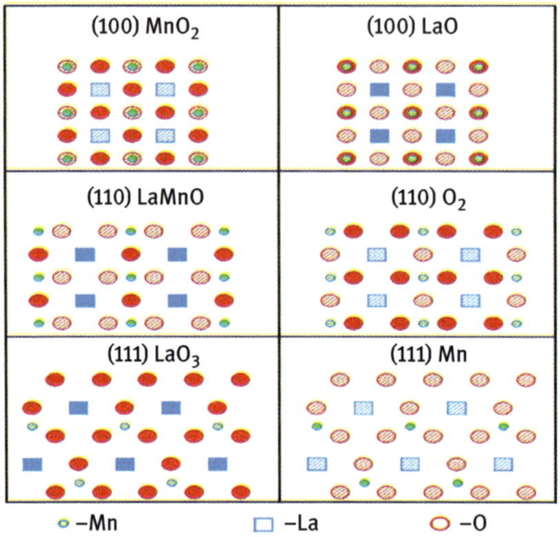

Fig. 2.10: Surface faces of lanthanum manganite. Filled symbols – in surface location; hatched – subsurface location. On-top and bridged oxygen forms can be adsorbed. Surface energies by Interacting Bonds Method (IBM) (kcal/mol): 70 (100) < 110 (111) < 127 (110).

Heat of oxygen adsorption on lanthanum manganite faces is shown in Table 2.3.

Considering the X-ray Photoelectron Spectroscopy (XPS) and Secondary Ions Mass Spectroscopy (SIMS) data [54, 58, 62, 64–66], one may conclude that in most cases the ratio between the lanthanum cations (or sum of lanthanum and alkaline-earth cations for samples with such a substitution in the lanthanum sublattice) and transition metals in the subsurface layer differs insignificantly from that in the bulk, though it may vary depending on the conditions of sample synthesis and pretreatment. Thus, a reductive treatment favors segregation of transition metal cations, whereas an oxidative treatment in humid atmospheres enriches the surface with lanthanum and alkaline-earth cations [62, 64–66]. Hence, the most developed faces of perovskites appear to be represented by all possible terminations with a nearly equal probability.

On any perovskite face, the oxygen atoms bound to lanthanum cations in the regular positions seem to be chemically inert at all temperatures of catalysis. The faces of the (110) type shall be mainly covered by bridged oxygen species bound to either two manganese cations or manganese and lanthanum cations. The (100) faces with MeO_2 composition may contain penta-coordinated cations of transitions metals, which are sites for oxygen adsorption in the terminal MO form. From stoichiometric lanthanum cobaltites, ferrites, nickelates and manganites, such oxygen species are usually desorbed at 700–800 °C [62, 67–70].

O_2 TPD studies (Fig. 2.11) revealed that for lanthanum manganites a small amount of the surface oxygen (up to 5–10% of the monolayer coverage, activation energy of

Table 2.3: Heat of oxygen adsorption (kcal/mol) on the lanthanum manganite faces [62].

Parameter	Face					
	(100)		(110)		(111)	
Face composition	MnO_2	LaO	LaMnO	O_2	LaO_3	Mn
qO_2 O–Mn	**52.6**		65.4			**68.0**
O–La		**71.7**	78.6	45.9	**57.8**	**63.6**
O–La_2 O–La_2Mn O–MnLa		177.4	234.1			160.4

Heat of oxygen adsorption on defect centers with clustered Mn cations is ~35 kcal/mol.

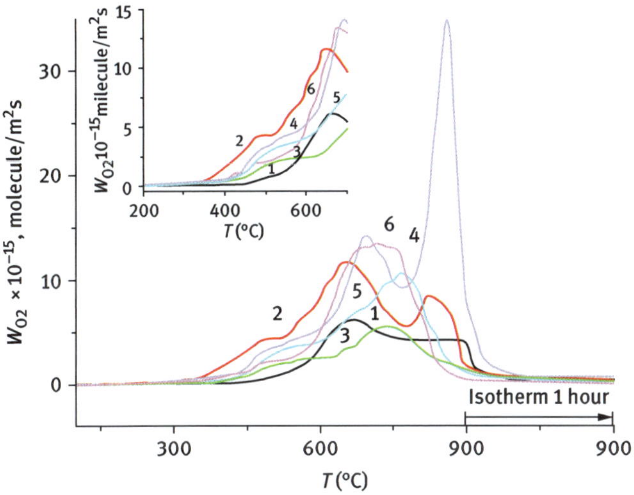

Fig. 2.11: Temperature dependence of the rates of O_2 desorption for LaMn700 (1), LaMn 500 (2), LaCeMn (3), LaFeMn (4), LaBiMn (5) and LaSrBiMn (6).

desorption, $E_a \le 40$ kcal/mol) is removed at temperatures up to 500 °C. A small amount of adsorption centers capable to retain this oxygen allows assigning them to defects (microstrains, intergrain boundaries, etc). For a pure lanthanum manganite sample prepared by the citrate method, the amount of oxygen desorbed at 913–1073 K was found to be ~25% of a monolayer. Hence, this high-temperature peak can be assigned to the oxygen desorption from the regular surface sites combined with simultaneous oxygen flux from the subsurface. Deconvolution of TPD profiles into four peaks allowed us to estimate amount of desorbed oxygen and E_d by analysis of ascending profiles: α form (E_d α 40 kJ/mol), β (80–120 kJ/mol), γ_1 (160–250 kJ/mol) and γ_2 (around 400 kJ/mol), respectively. Catalytic activity in CO oxidation correlates with the amount of weakly bound α oxygen form (Table 2.4).

Table 2.4: Different forms of oxygen (monolayers, *N*) removed by desorption in He or CO TPR and rates of CO oxidation reaction [69, 70].

Sample (*S*)*	By TPD (N)*					By CO TPR (*N*)		Rate of CO oxidation × 10^{-16} (molecule/m^2 s) 180 °C
	α	β	γ_1	γ_2	Total	≤300 °C	Total	
LaMn700 (20)	0.04	0.5	0.5	–	1.0	1.0	9.0	1.4
LaMn (53)	0.3	1.3	–	0.4	2.0	0.9	4.5	1.4
LaCeMn (100)	0.01	0.3	0.6	–	0.9	0.5	2.5	1.8
LaFeMn (48)	0.3	1.5	–	1.1	2.9	0.7	5.0	2.9
LaBiMn (49)	0.01	0.4	1.0	–	1.4	1.3	6.0	2.5
LaSrBiMn (26)	0.7	–	1.2	–	1.9	1.8	9.0	12.9

*S – specific surface area, m^2/g; N – surface coverage in monolayers.

In the perovskite structure, clustering becomes possible only in the vicinity of such extended defects as clusters of oxygen vacancies, twins or intergrowth structures where the local enrichment by transition metal cations occurs [61, 62, 68, 71].

If in perovskite-like oxides some lanthanum cations are replaced by the alkaline-earth cations, various defects are generated depending on the degree of substitution and the nature of transition metal cation. Among them, cation and/or anion vacancies, ordering of modifying cations in some planes, microstrains and other extended defects like microdomain interfaces, intergrowth of phases with different composition and cooperative distortion of lattice, which may even produce phases of other symmetry and/or structure, were revealed [61, 62, 64, 68, 71–74]. Oxygen vacancies appearing due to this partial substitution will affect the catalyst performance only if they emerge on the surface. It certainly takes place for partially substituted lanthanum manganites, ferrites, cobaltites and nickelates, where appearance of oxygen species desorbing at temperatures below 500 °C and probably related to ion-radical O^- species bound with the surface oxygen vacancies is well documented [67, 75]. These radical species can be associated with a new O1s peak in the X-ray photoelectron spectroscopy (XPS) spectra of substituted samples lying at ~2 eV above the level typical for oxides and nonpromoted perovskites [75]. However, at least in part, this peak may also correspond to surface hydroxyls bound to lanthanum cations [74].

If oxygen vacancies appear in the coordination sphere of transition metal cations in perovskites, the oxygen octahedron easily transforms into another polyhedron (tetrahedron, bipyramid, etc.), thus removing the vacancy [60, 61, 71]. Therefore, one may assume that relatively stable surface oxygen vacancies are most likely present in the lanthanum-containing faces. This means that oxygen ion radicals appear to be mainly linked to the subsurface transition metal cations. Hence, a cation substitution produces the active oxygen species on rather inert lanthanum-containing faces of perovskites.

Hence, analysis carried out in this chapter demonstrated that for simple transition metal oxides and complex oxides with perovskite structure, specificity of their particles' morphology as well as defect structure is reflected in a broad variation of the bonding strength of surface oxygen species and their coverage. As a result, specific catalytic activity in oxidation reactions of the same oxide phase can differ substantially in limits controlled by the specificity of its structure and composition leading to the so-called structure sensitivity. Reaction media effect can also vary in a broad range, depending upon the oxide structure rigidity and reaction conditions determining possibility of restructuring of the surface layer of oxides into the steady-state structure not depending on its initial state.

References

[1] Boreskov, G. K. Mechanism of solid catalysts action, Heterogeneous catalysis in chemical industry, Moscow, GNTICHL, 1955, 5–28.

[2] Haber, J. Catalysis and Surface Chemistry of Oxides. In: Proceedings, 6th International Congress on Catalysis, London, 1976 (G. C. Bond, P. B. Walls, and F. C. Tompkins, Eds.). The Chem. Soc., London, 1976, 1, 85–112.

[3] Gonzalez-Gruz, L., Joly, J.-P., & Germain, J.-E. Fast thermodesorption of oxygen from metallic oxides. J. Chim. Phys. Phys.-Chim. Biol. 1978, 75, 324–328.

[4] Halpern, B., & Germain, J. E. Thermodesorption of oxygen from powdered transition metal oxide catalysts. J. Catal. 1975, 37, 44–56.

[5] Sadykov, V. A., Razdobarov, V. A., Veniaminov, S. A., Bulgakov, N. N., Kovalenko, O. N., Pankratyev, Yu. D., Popovskii, V. V., Kryukova, G. N., & Tikhov, S. F. Nature of the active oxygen of Co_3O_4. React. Kinet. Catal. Lett. 1988, 37, 109–114.

[6] Joly, J.-P. Measurement of the heats of desorption of oxygen from metal oxides. Isochore method and Langmuir's isotherm. J. Chim. Phys. Phys.-Chim. Biol. 1975, 72, 1013–1018.

[7] Babenko, V. S., Aleksandrov, V. Yu., Buyanov, R. A., Popovslii, V. V., & Afanasiev, A. D. Energy state of surface oxygen in iron oxide catalysts promoted by potassium. Izv. SO AN SSSR. Ser. Khim. Nauk 1981, 3, 76–80.

[8] Sadykov, V. A., Tikhov, S. F., Popovskii, V. V., & Pankratyev, Yu. D. Binding energy and reactivity of surface oxygen of cobalt molybdates. React. Kinet. Catal. Lett. 1987, 34, 401–406.

[9] Sadykov, V. A., Tikhov, S. F., Tsybulya, S. V., Kryukova, G. N., Veniaminov, S. A., Kolomiichuk, V. N., Bulgakov, N. N., Isupova, L. A., Paukshtis, E. A., Zaikovskii, V. I., Kustova, G. N., & Burgina, E. B. Structural Sensitivity of the Oxidation Reactions Catalyzed by Dispersed

Transition Metal Oxides: Role of Defect Structure (R. K. Grasselli, S. T. Oyama, A. M. Gaffney, and J. E. Lyons, Eds.). 3rd World Congress Oxidation Catalysis, San Diego, CA, 21–26. September 1997 (Stud. Surf. Sci. Catal., v. 110), Elsevier, Amsterdam, 1997, 1155–1164.

[10] Sadykov, V. A., Isupova, L. A., & Bulgakov, N. N. Effect of mechanical activation on the bulk and surface defect structure and reactivity of some transition metal oxides. Chem. Sustainable Dev. 1998, 6, 215–222.

[11] Bulgakov, N. N., & Sadykov, V. A. Surface energies of hematite faces and heats of oxygen adsorption: calculations by modified semiempirical interacting bonds method. React. Kinet. Catal. Lett. 1996, 58, 397–402.

[12] Sadykov, V. A., Isupova, L. A., Tsybulya, S. V., Cherepanova, S. V., Litvak, G. S., Burgina, E. B., Kustova, G. N., Kolomiichuk, V. N., Ivanov, V. P., Paukshtis, E. A., Golovin, A. V., & Avvakumov, E. G. Effect of mechanical activation on the real structure and reactivity of iron (III) oxide with corundum-type structure. J. Solid State Chem. 1996, 123, 191–202.

[13] Tikhov, S. F., Sadykov, V. A., Kryukova, G. N., Paukshtis, E. A., Popovskii, V. V., Starostina, T. G., Anufrienko, V. F., Razdobarov, V. A., Bulgakov, N. N., & Kalinkin, A. V. Microstructural and spectroscopic investigations of the supported copper-alumina oxide system. Nature of aging in oxidizing reaction media. J. Catal. 1992, 134, 506–524.

[14] Henrich, V. E. Electronic and Geometric Structure of Defects on Oxides and Their Role in Chemisorptions (J. Novotny and L.-C. Dufour, Eds.) Surface and Near-Surface Chemistry of Oxide Materials, Materials Sci. Monographs,Amsterdam, Elsevier, 1988, 47, 23–60.

[15] Brown, G. E., Henrich, V. E., Casey, W. H., Clark, D. L., Eggleston, C., Felmy, A., Wayne Goodman, D., Grätzel, M., Maciel, G., McCarthy, M. I., Nealson, K. H., Sverensky, D. A., Toney, M. F., & Zachara, J. M. Metal oxide surfaces and their interactions with aqueous solutions and microbial organisms. Chem. Rev. 1999, 99, 77–174.

[16] Zecchina, A., Scarano, D., Bordiga, S., Ricchiardi, G., Spoto, G., & Geobaldo, F. IR studies of CO and NO adsorbed on well characterized oxide single microcrystals. Catal. Today 1996, 27, 403–435.

[17] Lunina, E. V., Markaryan, G. L., Parenago, O. O., & Fionov, A.V. Nitroxide complexes with the solid Lewis acids. Colloids Surf., A 1993, 72, 333–343.

[18] Bucket, M. I., & Marks, L. D. Electron irradiation damage in NiO. Surf. Sci. 1990, 232, 353–366.

[19] Smith, D. J., Bursill, L. F., & Jefferson, D. F. Atomic imaging of oxide surfaces: I. General features and surface rearrangements. Surf. Sci. 1986, 175, 673–683.

[20] Kryukova, G. N., Chuvilin, A. L., & Sadykov, V. A. Surface structure of the oxide catalyst microcrystals: high resolution electron microscopy study, in Atomic Scale Imaging of Surfaces and Interfaces. Mat. Res. Soc. Meet. Symp. Ser., Boston 1993, 295, 179–182.

[21] Jacobs, J.-P., Maltha, A., Reintjers, J. G. H., Drimal, J., Ponec, V., & Brongersma, H. H. The surface of catalytically active spinels. J. Catal. 1994, 147, 294–300.

[22] Beafils, J. P., & Barbaux, Y. Determination of the preferential exposed plane of powdered supports of catalysts by differential neutron diffraction. J. Chim. Phys. Phys.-Chim. Biol. 1981, 78, 347–352.

[23] Beafils, J. P., & Barbaux, Y. Study of adsorption on powders by surface differential diffraction measurements. Argon on cobalt oxide (Co3O4). J. Appl. Crystallogr. 1982, 15, 301–307.

[24] Ziolkowski, J., & Barbaux, Y. Identification of sites active in oxidation of 1-butene to butadiene and carbon dioxide on tricobalt tetroxide in terms of the crystallochemical model of solid surfaces. J. Molec. Catal. 1991, 67, 199–215.

[25] Catlow, C. R. A., Ackerman, L., Bell, R. G., Gay, D. H., Holt, S., Lewis, D. W., Nygren, M. A., Sastre, G., Sayle, D. C., & Sinclair, P. E. Modelling of structure, sorption, synthesis and reactivity in catalytic systems. J. Molec. Catal. A 1997, 115, 431–448.

[26] Van Santen, R. A., & Neurock, M. Concepts in theoretical heterogeneous catalytic reactivity. Catal. Rev.-Sci. Eng. 1995, 37, 557–698.
[27] Manassidis, I., & Gillan, M. J. Structure and energetics of alumina surfaces calculated from first principles. J. Am. Ceram. Soc. 1994, 77, 335–338.
[28] Gautier, M., Renaud, G., Pham Van, L., Vilette, B., Pollak, M., Thromat, N., Jolloet, F., & Duraud, J.-P. α-Al_2O_3 (0001) surfaces: atomic and electronic structure. J. Am. Ceram. Soc. 1994, 77, 323–334.
[29] Streitz, F. H., & Mintmire, J. W. Electrostatic potentials for metal-oxide surfaces and interfaces. Phys. Rev. B. 1994, 50, 11996.
[30] Ellis, D. E., Guo, J., & Lam, D. J. Cluster models of bulk, surface, and impurity structure in α-alumina. J. Am. Ceram. Soc. 1994, 77, 398–403.
[31] Guo, J., Ellis, D. E., & Lam, D. J. Electronic structure and energetics of sapphire (0001) and (11¯02) surfaces. Phys. Rev. B 1992, 45, 13647.
[32] Mackrodt, W. C., Davey, R. J., & Black, S. N. The morphology of α-alumina and α-ferric oxide: the importance of surface relaxation. J. Cryst. Growth 80, 1987, 441–446.
[33] Hass, K. C., Schneider, W. F., Curioni, A., & Andreoni, W. The chemistry of water on alumina surfaces: reaction dynamics from first principles. Science 1998, 282, 265–268.
[34] Wang, X.-G., Weiss, W., Shaikhutdinov, Sh. K., Ritter, M., Petersen, M., Wagner, F., Schlögel, R., & Scheffer, M. The Hematite (α-Fe_2O_3) (0001) surface: evidence for domains of distinct chemistry. Phys. Rev. Lett. 81, 1998, 1038–1041.
[35] Condon, N. G., Murray, P. W., Leibsle, F. M., Thornton, G., Lennie, A. R., & Vaughan, D. J. Fe_3O_4 (111) termination of α-Fe_2O_3 (0001). Surf. Sci. 1994, 310, L609–L613.
[36] Sadykov, V. A., Kryukova, G. N., Tikhov, S. F., Popovskii, V. V., & Bulgakov, N. N. On the structure of the surface active centers on CuO. React. Kinet. Catal. Lett. 1986, 32, 123–127.
[37] Kuznetsov, V. I., Sadykov, V. A., Razdobarov, V. A., & Klimenko, A. G. The structural features of cobalt oxides: Fe57 Mossbauer spectroscopy, TEM, and static magnetic susceptibility measurements. J. Solid State Chem. 1993, 104, 412–421.
[38] Kryukova, G. N., Tsybulya, S. V., Solovyeva, L. P., Sadykov, V. A., Litvak, G. S., & Andrianova, M. P. Effect of heat treatment on microstructural evolution of the haematite derived from synthetic goethite. Materials Sci. Eng. A 1991, 149, 121–127.
[39] Kryukova, G. N. Studies of the defect structure of oxide catalysts by transmission electron microscopy, Ph.D. Thesis, Boreskov Institute of Catalysis, Novosibirsk, 1990.
[40] Schönnenbeck, M., Cappus, D., Klinkmann, J., Freund, H.-J., Petterson, L. G. M., & Bagus, P. S. Adsorption of CO and NO on NiO and CoO: a comparison. Surf. Sci. 1996, 347, 337–345.
[41] Sadykov, V. A., Kryukova, G. N., Tsybulyev, P. N., & Voronin, P. N. Plasmochemical Method of Oxide Catalysts Preparation: New Morphological and Structural Properties, in CHISA-90. Proc. 10th Int. Congr. Chem. Eng., Chem. Equipment, Design and Autom., Czech., Praha, 1990, 695.
[42] Angelov, S., Zhecheva, E., & Mehandjiev, D. Dispersity, structure and magnetic properties of polycrystalline cobalt (II, III) oxide. Comm. Dept. Chem. 1980, 13, 369.
[43] Belova, D., Shalaginov, V. V., Galyamov, B. Sh., Roginskaya, Yu. E., & Shub, D. M. Study of the defect structure of nonstoichiometric cobalt oxide (Co_3O_4) films. Zh. Neorg. Khim. 1976, 23, 286.
[44] Tzibirova, F. K., Belova, I. D., Zaharin, D. S., Reinman, S. I., Roginskaya, Yu. E., & Venevtsev, Yu. I. High-spin cobalt (III) in cobalt oxide ($Co_{3-x}O_4$) defect films, Fiz. Tv. Tela (Solid State Physics) 1984, 26, 890–893.
[45] Jansen, R., Brabers, V. A. M., & van Kempen, H. One-dimensional reconstruction observed on Fe_3O_4 (110) by scanning tunneling microscopy. Surf. Sci. 1995, 328, 237.

[46] Wiesendanger, R., Shvets, I. V., Burgler, D., Tarrach, G., Guntherodt, H. J., Coey, J. M. D., & Graser, S. Topographic and magnetic-sensitive scanning tunneling microscope study of magnetite. Science 1992, 255, 583–586.

[47] Gaines, J. M., Bloemen, P. J. H., Kohlhepp, J. T., Bulle-Lieuwma, C. W. T., Wolf, R. M., Reinders, A., Jungblut, R. M., van der Heiden, P. A. A., van Eemeren, J. T. W. M., van de Stegge, J., & de Jonge, W. J. M. An STM study of Fe_3O_4(100) grown by molecular beam epitaxy. Surf. Sci. 1997, 373, 85–94.

[48] Tarrach, G., Burgler, D., Schaub, T., Wiesendanger, R., & Guntherodt, H. J. Atomic surface structure of magnetite (Fe_3O_4) (001) in different preparation stages studied by scanning tunneling microscopy. Surf. Sci. 1993, 285, 1–14.

[49] Isupova, L. A., Sadykov, V. A., Avvakumov, E. G., & Kosova, N. V. Mechanochemical activation in technology of high-temperature oxide catalysts. Chem. Sustainable Dev. 1998, 6, 207.

[50] Pauli, I. A., Avvakumov, E. G., Isupova, L. A., Sadykov, V. A., & Poluboyarov, V. A. Effect of mechanochemical activation on synthesis and catalytic properties of lanthanum cobaltite. Sib. Khim. Zh. 1992, 3, 133.

[51] Sadykov, V. A., Isupova, L. A., Tikhov, S. F., & Kimkhai, O. N. Perovskite catalysts: high-surface-area powders synthesis, monolith shaping and high-temperature application, in Synthesis and Properties of Advanced Ceramic Materials. Mat. Res. Soc. Symp. Ser., MRS Fall Meeting, 1994, Boston, 1995, 386, 293.

[52] Isupova, L. A., Sadykov, V. A., Tikhov, S. F., Kimkhai, O. N., Ya, A., Rozovskii, V. F., Tretyakov, V. V., & Lunin, Monolith. Perovskite Catalysts for Environmentally Benign Fuels Combustion and Toxic Wastes Incineration (G. Centy, C. Cristiani, P. Forzatti, and S. Perathoner, Eds.). Environmental Catalysis for a Better World and Life, Proceedings 1st World Congress on Envir. Catal., Pisa, Italy, 1995, Societa, Chimica Italiana, Rome, Italy, 167.

[53] Isupova, L. A., Sadykov, V. A., Solovyeva, L. P., Andrianova, M. P., Ivanov, V. P., Kryukova, G. N., Kolomiichuk, V. N., Avvakumov, E. G., Pauli, I. A., Andryushkova, O. V., Poluboyarov, V. A., Rozovskii A. Ya., Tretyakov V. F. Monolith perovskite catalysts of honeycomb structure for fuel combustion (G. Poncelet, J. Martens, B. Delmon, P. A. Jacobs, and P. Grange, Eds.). Preparation of Catalysts VI, Stud. Surf. Sci. Catal., Amsterdam, Elsevier, 1995, 91, 637–645.

[54] Isupova, L. A., Sadykov, V. A., Ivanov, V. P., Rar, A. A., Tsybulya, S. V., Andrianova, M. P., Kolomiichuk, V. N., Petrov, A. N., & Kononchuk, O. F. Catalytic and physico-chemical properties of $La_{1-x}SrxCoO_3$ perovskites. React. Kinet. Catal. Lett. 1994, 53, 223–229.

[55] Sadykov, V. A., Tsyrulnikov, P. G., Popovskii, V. V., & Tikhov, S. F. Interaction of manganese oxides with reaction media during heterogeneous-catalytic oxidation of carbon monoxide by molecular oxygen. II. Electrode potentials of manganese oxides. Kinet. Catal. 1981, 22, 722–731.

[56] Sadykov, V. A., Tsyrulnikov, P. G., Popovskii, V. V., & Tikhov, S. F. Interaction of manganese oxides with reaction media during heterogeneous-catalytic oxidation of carbon monoxide by molecular oxygen. III. Studies of the steady-state surface layer formation. Kinet. Catal. 1981, 22, 1219–1226.

[57] Sadykov, V. A., Tikhov, S. F., & Popovskii, V. V. Interaction of copper oxides with reaction media during heterogeneous-catalytic oxidation of carbon monoxide by molecular oxygen. V. Formation of the stationary state of the surface layer of copper oxides. Kinet. Catal. 1986, 27, 147–155.

[58] Tikhov, S. F., Sadykov, V. A., Pak, E. A., Kimkhai, O. N., Moroz, E. M., Ivanov, V. P., Kustova, G. N., & Alikina, G. M. The Influence of the Genesis on Catalytic and Physico-chemical Properties of Mn Lanthanides (L. Petrov and G. Kadinov, Eds.). Heterog. Catalysis, Proc. 7th Int. Symp., Bourgas, 29 Sept.–3 Oct. 1991, Vratza, Sofia, 1991, part I, 423.

[59] Traversa, E., Nunziante, P., Sakamoto, M., Sadaoka, Y., Carotta, M. C., & Martinelli, G. Thermal evolution of the microstructure of nanosized $LaFeO_3$ powders from the thermal decomposition of a heteronuclear complex, $La[Fe(CN)6]·5H_2O$. J. Mater. Res. 1998, 13, 1335–1344.

[60] Faaland, S., Knudsen, K. D., Einarsrud, M.-A., Rørmark, L., Høier, R., & Grande, T. Structure, stoichiometry, and phase purity of calcium substituted lanthanum manganite powders. J. Solid State Chem. 1998, 140, 320–330.

[61] Grenier, J.-C., Pouchard, M., & Hagenmuller, P. Vacancy Ordering in Oxygen-Deficient Perovskite-Related Ferrites, in Structure and Bonding, Berlin–Heidelberg–New York, Springer-Verlag, 1981, 47, 2 (1–25).

[62] Sadykov, V. A., Bulgakov, N. N., Muzykantov, V. S., Kuznetsova, T. G., Alikina, G. M., Lukashevich, A. I., Potapova, Yu. V., Rogov, V. A., Burgina, E. B., Zaikovskii, V. I., Moroz, E. M., Litvak, G. S., Yakovleva, I. S., Isupova, L. A., Zyryanov, V. V., Kemnitz, E., & Neophytides, S. Mobility and Reactivity of the Surface and Lattice Oxygen of Some Complex Oxides with Perovskite Structure, "Mixed Ionic Electronic Conducting Perovskites for Advanced Energy Systems" (N. Orlovskaya and N. Browning, Eds.). Boston/Dordrecht/London, Kluwer Academic Publishers, 2004, 49–70.

[63] Sadykov, V. A., Tikhov, S. F., Tsybulya, S. V., Kryukova, G. N., Veniaminov, S. A., Kolomiichuk, V. N., Bulgakov, N. N., Paukshtos, E. A., Ivanov, V. P., Koshcheev, S. V., Zaikovskii, V. I., Isupova, L. A., & Burgina, E. B. Role of defect structure in structural sensitivity of the oxidation reactions catalyzed by dispersed transition metal oxides. J. Molec. Catalysis: A: Chem. 2000, 158, 361–365.

[64] Isupova, L. A., Tsybulya, S. V., Kryukova, G. N., Rogov, V. A., Yakovleva, I. S., & Sadykov, V. A. Microheterogeneous Solid Solutions in Perovskites: Formation, Microstructure, and Catalytic Properties, "Mixed Ionic Electronic Conducting Perovskites for Advanced Energy Systems" (N. Orlovskaya and N. Browning Eds.). Boston/Dordrecht/London, Kluwer Academic Publishers, 2004, 143–162.

[65] Tabata, K., Hirano, Y., & Suzuki, E. XPS studies on the oxygen species of $LaMn_{1-x}CuxO_{3+\lambda}$. Appl. Catal. A: General 1998, 170, 245–254.

[66] Lin, J., Wee, A. T. S., Tan, K. L., Neoh, Koon Gee., & Teo, Wah Koon. XPS/FTIR study of the interaction between nitric oxide and yttrium barium copper oxide ($YBa_2Cu_3O_7$). Inorg. Chem. 1993, 32, 5522.

[67] Misono, M., & Nitadori, T. Redox and Catalytic Properties of Perovskite-Type Mixed Oxides. A Comparative Study of $Ln_{1-x}SrxBO_3$ (Ln=Rare earth, B=Co, Fe, Mn), Adsorption and Catalysis on Oxide Surfaces, Studies in Surface Science and Catalysis, Amsterdam,Elsevier, 1985, 21, 409–419.

[68] Yamazoe, N., & Teraoka, Y. Oxidation catalysis of perovskites – relationships to bulk structure and composition (valency, defect, etc.). Catal. Today 1990, 8, 175–199.

[69] Kuznetsova, T., Sadykov, V., Batuev, L., Moroz, E., Burgina, E., Rogov, V., Kurina, L., & Neophytides, S. Structural features and the lattice oxygen reactivity of low-temperature lanthanum manganites doped with different cations. React. Kinet. Catal. Lett. 2005, 86, 249–256.

[70] Kuznetsova, T., Sadykov, V., Batuev, L., Kurina, L., & Neophytides, S. Effect of the surface/bulk doping of lanthanum manganite on the oxygen mobility, reactivity and catalytic activity in the CO oxidation. React. Kinet. Catal. Lett. 2005, 86, 257–266.

[71] van Roosmalen, J. A. M., & Cordfunke, E. H. P. A new defect model to describe the oxygen deficiency in perovskite-type oxides. J. Solid State Chem. 1991, 93, 212–219.

[72] Petunchi, J. O., & Lombardo, E. A. The effect of bulk and surface reduction upon the catalytic behavior of perovskite oxides. Catal. Today 1990, 8, 201–219.

[73] Karppinen, M., Yamaguchi, H., Suematsu, H., Isawa, K., Nagano, M., Itti, R., & Fukunaga, O. Control on the copper valence and properties by oxygen content adjustment in the $LaCuO_{3-y}$ System (0≤y≤0.5). J. Solid State Chem. 1997, 130, 213–222.
[74] Fierro, J. L. G. Structure and composition of perovskite surface in relation to adsorption and catalytic properties. Catal. Today 1990, 8, 153–174.
[75] O'Connell, M., Norman, A. K., Hüttermann, C. F., & Morris, M. A. Catalytic oxidation over lanthanum-transition metal perovskite materials. Catal. Today 1999, 47, 123–132.

3 Structure of the nearest environment, spatial distribution and chemical properties of coordinatively unsaturated cations on the surface of transition metal oxides

Specific activity of transition metal oxides in catalysis of redox reactions and even reaction mechanism could strongly depend on their real structure and surface composition. The literature provides various descriptions of the possible surface structure of such oxides, which are summarized in some reviews [1–3]. The idealized description often includes a cross-section of the oxide lattice along the most developed faces of oxides according to the principal component analysis (PCA) [1, 2, 4]. In the most general form, irrespective of the stoichiometry and lattice type, a model of the surface and the subsurface layer is displayed in Fig. 3.1, similar to [1, 3, 5]. In this model, nonstoichiometry of the oxide manifests itself as the presence of anionic vacancies uniformly distributed over the oxide surface and in its bulk. The oxygen coordination of cations is determined by the cross-section and the type of idealized lattice. In some cases, the presence of many coordinatively unsaturated "half-naked" cations on the surface is assumed [1, 3, 4]. In addition to PCA data, these models are based on spectral data (low-energy electron diffraction (LEED), X-ray photoelectron spectroscopy (XPS), low energy electron loss spectroscopy (LEELS), Auger spectroscopy) as well as scanning tunneling microscopy [1–4]. In spite of their high resolution, the indicated methods perform mostly the averaged analysis of the oxide surface. All the effects related to nonstoichiometry and coordinatively unsaturated cations of the oxide surface show up as relatively small deviations from the principal spectra and microimages.

In this respect, it seems promising to use the infrared (IR) spectroscopy of adsorbed molecules. This method is widely employed to investigate features of intermediates on the surface of oxides and analyze mechanisms of catalytic reactions [6, 7]. A model of atomic structure of the surface on which these intermediates are coordinated is commonly oversimplified. Therewith, the surface structure (e.g., the distribution of coordinatively unsaturated cations) can play an important role in determining a possible mechanism of redox reactions. Thus, the redox Mars–van Krevelen mechanism implies a uniform distribution of anionic vacancies (coordinatively unsaturated cations) over the catalyst surface. For the Langmuir–Hinshelwood adsorption mechanism, such cations should be in close vicinity to each other, as in the case of metals [5].

The spectroscopy of simple CO and NO probe molecules is able to selectively acquire information concerning exactly the coordinatively unsaturated cations on the surface: their oxidation state, electronic state and composition of the nearest environment [8, 9]. The indicated approach has been widely applied in the 1980s, but later became less popular. This is related to the fact that, in distinction to

https://doi.org/10.1515/9783110587777-003

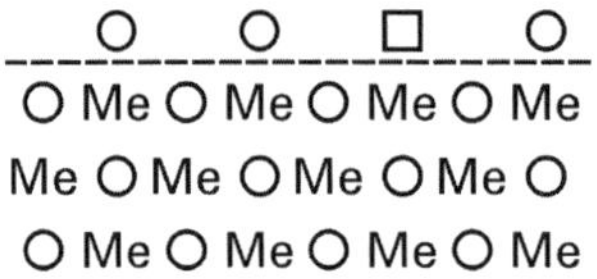

Fig. 3.1: A typical scheme of the oxide surface. Side view: O – oxygen anion, □ – oxygen vacancy, – – – – the surface of solid oxide.

nuclear magnetic resonance (NMR) or X-ray absorption fine structure (EXAFS), the vibrations of chemosorbed CO and NO molecules per se cannot directly provide information on the structure of the nearest environment. For reliable interpretation of spectra in a narrow range of absorption band frequencies, it is necessary to have a large body of data on both the spectra and the properties of oxides themselves with a wide variation of conditions of their mild treatment, which allows changing the reduction degree (the concentration of anionic vacancies) of the oxide surface and conditions of spectra recording (CO pressure and temperature). To reveal a relation of these sites with the activity, the activities should be compared for similar treatments under steady and unsteady conditions. In this section, analysis of such information for some simple transition metal oxides is carried out.

3.1 Copper oxide

Although CuO (tenorite) has the monoclinic lattice whose structure differs from lattices of the majority of oxides, the use of CO molecule for surface probing is very popular because a wide set of peaks is observed in the region of ~2,100–2,150 cm^{-1}, which is typical for Cu^{1+} –CO complexes [8, 10–12]. For these complexes, ν_{CO} and ΔH_{ads} change symbately owing to the mutual enhancement of π–σ binding [11, 13]. Cu^0CO complexes are characterized by the bands with ν_{CO} < 2,100 cm^{-1} and a lower adsorption heat. For Cu^{2+}CO complexes with ν_{CO} > 2,180 cm^{-1}, the adsorption heats are even lower. In addition, a different dependence is typical for such complexes. Exactly the difference in adsorption heats allows separating the indicated complexes in the boundary spectral regions [14]. The presence of many bands in the region of ~2,100–2,150 cm^{-1} testifies to high differentiation between types of coordinatively unsaturated copper cations on the oxide surface, which is caused by a change in the effective charge from 1 – δ to 1 + δ. In its turn, such changes may be caused by the nearest environment (coordination) of copper cations [11]. Any reliable quantum-chemical models able to strictly describe relations between ν_{CO}, effective charge of cation and the structure of its nearest environment are still absent [8]. More real is the approach with the analysis of changes in the spectra upon various mild redox treatments of the oxide surface, which takes into account the energy of the surface oxygen (see Chapter 2) and the surface chemistry of copper oxides [4, 15, 16]. In principle, the approach based on a detailed analysis of chemical properties has allowed chemists in the nineteenth century to reveal quite successfully the structure of methane and benzene without direct spectral information about these molecules.

The maximum oxidation state of the surface (the minimum concentration of coordinatively unsaturated cations) is provided by high-temperature (h.t.) treatment in oxygen with subsequent cooling to room temperature (r.t.) and evacuation. In this case, IR spectra of CO adsorbed at 300 K have only one band in the region of ~2,120 cm^{-1} (Fig. 3.2) (see also [17]), which corresponds to reduced Cu^{1+} cations [11, 17]. For such oxidized samples, the bands corresponding to Cu^{2+} were not observed even at 163 K, where they could be expected [17] taking into account their low adsorption heat. Such sites appeared exactly upon desorption of weakly bound oxygen at r.t. but not upon interaction with CO, because the bands were not detected in the absorption region of carbonate–carboxylate complexes (CCC) [11, 17]. The occurrence of reduced sites is especially remarkable because copper oxide is a p-semiconductor with the excess oxygen stoichiometry $CuO_{1+\delta}$. This experimental fact, which was obtained with two oxides produced by different methods [11, 17], allows

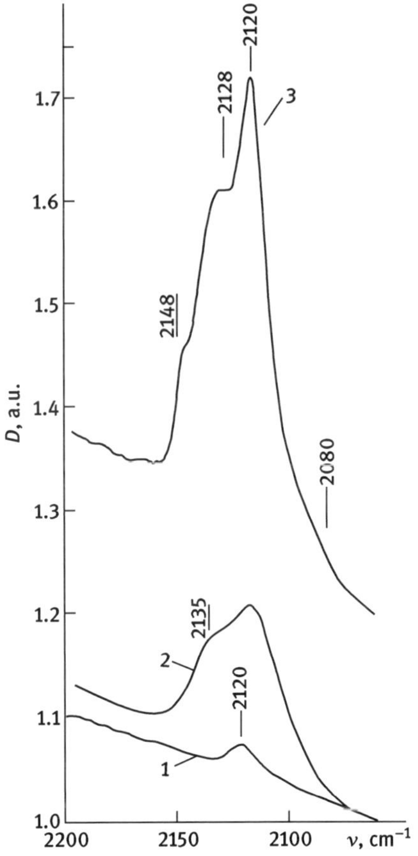

Fig. 3.2: IR spectra of adsorbed CO (10 Torr, 300 K) after different types of high-temperature treatment: (1) O_2 (573 K); (2) vacuum (573 K); (3) CO (573 K, ~ 60 s).

concluding that the model of oxide displayed in Fig. 3.1 is unreliable because it excludes the existence of such sites after h.t. oxidative treatment.

Most researchers analyze IR spectra of adsorbed molecules after h.t. treatment in vacuum [8–11, 17], which enhances oxygen desorption from the surface of oxides. According to the model in Fig. 3.1, such treatment should increase the concentration of reduced sites and decrease the effective charge of cations. Indeed, the intensity of the 2,120 cm^{-1} band substantially increases after vacuum treatment. However, additionally there appear not the more reduced sites with a weaker band but, on the contrary, the more oxidized sites of CO adsorption with the band at ~2,135 cm^{-1} (Fig. 3.2) [17]). At 163 K, the bands at 2,180–2,190 cm^{-1} are observed, which are typical of Cu^{2+} [17]. It seems even more remarkable that evacuation activates (!) the oxide surface in comparison with treatment in oxygen. As a result, CO can be oxidized by the surface oxygen already at r.t. This shows up as the appearance of CO oxidation products, which is reflected by the bands corresponding to CCC [10, 11, 17].

It should be noted that some details of the spectra may differ depending on the oxide origin and conditions of its treatment and spectra recording. For example, in [16, 17] evacuation was carried out at 573 K and spectra were recorded at a CO pressure of 12–14 Torr. In [10], after vacuum treatment at 710 K and CO pressure 100 Torr, the band at 2,115 cm^{-1} was observed.

A short-term h.t. treatment in CO results in appearance of sites more reduced than those with the band at 2,120 cm^{-1} (Fig. 3.2, shoulder at 2,080 cm^{-1}). However, there appears also the band at ~2,148 cm^{-1} corresponding to even more oxidized Cu^{1+} ions (Fig. 3.2). The oxidized sites were observed also in [10] after strong vacuum treatment. Thus, the surface sites of copper oxide differ by their adsorption features, which is not reflected in the model of Fig. 3.1. Such nonuniformity may be caused only by the structural features of surface sites, which determine coordination of copper cations by oxygen ions. Sites differing from the structure of ideal faces of copper oxide may be represented by the step formed on high-index faces, joints of different faces of crystallites having finite sizes, and steps at the intersection of dislocations, twins and other extended defects with the surface [1, 18]. The concentration of such coordinatively unsaturated cations on the surface is not high (~10^{11}–10^{13} site/cm^2) [11, 19]. This confirms essential difference of coordinatively unsaturated cations observed in IR spectra of adsorbed CO from the cations of ideal faces [4] for which the concentration of copper ions should be comparable with the total concentration of copper on the surface (~10^{15} at/cm^2). The results are summarized in Table 3.1.

The chemical behavior of the copper oxide surface after treatments 3 and 2 differs from that after treatment 1. After treatment 1, the band at 2,120 cm^{-1} does not change with time at r.t. (Fig. 3.2). After treatment 2, holding in CO at r.t. for 12 h decreases intensity of the band at 2,125–2,135 cm^{-1}, whereas intensity of more reduced sites at 2,120 cm^{-1} increases (Fig. 3.3a). After treatment in CO, all the bands increase in intensity, including bands corresponding to Cu^0CO complexes

Table 3.1: Main characteristics of CO adsorption sites on the CuO surface.

Type of site	ν_{CO}, cm^{-1}	ΔH_{ads}, kJ/mol	Charge of Cu^{n+} adsorption site	Ref.
1	~2,115–2,120	~90	1	10–12
2	~2,125–2,130	~100	1	10–12
3	~2,135–2,148	~110	1	10–12
4	~2,110	~80	1	11
5	~2,085–2,105	<80	0	10–12
6	~2,180–2,190	$\ll 80$	2	10,17

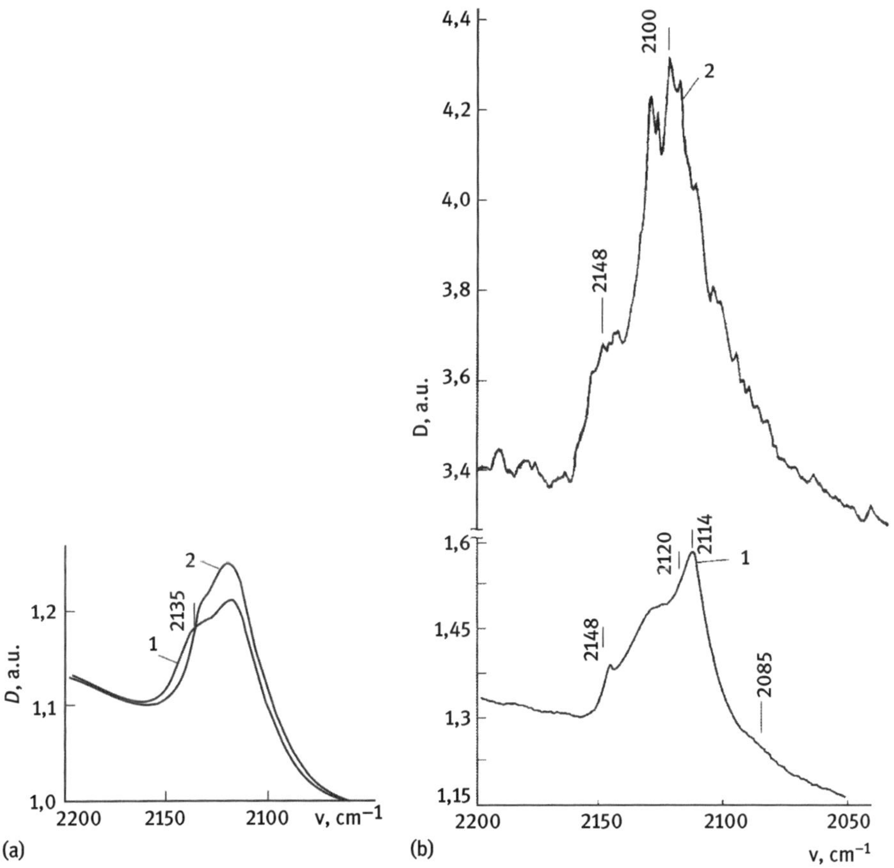

Fig. 3.3: IR spectra of adsorbed CO (10 Torr, 300 K) after different types of h.t. reduction treatment: 1– 10–15 min of exposure; 2–12 h of exposure. (a) vacuum, (b) CO.

(ν_{CO} 2,080, 2,105) (Fig. 3.3b). It means that in comparison with h.t. oxygen treatment, vacuum and CO treatments essentially activate the surface for its further reduction. Therewith, as shown in [18], h.t. treatment in helium (an analog of vacuum treatment) increases the unsteady activity in the pulse mode toward CO oxidation by more than an order of magnitude. The reduction in CO after removal of a certain oxygen amount results in a topochemical growth of the surface reduction rate caused by the growing size of the new phase nuclei (Cu_2O or Cu^0) [4, 18].

Of particular interest is activation of the copper oxide surface after vacuum treatment. Evidently, it is related not only to the increased degree of surface reduction within the homogeneous model (Fig. 3.1). If the surface reduction and analysis of changes in specific catalytic activity are performed at the same temperature, the activity will decrease. However, if the CuO surface is reduced at 140 °C and activity is measured at r.t., the activity will increase in the initial period [18]. Hence, h.t. vacuum treatment is accompanied by the formation of reduced metastable structures on the surface, which include also the subsurface layers and freeze when the temperature is rapidly lowered. Exactly these structures provide the pronounced low-temperature activity of copper oxide. They are gradually destroyed upon contact with the reaction medium, thus decreasing the specific activity [20]. Taking into account a low total concentration of the involved cations, the indicated structures are very small. Most probably, they have the subnano (cluster) size and involve several cations. These structures serve as the nucleation sites for new phases during topochemical reduction. This is corroborated by the fact that oxygen admission decreases the intensity of more reduced sites with the bands at 2,114 and 2,085 cm^{-1} but increases the intensity of sites with 2,120 and 2,130 cm^{-1} bands (Fig. 3.4), thus indicating a genetic relation between them.

It should be noted that this pattern is qualitatively similar for copper oxides obtained by different methods [16, 17]. Thus, copper oxide used above was obtained from copper hydroxide (specific surface area S_{sp} 8.6 m^2/g) and is represented by well-crystallized spherical particles, whereas in [17] copper oxide obtained by decomposition of basic copper carbonate was represented by mosaic aggregates of particles (S_{sp} 22 m^2/g). It means that the revealed regularities are quite general.

The results obtained allow a significant refinement of the general model of oxide surface (Fig. 3.5). For the oxidized surface, a decrease in effective charge of the coordinatively unsaturated cation to Cu^{1+} may be caused by lowering the total oxygen concentration in the nearest environment or decreasing the effective charge of the metal in the second coordination sphere. This is possible for the steps, at the sites where extended defects terminate on the surface or at the intersection of faces (edges) at a finite size of oxide particles (Fig. 3.5a). The latter is most probable taking into account relative stability of these adsorption sites upon prolonged contact with CO (Fig. 3.2). The oxide reduced in vacuum is characterized by the increased concentration of type I sites and by the appearance of CO

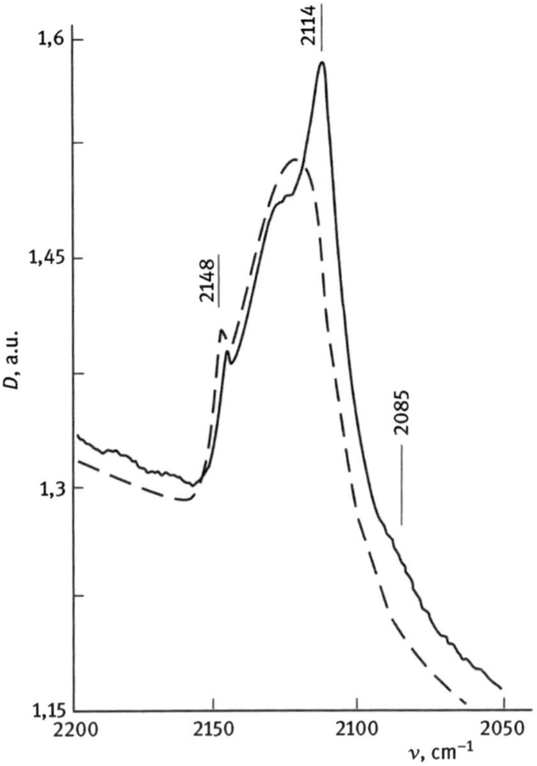

Fig. 3.4: IR spectra of adsorbed CO (300 K, 13 Torr) after CO treatment (573 K): 1 – before (solid line), 2 – after O_2 admission (dashed line).

adsorption sites of type II having a high effective charge. In addition, changes occur in the composition and structure of the subsurface layer, facilitating activation of the surface for further reduction and catalytic reaction. This is reflected in Fig. 3.5b as the appearance of oxygen vacancies in subsurface layer and the appearance of weak Me–Me bonds within cluster subnanostructures. Treatment in CO leads to the formation of the new phase nuclei, which can spontaneously increase in the presence of CO (Fig. 3.5c) because cleavage of the oxygen–metal bond in the oxide is compensated not only by the formation of CO_2 but also by strengthening of the Me–Me bond in the growing nucleus [21]. The fact of close mutual arrangement of all coordinatively unsaturated sites is confirmed by their low total concentration and interconversion.

Thus, coordinatively unsaturated copper cations on the surface of bulk copper oxides are the metastable cluster structures occupying a small part of the oxide surface. The concentration of such sites clearly correlates with the unsteady low-temperature activity in the complete oxidation of CO.

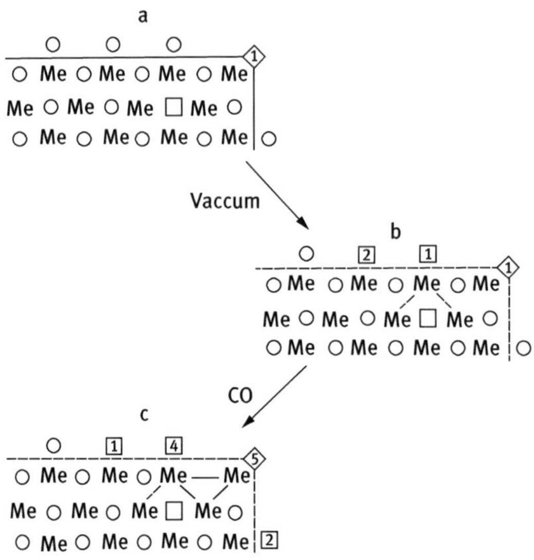

Fig. 3.5: Simple models of oxide surfaces and subsurface layer after different h.t. treatments (side view): (a) O_2; (b) vacuum; (c) CO. Designation as in Fig. 3.1. Numbered: possible sites of CO adsorption (Table 3.1).

3.2 Cobalt oxides

Data on interpretation of IR spectra of adsorbed CO on cobalt oxides are quite contradictory [8, 22–27]. This occurs mostly because the properties of coordinatively unsaturated cations on the surface are associated with the properties of idealized cobalt oxide structures: NaCl for CoO and spinel for Co_3O_4. To change the surface composition, the degree of surface reduction is varied under harsh conditions of the bulk solid-phase reactions or upon prolonged interaction with the reaction medium [22, 24]. The simplest approach to distinguish between types of surface sites is the differential reduction of oxide with retaining its phase composition in the bulk. For Co_3O_4, prepared from basic cobalt carbonate (BCC), h.t. treatment in oxygen, cooling to r.t., and evacuation at r.t. followed by cooling to 80 K and CO admission led to appearance of the band at ~2,133 cm^{-1} (Fig. 3.6, ~80 K) and weak bands in the adsorption region of CCC.

Upon heating to 300 K, the band shifted to the region of ~2,140 cm^{-1}; the intensity of CCC slightly increased (Fig. 3.7) and then remained virtually constant for 30 min.

After vacuum treatment, intensity of the ~ 2,140 cm^{-1} band increased and additional bands appeared in the low-frequency region at 2,066 and 1,840–1,990 cm^{-1}. These bands increased with time and the intensity of CCC increased slightly (Fig. 3.7). After oxygen admission, all the bands disappear and the 2,164 cm^{-1} band appears, which indicate a genetic relation between reduced and oxidized sites of

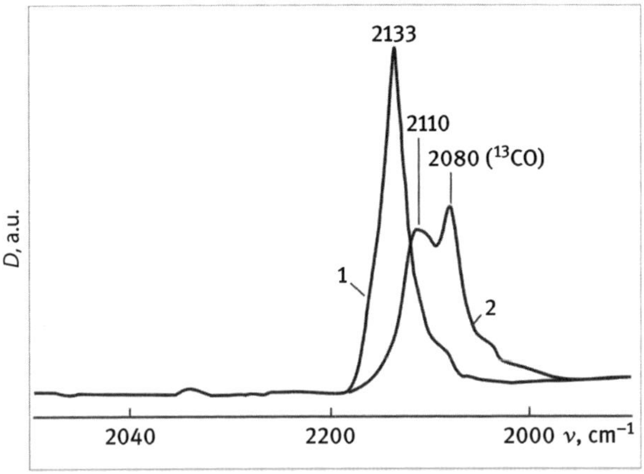

Fig. 3.6: Fourier-transform infrared (FTIR) spectra of CO adsorbed on Co_3O_4 (BCC, 673 K, O_2 h.t. treatment) at ~80 K, P_{CO} ~ 10^{-1} Torr. (1) ^{12}CO; (2) ^{12}CO + ^{13}CO (1:1).

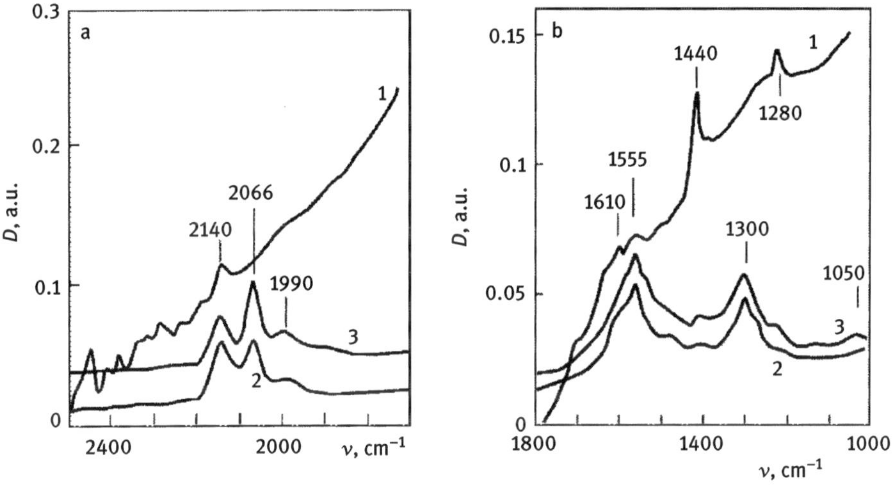

Fig. 3.7: FTIR spectra of CO adsorbed on Co_3O_4 (BCC, 673 K) at 300 K: a – carbonyl region; b – CCC region after different h.t. treatments: (1) O_2; (2) vacuum 1 min after, (3) vacuum 30 min after.

CO adsorption. As for the band at 2,133 cm^{-1}, dilution with ^{13}CO isotope shifts it to the low-frequency region [8] (Fig. 3.6). Another high-frequency shift from 2,048 to 2,064 cm^{-1} was observed with the increase of partial pressure of CO (Fig. 3.8).

The second approach to analyzing the spectra of adsorbed CO was related to variation of the phase composition. The bands for CoO treated in vacuum were similar to those observed for Co_3O_4, except the 2,140–2,133 cm^{-1} band (Fig. 3.9). H.t.

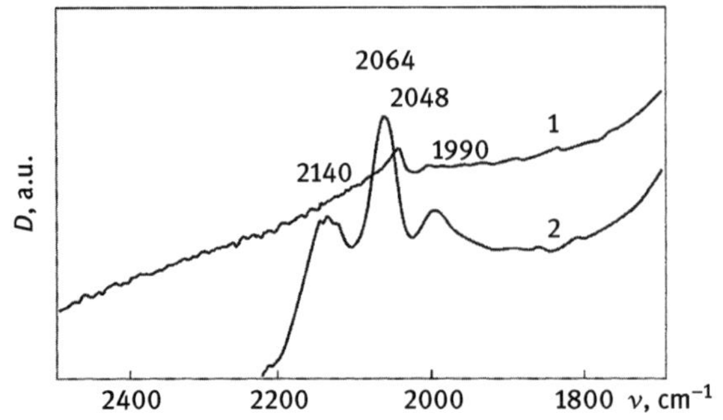

Fig. 3.8: FTIR spectra of CO adsorbed on Co_3O_4 (BCC, 673 K) at 300 K after h.t. vacuum treatment: (1) $P_{CO} \sim 10^{-3}$ Torr; (2) $P_{CO} \sim 3$ Torr.

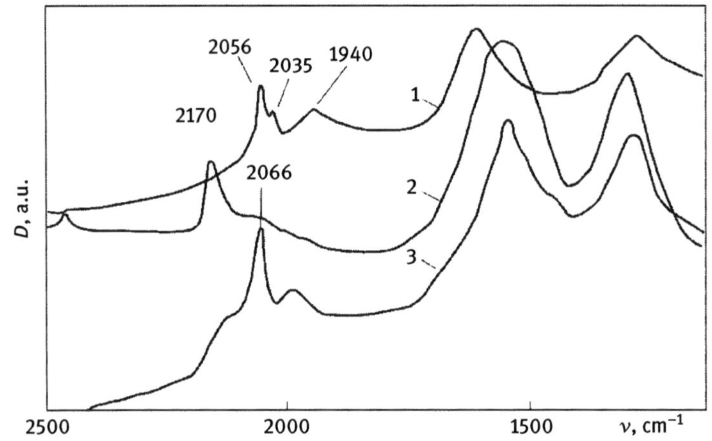

Fig. 3.9: FTIR spectra of CO adsorbed on CoO at 300 K after different h.t. treatments: (1) vacuum, 673 K; (2) O_2, 573 K; (3) O_2, 573 K; vacuum, 673 K.

treatment in oxygen with evacuation at 300 K resulted in appearance of the band at 2,170 cm^{-1}; a weak shoulder in the region of 2,160 cm^{-1} was retained (Fig. 3.9). The subsequent h.t. vacuum treatment led to the bands reproducing virtually completely the spectra of CO on Co_3O_4 (compare Figs. 3.7 and 3.9).

In the high-frequency region, the 2,170–2,190 cm^{-1} bands are assigned to CO complexes with Co^{3+} and Co^{2+} [22–24, 28] (Table 3.2). Some differences in the spectra at similar treatments are caused by the synthesis conditions of oxide samples and conditions of spectra recording. The bands at 2,120–2,150 cm^{-1} correspond most likely to CO complexes with Co^{1+} and Co^{2+} [23, 25–27] (Table 3.2).

Table 3.2: Some carbonyl complexes with the surface sites of cobalt oxides.

No	ν_{CO}, cm^{-1}	ΔH_{ads}, kJ/mol	Sample, pretreatment, conditions; stability	Proposed type of adsorption site	Ref.
1	2,170–2,195	-	Co_3O_4; h.t. vacuum treat., 150–300 K, 100 Torr CO; desorption under vacuum at 300 K	$Co^{3+}CO$	[22]
2	2,170–2,180	80–84	Co_3O_4; h.t. vacuum treat., 300 K, 10 Torr CO; CoO, h.t. oxygen treat., 300 K, 10 Torr CO	$Co^{2+}CO$	[23, 24, 33]
3	2,140–2,145 2,136–2,122 2,120–2,150	- - 20–30	CoO (100); h.t. vacuum treat. at 80 K; CoO, 77 K (Θ = 0–1). Co_3O_4; h.t. vacuum treat.; 300 K, 10 Torr CO, desorption at vacuum r.t.	$Co^{2+}CO$ $Co^{2+}CO$ $Co^{1+(2+)}{}_{clust}CO$	[25, 26, 23, 27]
4	2,000–2,100	-	Co_3O_4; h.t. vacuum treat., 150–300 K, 100 Torr CO; stable under vacuum, 300 K	$Co^{1+}CO$	[22]
5	2,040–2,080 2,060	105–125 -	CoO, Co_3O_4; h.t. vacuum treat., 300 K, 10 Torr CO, stable at r.t. Co_3O_4; 300 K, 30 Torr CO	Co^0CO Co^0CO	[23, 30] [24, 32]
6	1,840–1,980	-	CoO, Co_3O_4; h.t. vacuum treat., 300 K, 10 Torr CO	$Co^0{}_2CO$; $Co^0(CO)_x$	[22, 30, 31]

It should be noted that attribution of the 2,137–2,141 cm^{-1} band [29] to vibrations of "free CO trapped in photochemical matrices" is incorrect under our experimental conditions because this band was observed for nonporous CoO formed upon oxidation of metallic cobalt [25]. In [26], the band at 2,136 cm^{-1} was also observed at low coverages with CO adsorbed on CoO.

In our study, the band with ν_{CO} < 2,100 cm^{-1} was assigned to CO complexes with cobalt in the oxidation state Co^0 because such bands were observed upon thermal decomposition of cobalt polycarbonyls $Co_2(CO)_8$ in the oxygen-free medium on the supports containing titanium hydride and silicon dioxide. The presence of metallic cobalt was verified by magnetic methods [30]. Similarly, according to [31], bands in the region of 2,100–2,000 cm^{-1} correspond to monocarbonyl complexes Co^0CO, whereas bands at 1,800–1,980 cm^{-1} correspond to metal carbonyls with the bridging structure $(Co)_2CO$ and also to subcarbonyl complexes of the $Co^0(CO)_x$ (x = 1–4) type, which were obtained in the matrices of noble gases [32]. For the bands at ~2,040–2,060 cm^{-1}, a high-frequency shift 2,048→2,064 cm^{-1} took place with a growth of CO pressure (coverage) (Fig. 3.8), which is typical for metals [8, 33].

Thus, the general pattern of CO complexes with cobalt closely resembles that of CO complexes with copper. The difference consists in a higher reactivity of cobalt oxides compared with copper oxides. This leads to the interaction of surface oxygen with CO and formation of the oxidation products even on the low-reactive surface formed upon h.t. treatment in oxygen at r.t. (Figs. 3.2 and 3.7). Even more remarkable is the formation of $Co^0(CO)_x$ complexes after vacuum treatment. The indicated complexes are produced by the interaction with CO because their intensity after vacuum treatment of CCC strongly increases (Fig. 3.7). It means that the chemistry of processes on the surface of cobalt oxides essentially differs from that of classical solid-phase processes accompanying the solid-phase reduction of cobalt oxides.

Coordinatively unsaturated cations on the surface, which are observed in the IR spectra of adsorbed CO, are much more reactive sites than those present on the surface of deeply oxidized oxides. This is illustrated by changes of the unsteady catalytic activity measured in pulse mode and changes of the phase composition of cobalt oxides upon h.t. treatments for CoO and Co_3O_4 produced by h.t. oxidation of CoO followed by its h.t. evacuation (Fig. 3.9 and Table 3.3) [23]. According to these data, formation of oxidized Co_3O_4 does not increase the rate of catalytic oxidation of CO, and only the substantial vacuum treatment activates the oxide. Such a comparison allows a conclusion that in the first approximation, the unsteady specific activity of cobalt oxides correlates with the intensity of the ~2,060 cm^{-1} band (Table 3.3), which we attributed to metallic cobalt monocarbonyls. It does not mean that exactly the Co^0 sites provide a high unsteady activity of cobalt oxides in the reaction medium. They are evidently oxidized in the reaction medium. But it means that these highly active sites have very labile chemical and structural properties, which are not typical of the ideal faces of cobalt oxides.

In [28], a comparison of activities after holding in the reaction medium (time on run 70 min) and subsequent treatment in oxygen (10 vol.%) or argon allowed a conclusion that exactly the oxidative treatment activates the Co_3O_4 surface. In such cases, it is necessary to analyze in detail the conditions of experiment. Under our conditions, we compared the effect of treatments of the initial samples in oxygen and helium on the unsteady activity (time on run ~10–20 s). If the authors of [28] performed h.t. treatment consecutively in oxygen and argon, the subsequent unsteady activity would be higher exactly in this case as compared to treatment in oxygen.

The real structure of the initial cobalt oxide (CoO) obtained by thermal decomposition of basic carbonate in helium is demonstrated in Fig. 3.10. Transmission electron microscopy (TEM) data clearly show the surface steps, twin (arrowed) (Fig. 3.10a), intergrowth boundaries of crystallites (Fig. 3.10b) and dislocations (Fig. 3.10c) occupying a small part of the particle surface.

Table 3.3: Relation between h.t. treatment of different cobalt oxides, their unsteady catalytic activity and FTIR spectra of adsorbed CO.

Sample	h.t. treatment	Specific activity, 10^{15} molec. CO/ (m^2 s), 300 K	FTIR spectra in figures
CoO	He (~10^{-6} bar O_2), 623 K	~230×	Fig. 3.9, spectrum 1
	10% O_2, 623 K	~80×	Fig. 3.9, spectrum 2
	10% O_2, 623 K He (~10^{-6} bar O_2), 623 K	~2,200×	Fig. 3.9, spectrum 3
Co_3O_4	100% O_2, 623 K	~180×	Fig. 3.7, spectrum 1
	He (~10^{-6} bar O_2), 623 K	~750×	Fig. 3.7, spectrum 2

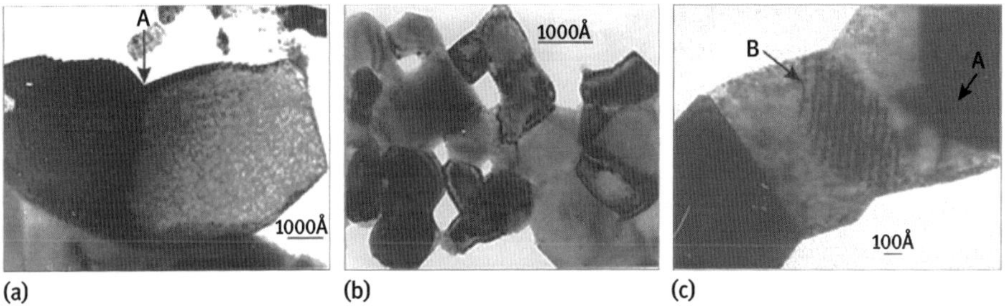

Fig. 3.10: Real defect structure of CoO: (a) twin (arrowed); (b) different grain boundaries of CoO crystallites; (c) twin (111) (arrowed A) and dislocation dipoles (arrowed B).

H.t. oxidation of CoO leads to the formation of cobalt oxide with the spinel structure; it has the developed dislocation network (Fig. 3.11a), which becomes more pronounced after h.t. treatment in helium (Fig. 3.11b) and acquires distinct planar defects (Fig. 3.11c).

Hence, significant changes in the defect structure of cobalt oxides are accompanied by the changes in the state of coordinatively unsaturated sites on the surface revealed by FTIR and unsteady catalytic activity.

The anomalously high activity of coordinatively unsaturated sites on the surface of cobalt oxides was verified by experiments at ~ 80 K [27]. The adsorption of CO and CO + O_2 at the liquid nitrogen temperature on Co_3O_4 treated in vacuum revealed increase in the intensity of CO_2 band (~2,345 cm^{-1}) (Fig. 3.12). Therewith, the intensity of the CCC band virtually did not change with time, while the bands corresponding to Co^0CO complexes appeared in the spectra (Fig. 3.12 (1)) and disappeared in the presence of oxygen (Fig. 3.12 (2)). Thus, CO oxidation proceeds via the primary formation of

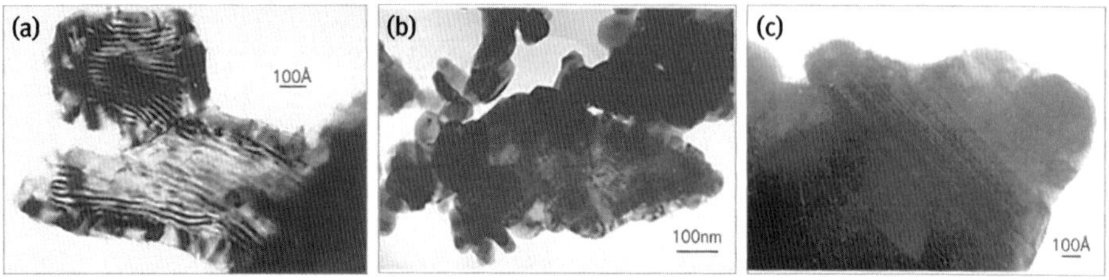

Fig. 3.11: CoO after h.t. oxidation: (a) dislocation network in Co_3O_4 after h.t. oxidation of CoO; (b) after h.t. oxidation followed by h.t. vacuum treatment; (c) planar defects in Co_3O_4 after oxidation of CoO.

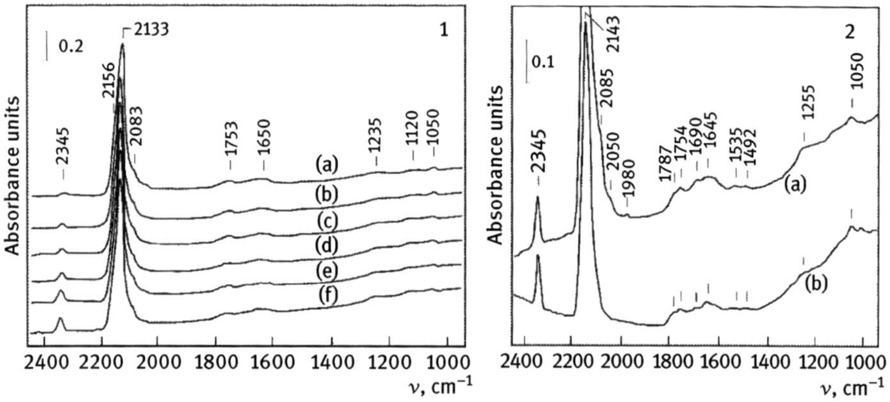

Fig. 3.12: FTIR spectra of CO adsorbed on Co_3O_4 at 80 K after vacuum treatment: (1) (P_{CO} = 0.1 Torr); (2) (P_{CO} = 0.1 Torr, P_{O2} = 0.1 Torr) after (a) 1, (b) 8, (c) 20, (d) 45, (e) 100, (f) 170 min [27].

CO_2, which is consistent with the data of [28, 34–36], and this process occurs virtually without activation.

Another important feature of some coordinatively unsaturated sites on the surface of cobalt oxides is their aggregation into cluster structures on which the adsorbed complexes demonstrate the so-called dynamic shift – the high-frequency band shift upon increase in coverage (growth of CO partial pressure in IR cell). This effect disappears when ^{12}CO is diluted with ^{13}CO isotope (Figs. 3.6 and 3.13). According to [8], the indicated effect is caused by the lateral dipole–dipole interaction of adsorbed molecules under the following conditions: "the adsorbed molecules must have the same intrinsic frequency; the adsorption plane must be flat; the adsorbed molecules must be in close proximity and oriented parallel to each other" [8] (Fig. 3.14).

This effect is well known for supported metals [37]. For oxides with high ionicity of the metal–oxygen bond, this effect, which was first discovered by A. Tsyganenko [38, 39], is much lower. It was attributed mostly to the so-called static

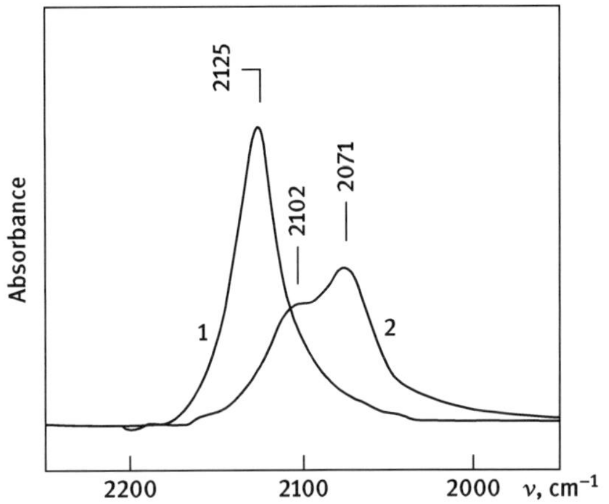

Fig. 3.13: FTIR spectra of CO adsorbed on CoO at 80 K: (1) ^{12}CO, (2) $^{12}CO + ^{13}CO$ (1:7) [27].

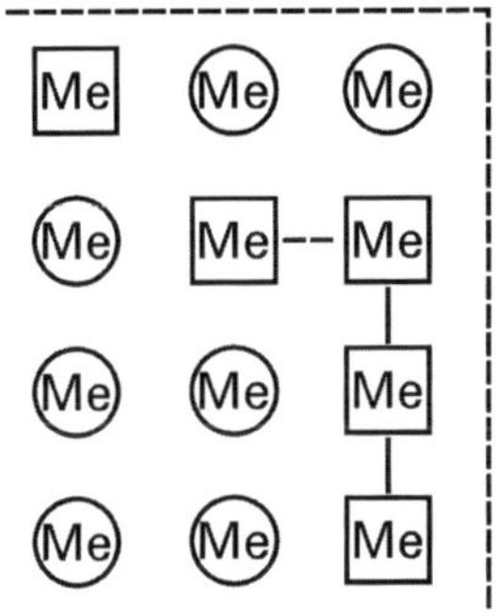

Fig. 3.14: Simple model of oxide surface after mild reduction (top view): - - - - - edge of the surface step; [Me] – surface anion vacancy over Me cation; (Me) – surface oxygen anion over Me cation.

effect [8, 39, 40]. In our case, the shift reaches 23 cm^{-1} (Figs. 3.6. and 3.13). In our opinion, taking into account chemical properties of these adsorption sites, this effect is related to the presence of cobalt cations (precursors of the reduced sites) in close vicinity to each other as clusters on the surface of cobalt oxides. As shown by quantitative estimates of changes in pressure upon admission of similar CO portions into the empty cell and the cell with cobalt oxide, the total number of such sites is not large (5–10% of a monolayer coverage). Theoretical calculations for small Co_xO_y clusters confirm that the clusters can be highly active in oxidation reactions [35, 36].

A decrease in the 2,164 cm^{-1} band intensity with the time of holding in the reaction medium and catalyst deactivation can be interpreted in terms of clusters as the high-active sites of low-temperature CO oxidation on the surface of cobalt oxides [33]. The indicated band was attributed to clusters in the oxidized state. We think that the catalyst deactivation is associated not only with the surface reduction but, to a greater extent, with a slow destruction of clusters under the action of reaction medium.

Thus, for cobalt oxides, the qualitative pattern showing the dependence of IR spectra of adsorbed CO on various treatments is similar to the pattern for copper oxide. Various coordinatively unsaturated cations in the oxidation state lower than Co^{3+} and Co^{2+}, which are typical of cobalt spinel, are observed. Cobalt oxides are much more reactive, which leads even at 80 K to the appearance of reduced sites and products of their reduction – CCC and CO_2.

3.3 Iron oxides

The most abundant iron oxide is hematite having the corundum structure. In chemical terms, iron oxides have a lower reactivity; so, in distinction to cobalt oxides, their surface properties can be analyzed at 77–80 K by considering only the adsorption and excluding the chemical interaction of CO with oxygen of the oxide surface. In some works comparing the data obtained by TEM and IR spectroscopy of adsorbed CO for the oxides with corundum structure, different types of coordinatively unsaturated cations on the surface were attributed to basal or prismatic faces [41, 42]. We have compared IR spectra of adsorbed CO at 80 K for the iron oxide samples synthesized by different methods and having different sets of the most developed faces according to TEM (Table 3.4 and Fig. 3.15). For samples 1 and 3 with the most developed prismatic (0001) faces, a set of bands coincides: ~2,190, ~2,170, ~2,164, and ~2,150 cm^{-1} (Fig. 3.15). Changes are observed only in the intensity ratio of different bands. For samples 4 and 7, which have (1$\bar{1}$00) and (0001) faces, the spectra are also quite similar. In addition, the band at 2,203 cm^{-1} (Fig. 3.15) is observed. For γ-Fe_2O_3 sample 5, high-frequency bands are completely absent. Only for the oxide sample obtained by thermal decomposition of iron hydroxide and having the most developed prismatic faces, the spectrum of adsorbed CO is close to the spectrum [42] with one band at ~2,166 cm^{-1}.

The adsorption heats of NO are much higher than those of CO (Table 3.5). So the spectral pattern is less differentiated. For sample 3, a slightly asymmetric band observed at 1,809 cm^{-1} (Fig. 3.16) is typical of Fe^{2+} complexes [9]. For sample 2, an additional band appears at 1,888 cm^{-1}, which corresponds to iron cations with a high charge [9]. Hence, the spectral pattern for complexes of adsorbed CO and NO is quite complicated. It means that types of surface coordinatively unsaturated cations are more diverse than it could be expected from the morphological analysis of iron oxide particles.

As earlier, we have modified the surface state of iron oxides using h.t. pretreatments in oxygen and vacuum. Similar to copper and cobalt oxides, one high-frequency band at the minimum intensity of CCC is observed at 300 K after pretreatment in oxygen (Fig. 3.17). After vacuum pretreatment, many bands appear in the region of

Table 3.4: Some properties of iron oxides.

No.	Method of preparation	Phase	Specific surface area, m^2/g	Most developed faces (TEM)	Wafer density, g/cm^2 (in IR cell)
1	Thermal decomposition of oxalate, 673 K, air	α-Fe_2O_3	14	(0001)	0.018
2	Thermal decomposition of goethite (1), 673 K, air	α-Fe_2O_3	64	$(01\bar{1}0)$, (0001)	0.015
3	Thermal decomposition of ammonium oxalatoferriate, 773 K, air	α-Fe_2O_3	18	(0001)	0.017
4	Thermal decomposition of goethite (2), 673 K, air	α-Fe_2O_3	90	$(1\bar{1}00)$, (0001)	0.029
5	Reduction of sample 3 by CO in He at 673 K, oxidation by O_2 in He at 500 K	γ-Fe_2O_3	27	(100), (110)	0.024
6	Aging of iron hydroxide at pH ~ 3, washing, calcination at 673 K, air	α-Fe_2O_3	27	(0001)	0.039
7	Thermal decomposition of goethite (2), 873 K, air	α-Fe_2O_3	20	$(1\bar{1}00)$, (0001)	0.024

$\nu_{CO} \leq 2{,}140\ cm^{-1}$ (Fig. 3.17a). Simultaneously, the band intensity strongly increases in the CCC region (Fig. 3.17b). Special experiments revealed a weak effect of CCC on IR spectra of adsorbed CO and reactivity of the oxide surface toward reduction. For nitrosyl complexes, which do not interact with the surface oxygen, vacuum pretreatment only increases the intensity of asymmetric band at 1,809 cm^{-1} as compared to the oxygen pretreatment (Fig. 3.18a). This demonstrates that vacuum pretreatment initially produces no significant changes in the oxidation state of coordinatively unsaturated Fe^{2+} cations on the oxide surface but alters their concentration and reactivity toward further reduction upon interaction with CO. It means that, similar to the case of copper and cobalt oxides, vacuum treatment changes the structure of the subsurface layer facilitating its activation (see Fig. 3.5). Upon reduction of the iron oxide surface by carbon monoxide, for adsorbed nitrosyl complexes a shoulder at 1,740 cm^{-1} appears, while intensity of the 1,809 cm^{-1} band declines (Fig. 3.18a). For complexes of CO adsorbed at 80 K,

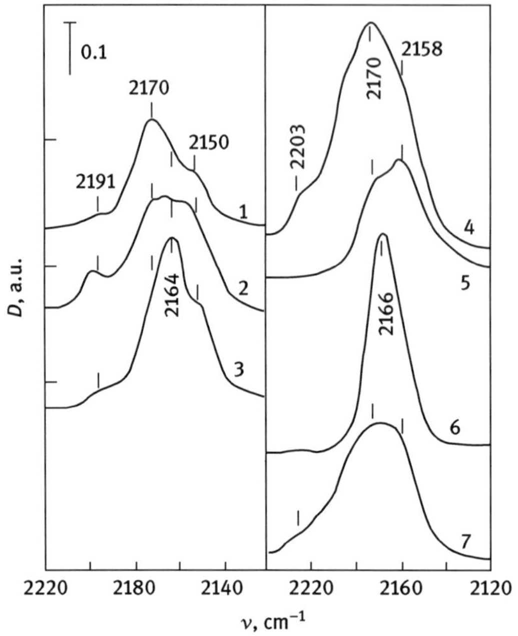

Fig. 3.15: IR spectra of CO adsorbed on various Fe_2O_3 samples at 80 K (P_{CO} = 0.5 Torr) after vacuum treatment (sample numbers as in Table 3.4) [43].

Table 3.5: The main types of coordinatively unsaturated iron cations on the surface of iron oxides.

No.	ν_{CO}, cm^{-1}	$\Delta H_{ads\ CO}$, kJ/mol	ν_{NO}, cm^{-1}	$\Delta H_{ads\ NO}$, kJ/mol	Sample (No. in Table 4), treatment conditions	Ref.
1	2,195–2,205	–	1,888		$Fe^{(2+\delta)+}$	[43], [9]
2	2,170–2,195	~21	1,809	90–110	Fe^{2+}	[43], [9]
3	~2,160–2,165	–	–	–	OH	–
4	2,140–2,100	76–84	1,809	90–110	Fe^{1+}	[43], [9]
5	~2,060	>84			Fe^0	[43]
6	2,064, 2,040, 1,980	~20	~1,740	–	$Fe^0_x(CO)_y$	[43]

a similar procedure leads to a strong decrease in intensity of bands at $\nu_{CO} \geq 2{,}170\ cm^{-1}$ and an increase in intensity of low-frequency bands (Fig. 3.18b). This testifies to a genetic relation between oxidized and reduced cations on the surface.

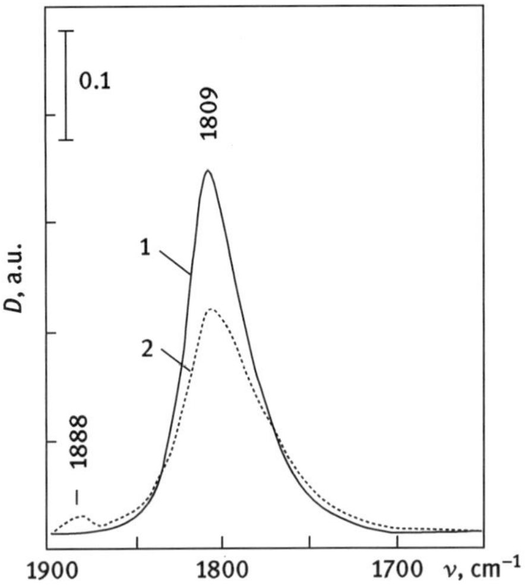

Fig. 3.16: IR spectra of NO adsorbed at 300 K on sample (3) spectrum 1 and sample (2) spectrum 2 [43].

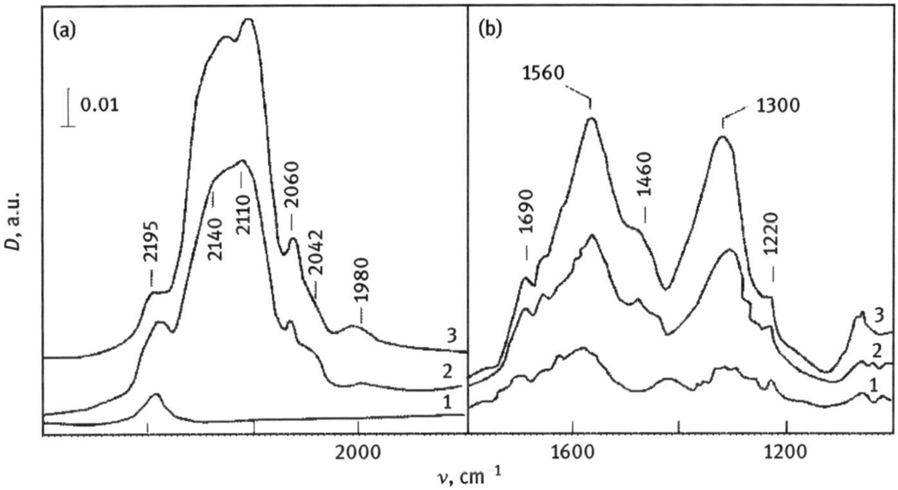

Fig. 3.17: IR spectra of CO adsorbed on sample (3) at 300 K: (a) carbonyl region; (b) carbonate region: (1) after treatment in O_2 at 673 K; (2) after vacuum treatment at 673 K. P_{CO} = 0.01 Torr [43].

The opposite experiment – preadsorption of NO – results in nearly a complete disappearance of the spectra of adsorbed CO at 80 K, except the band at ~2,166 cm^{-1}, which was assigned to CO adsorbed on the surface hydroxyl groups [14, 17]. The assignment of different bands of CO and NO adsorbed on iron oxides is shown in Table 3.5.

One of the most interesting and important results is the "dynamic effect" of dipole–dipole interaction for CO and NO molecules adsorbed on coordinatively

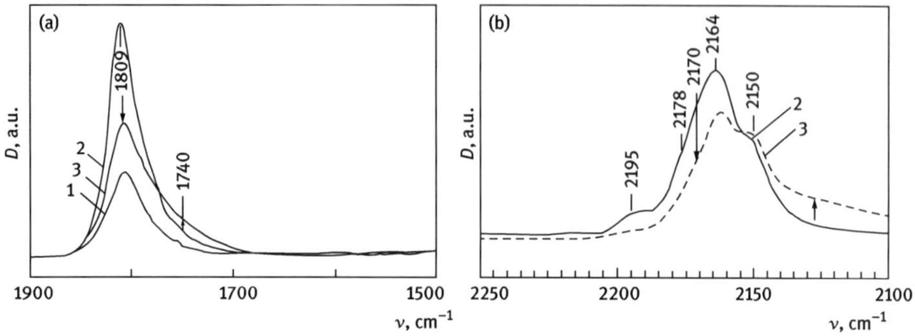

Fig. 3.18: IR spectra of NO adsorbed on sample (3) at 300 K (a) and CO adsorbed at 80 K (b): (1) after treatment in O_2 at 673 K; (2) after vacuum treatment at 673 K; (3) after vacuum treatment at 673 K, admission of CO at 300 K followed by vacuum treatment. Arrowed is the difference of spectra after reduction [43].

unsaturated iron cations. This effect emerges due to lateral interaction and disappears upon dilution of adsorbed molecules with other isotope, which leads to the low-frequency shift of the band maximum [8, 40, 42]. For iron oxides, we have observed a shift of the band of adsorbed nitrosyl complexes even at a relatively small (1:1) dilution with ^{15}NO isotope (Fig.3.19a). For complexes of CO adsorbed on iron oxide, this effect was reported for the first time in [42]. We have also observed it even at a dilution with 20% ^{13}CO (Fig. 3.19b).

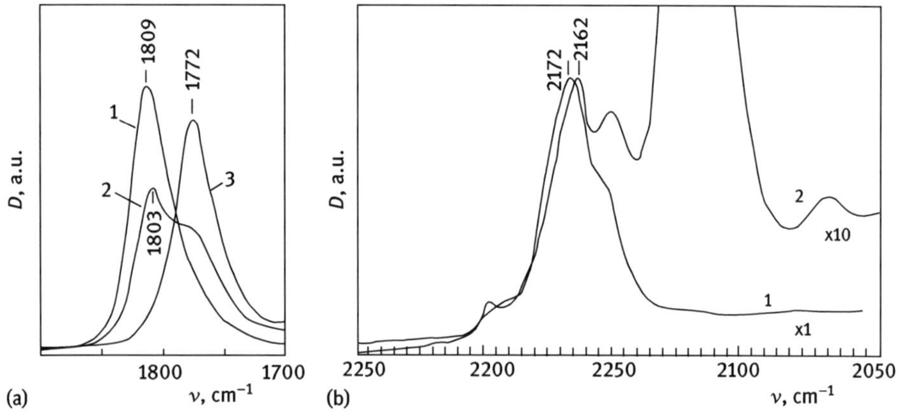

Fig. 3.19: IR spectra: (a) of NO adsorbed at 300 K after vacuum treatment at 673 K: (1) ^{14}NO, (2) $^{14}NO + {}^{15}NO$ (1:1), (3) ^{15}NO, $P_{NO} = 1$ Torr; (b) of CO adsorbed at 80 K: (1) ^{12}CO, (2) $^{12}CO + {}^{13}CO$ (1:7), $P_{CO} = 0.5$ Torr [43].

One more property, which is unusual for iron oxide, shows up after prolonged contact with CO at 300 K and evacuation at 673 K: readsorption of CO at 80 K leads to the appearance of three low-frequency bands at 2,060, 2,030, and 1,980 cm^{-1}, the intensity of which increases simultaneously with raising the CO pressure in IR cell (Fig. 3.20). These spectra resemble the spectra of iron subcarbonyls $Fe_x(CO)_y$ adsorbed on an inert support in the oxygen-free medium but not in the spectra of CO adsorbed on metallic iron [8, 44]. In our case, it means that reduction of oxidized iron oxide can be accompanied by the formation of at least several iron atoms in the oxidation state Fe^0, which have in the nearest coordination sphere not oxygen anions but iron cations that do not create steric hindrance for the adsorption of two or more CO molecules on one iron atom. These are either the clusters rising above the surface or the steps. Evidently, this completely contradicts the conventional surface models for oxides. Special adsorption experiments demonstrated that the total amount of CO adsorbed at 80 K, including adsorption on hydroxyls and physical adsorption, does not exceed 15% of a monolayer coverage of the oxide.

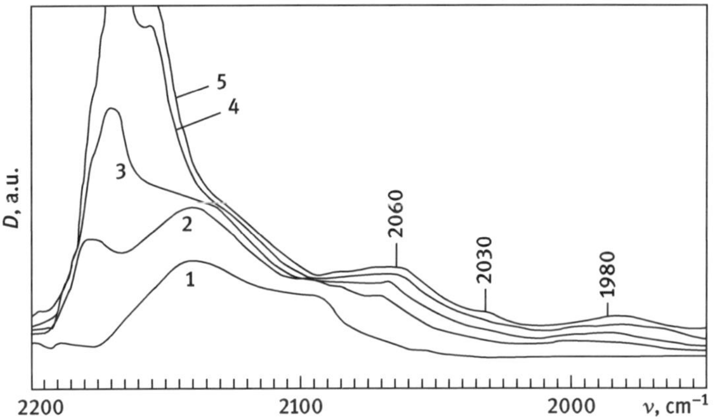

Fig. 3.20: IR spectra of CO adsorbed at 80 K after vacuum treatment at 673 K, CO adsorption at 300 K during 30 min followed by vacuum treatment at 623 K: (1) $P_{CO} = 10^{-3}$ Torr, (2) $P_{CO} = 10^{-2}$ Torr, (3) $P_{CO} = 5 \times 10^{-2}$ Torr, (4) $P_{CO} = 0.25$ Torr, (5) $P_{CO} = 0.5$ Torr.

A correlation (more accurately, an analogous change) of the steady-state specific activity and concentration of partially reduced surface sites shows that at least in the low-temperature region these sites can determine the activity of iron oxides in oxidation of CO (413 K, 1% CO + 1% O_2 in helium). The band at ~ 2,110 cm^{-1}, which forms upon interaction with CO at 300 K (Fig. 3.17), was chosen as the characteristic of reduced site. It is clearly seen in Fig. 3.21 that the activity tends to increase with the band intensity. A possible cause of the hyperbolic form of

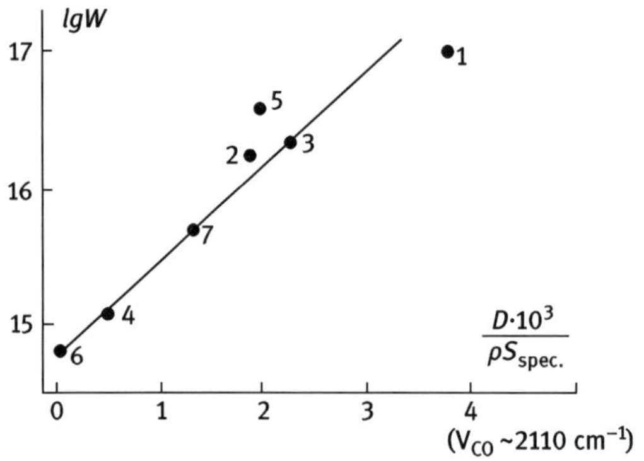

Fig. 3.21: *LgW* versus the normalized intensity of the adsorption band at ~2110 cm^{-1} ($D/\rho S_{spec}$) of different iron oxides, where *W* is the velocity (molecules CO/m^2 s), *D* the optical density, ρ the wafer density (g/cm^2) and S_{spec} the specific surface area. Numbering as in Table 3.4.

this dependence is that the concentration of sites estimated by IR spectroscopy may be underrated as compared to conditions in the reaction medium. In the oxygen-free medium created in a static setup, a part of the sites was reduced to metallic state, whereas another part remained oxidized (Fe^{2+}). In the reaction medium, the active sites – coordinatively unsaturated iron cations – are likely to have an intermediate oxidation state.

3.4 Conclusion

The study on the properties of coordinatively unsaturated cations on the surface of copper, cobalt and iron oxides demonstrates that their chemical properties under different pretreatment and synthesis conditions are not definitely related with their bulk properties – phase composition and stoichiometry. Such cations determine the low-temperature activity in the complete oxidation of CO under steady and unsteady conditions. Their concentration can change in a very wide range, but their total amount is much smaller than a monolayer coverage. The presence of dipole–dipole interaction can be attributed only to the arrangement of cations in close vicinity to each other. This is impossible for atomic structures within the ideal faces on the oxide surface, where cations are separated from each other by oxygen anions. Close location of coordinatively unsaturated cations should facilitate the dissociative adsorption of oxygen and the interaction of adsorbed CO and oxygen, similar to the models reported in [36, 45, 46]. Most likely, a high lability of chemical and structural properties of such cluster sites is the main cause of their high catalytic activity.

References

[1] Henrich, V. Electronic and geometric structure of defects on oxides and their role on chemisorption, Nowotny J., and Dufour L-C., eds., Surface and Near-Surface Chemistry of Oxide Materials, Amsterdam, Elsevier, 1988, 23–60.
[2] Fierro JLG, ed, Metal Oxides. Chemistry and applications, USA, Taylor & Francis Group, 2006.
[3] Royer, S., & Duprez, D. Catalytic oxidation of carbon monoxide over transition metal oxides. Chem. Cat. Chem. 2011, 3, 24–65.
[4] Svintsitskiy, DA., Kardash, TYu., Stonkus, OA., Slavinskaya, EM., Stadnichenko, AI., Koscheev, SV., Chupakhin, AP., & Boronin, AI. In situ XRD, XPS, TEM, and TPR study of highly active in CO oxidation CuO nanopowders. J. Phys. Chem. C 2013, 117, 14588–14599.
[5] Murzin, DYu. Engineering Catalysis, Berlin/Boston, Germany, Walter de Gruyter GmbH, 2013.
[6] Kydd, R., Ferri, D., Hug, P., Scott, J., Yang, W., & Amal, R. Temperature-induced evolution of reaction sites and mechanisms during preferential oxidation of CO. J. Catal. 2011, 277, 64–71.
[7] Jung, KT., & Bell, AT. An in situ infrared study of dimethyl carbonate synthesis from carbon dioxide and methanol over zirconia. J. Catal. 2001, 204, 339–347.
[8] Hadjiivanov, KI., & Vayssilov, GN. Characterization of oxide surfaces and zeolites by carbon monoxide as IR probe molecule. Adv. Catal. 2002, 47, 307–511.
[9] Davydov, A. Infrared Spectroscopy of Adsorbed Species on the Surface of Transition Metal Oxides, John Willey & Sons, Chichester, 1990.
[10] Busca, G. FT-IR study of the surface of copper oxide. J. Molec. Catal. 1987, 43, 225–236.
[11] Lokhov, YuA., Sadykov, VA., Tikhov, SF., & Popovskii, VV. Interaction of copper oxides with reaction medium in heterogeneous catalytic oxidation of carbon monoxide by molecular oxygen. 4. IR spectroscopic study of carbon monoxide interaction with copper oxide surface. Kinet. Catal. 1985, 26, 152–158.
[12] Lokhov, YuA., & Davydov, AA. Study of the state of transition metal cations on the surface of catalysts by the method of IR spectroscopy of adsorbed test molecules (CO,NO) .5. Influence of the state of adsorption sites on the frequency of the stretching vibration of adsorbed carbon monoxide. Kinet. Catal. 1980, 21, 1093–1098.
[13] London, JW., & Bell, AT. Infrared spectra of carbon monoxide, carbon dioxide, nitric oxide, nitrogen dioxide, nitrous oxide, and nitrogen adsorbed on copper oxide. J. Catal. 1973, 31, 32–40.
[14] Soltanov, RI., Paukshtis, EA., & Yurchenko, EN. IR spectroscopic investigation of the thermodynamics of the reaction of carbon monoxide with the surface of several oxide adsorbents. Kinet. Catal. 1982, 23, 135–141.
[15] Sadykov, VA., Tikhov, SF., Popovskii, VV., & Kryukova, GN. Reaction of copper oxides with the reaction medium in heterogeneous catalytic oxidation of carbon monoxide with molecular oxygen. 1. Nature of the new phases formed in topochemical reduction of copper oxides by carbon monoxide. Kinet. Catal. 1983, 24, 472–479.
[16] Sadykov, VA., Tikhov, SF., Popovskii, VV., & Kryukova, GN. Interaction of copper oxides with the reaction medium in heterogeneous catalytic oxidation of carbon monoxide with molecular oxygen. 3. Nature of reactive centers on copper oxide surface. Kinet. Catal. 1985, 26, 144–152.
[17] Tikhov, SF., Sadykov, VA., Kryukova, GN., Paukshtis, EA., Popovskii, VV., Starostina, TG., Kharlamov, GV., Anufrienko, VF., Poluboyarov, VF., Razdobarov, VA., Bulgakov, NN., & Kalinkin, AV. Microstructural and spectroscopic investigations of the supported copper-alumina oxide system: nature of aging in oxidizing reaction media. J. Catal. 1992, 134, 506–524.

[18] Sadykov, V. A., Tikhov, S. F., Bulgakov, N. N., & Gerasev, A. P. Catalytic oxidation of CO on CuO_x revisited: impact of the surface state on the apparent kinetic parameters. Catal. Today 2009, 144, 324–333.
[19] Senor, DA., & Amberg, CH. Infrared-band intensities of adsorbed carbon monoxide. J. Chem. Phys. 1965, 42, 1523–1529.
[20] Sadykov, VA., Tikhov, SF., & Popovskii, VV. Reaction of copper oxides with the reaction medium in heterogeneous catalytic oxidation of carbon monoxide with molecular oxygen. 5. Formation of the steady state on the surface of copper oxides. Kinet. Catal. 1986, 27, 133–140.
[21] Sadykov, VA., Tikhov, SF., Popovskii, VV., & Bulgakov, NN. Interaction of copper oxides with the reaction medium in heterogeneous catalytic oxidation of carbon monoxide with molecular oxygen. 2. Investigation of the kinetics and thermodynamics of topochemical transformations in reduction of cuprous oxide and copper–oxygen solid solution. Kinet. Catal. 1983, 24, 665–671.
[22] Busca, G., Guidetti, R., & Lorenzelli, V. Fourier-transform infrared study of the surface properties of cobalt oxides. J. Chem. Soc. Faraday Trans. 1990, 86, 989–994.
[23] Sadykov, VA., Lokhov, YuA., Tikhov, SF., Krjukova, GN., Bregikhin, MN., Popovskii, VV., Bulgakov, NN., Razdobarov, VA., Solovjeva, IP., Olenkova, IP., & Golovin, AV. The nature of the surface active centers and mechanism of CO catalytic oxidation on cobalt oxide catalysts. In: Heterogeneous Catalysis (Proc.6-th Intern. Symp, Jul.13–18). Publ. House Bulg. Acad. Sci., Sofia, Bulgaria 1987, Part 1, 359–364.
[24] Lin, H-K., Wang, C-B., Chiu, H-C., & Chien, S-H. In situ FTIR study of cobalt oxides for the oxidation of carbon monoxide. Catal. Lett. 2003, 86, 63–68.
[25] Schönnenbeck, M., Cappus, D., Klinkmann, J., Freand, H-J., M.Petterson, LG., & Bagus, PS. Adsorption of CO and NO on NiO and CoO: a comparison. Surf. Sci. 1996, 347, 337–345.
[26] Scarano, D., Zecchina, A., Spoto, G., & Geobaldo, F. Co–ZnO solid solution as a model to investigate the CO–cation interaction: An FTIR and HRTEM study. J. Chem. Soc. Faraday Trans. 1995, 91, 4445–4450.
[27] Lokhov, YuA., Tikhov, SF., Bredikhin, MN., Zhirnyagin, AG., & Sadykov, VA. Carbon monoxide oxidation on cobalt oxides at 80 K: an FT-IR study. Mendeleev. Commun. 1992, 1, 10–11.
[28] Jansson, J., Palmqvist, AEC., Fridell, E., Skoglundh, M., Österlund, L., Thormählen, P., & Langer, V. On the catalytic activity of Co_3O_4 in low-temperature CO oxidation. J. Catal. 2002, 211, 387–397.
[29] Pollard, MJ., Weinstock, BA., Butterwolf, TE., Griffiths, PR., Rewbery, AP., & Pain, JB. A mechanistic study of the low-temperature conversion of carbon monoxide to carbon dioxide over a cobalt oxide catalyst. J. Catal. 2008, 254, 218–225.
[30] Lisitsyn, AS., Golovin, AB., Kuznetsov, VL., & Ermakov, YuI. CO hydrogenation on cobalt catalysts. Effect of the conditions of $Co_2(CO)_8$ pyrolysis on the surface of TiO_2 on the catalytic and magnetic properties of the catalysts. J. Catal. 1985, 95, 527–538.
[31] Sheppard, N., & Nguen, TT. Advances in Infrared and Raman Spectroscopy, Vol. 5, London, Philadelphia, Rhein, 1979.
[32] Hanlan, LA., Huber, H., Kűndig, EP., McGarvey, BR., & Ozin, GA. Chemical synthesis using metal atoms – matrix infrared, Raman, ultraviolet-visible, and electron-spin resonance studies of binary carbonyls of cobalt, $Co(CO)_N$ (where N=1-4), and distortion problem in $C_O(CO)_4$. J. Amer. Chem. Soc. 1975, 97, 7054–7068.
[33] Crossley, A., & King, DA. Adsorbate island dimensions and interaction energies from vibrational spectra: CO on Pt {001} and Pt {111}. Surf. Sci. 1986, 95, 131–155.
[34] Jansson, J. Low-temperature CO oxidation over Co_3O_4/Al_2O_3. J. Catal. 2000, 194, 55–60.

[35] Xu, X-L., Yang, E., Li, J-Q., Li, Y., & Chen, W-K. A DFT study of CO catalytic oxidation by N_2O or O_2 on the Co_3O_4 (110) surface. Chem. Cat. Chem. 2009, 1, 384–392.
[36] Xie, Y., Dong, F., Heinbuch, S., Roccab, JJ., & Bernstein, ER. Oxidation reactions on neutral cobalt oxide clusters: experimental and theoretical studies. Phys. Chem. Chem. Phys. 2010, 12, 947–959.
[37] Eischens, RP., & Francis A. The effect of surface coverage on the spectra of chemisorbed CO. J. Phys. Chem. 1956, 60, 194–201.
[38] Tsyganenko, AA., Denisenko, LA., Zverev, SM., & Filimonov, VN. Infrared study of lateral interactions between carbon monoxide molecules adsorbed on oxide catalysts. J. Catal. 1985, 94, 10–15.
[39] Tsyganenko, AA., & Zverev, SM. Mechanism of lateral interactions between molecules adsorbed on oxide surfaces. React. Kinet. Catal. Lett. 1988, 36, 269–274.
[40] Lamberti, C., Zecchina, A., Groppo, E., & Bordiga, S. Probing the surfaces of heterogeneous catalysts by in situ IR spectroscopy. Chem. Soc. Rev. 2010, 39, 4951–5001.
[41] Scarano, D., Zecchina, A., Bordiga, S., Ricchiardi, G., & Spoto, G. Interaction of CO with α-Cr_2O_3 surface: a FTIR and HRTEM study. Chem. Phys. 1993, 177, 547–560.
[42] Zecchina, A., Scarano, D., & Reller, A. Infrared spectra of CO adsorbed on prismatic faces of a-Fe,O. J. Chem. Soc. Faraday Trans. 1, 1988(84), 2327–2333.
[43] Tikhov, SF., Sadykov, VA., Kryukova, GN., & Razdobarov, VA. Dipole coupling of CO(NO) adsorbed on Iron surface centres: a typical feature of surface active sites. Mendeleev. Commun. 1994, 2, 69–71.
[44] Peden, CHF., Parker, SF., Burrett, PH., Pearson, RG. Moessbauer and infrared studies of matrix-isolated iron-carbonyl complexes. J. Phys. Chem. 1983, 87, 2329–2336.
[45] Broqvist, P., Panas, I., & Persson, H. A DFT study on CO oxidation over Co_3O_4. J. Catal. 2002, 210, 198–206.
[46] Pollard, MJ., Weinstock, BA., Bitterwolf, TE., Griffiths, PR., Newbery, PA., & Paine, JB III. A mechanistic study of the low-temperature conversion of carbon monoxide to carbon dioxide over a cobalt oxide catalyst. J. Catal. 2008, 254, 218–225.

4 Mechanism of methane dry reforming over nanocomposite catalysts

4.1 Introduction

Basic knowledge of the mechanism of red–ox reactions catalyzed by oxide and nanocomposite catalysts is required to develop scientific bases of design of catalysts with high activity, selectivity and stability. Reliable elucidation of mechanistic features including detailed kinetic schemes is possible by application of unsteady-kinetic methods (including temporal analysis of products (TAP), isotope transition experiments Steady-State Isotope Transient Kinetics Analysis (SSITKA), etc.), in situ spectroscopic methods (Fourier-transform infrared spectroscopy (FTIRS), etc.) and microcalorimetry combined with detailed mathematical modeling. In this chapter efficiency of such approach is illustrated for the case of mechanism of methane dry reforming over nanocomposite catalysts comprised of metal alloy nanoparticles supported on mixed oxides with a high mobility and reactivity of the surface/near surface oxygen species.

4.2 General schemes

The main factor controlling the catalytic activity during methane dry reforming (MDR) is believed to be the reactant activation. The dissociative adsorption and activation of both CH_4 and CO_2 are structure sensitive, that is, they depend on the structure of nanocatalysts based on fluorite-like/perovskite-like oxides, as far as they depend on electronic as well as geometric factors [1, 2]. The existence of the strong metal–support interaction between active metal particles and support in catalysts drastically suppresses chemisorption of both H_2 and CO facilitating synthesis gas production [3].

It is generally accepted that methane is dissociatively adsorbed on the metal species to form CH_x fragments, whereas CO_2 activation may depend on the type of support used (Fig. 4.1).

If CO_2 is adsorbed on the support and activated at the interface between the metal particle and support (path I), rapid oxidation of CH_x on metal surface occurs. Alternatively, CO_2 can be activated on the metal surface (path II). It was suggested that path I was more effective for the inhibition of carbon formation than path II. For nanocatalysts, oxygen vacancies formed on the support during pre-reduction can act as active sites for dissociative adsorption of CO_2. For the irreducible oxide supports, CO_2 dissociation is supposed to be promoted by the H (ads) originating from the CH_4 dissociation, which can be assisted by oxygen atoms on the support. The reactive intermediates in the reaction mechanism are mostly considered to be support-related species [5–9]. Thus, Bradford and Vannice [7] have suggested for nickel supported catalysts that CO_2 participates in the reaction mechanism through the reverse water-gas

https://doi.org/10.1515/9783110587777-004

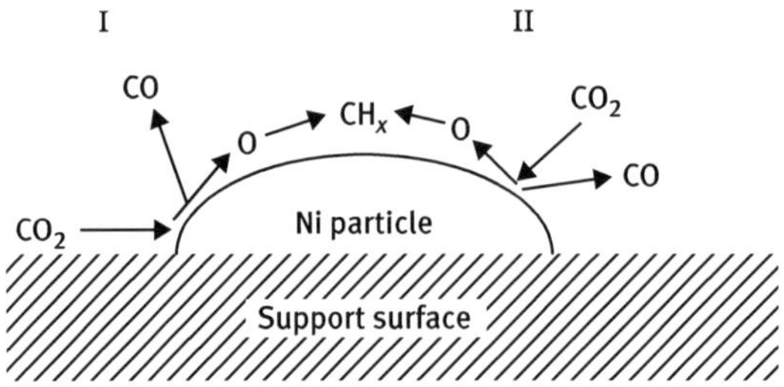

Fig. 4.1: Scheme of the MDR reaction [4].

shift to produce surface OH groups. The surface OH groups react with adsorbed CH_x intermediates being formed through CH_4 decomposition, yielding a formate-type intermediate, CH_xO. CH_xO decomposition leads to the principal reforming products, that is, H_2 and CO. Decomposition of both CH_4 and CH_xO are slow kinetic steps. According to observations by O'Connor et al. [8] over both Pt/Al_2O_3 and Pt/ZrO_2 catalysts, methane decomposition takes place over platinum. The main difference between the two catalysts concerns the carbon dioxide dissociation. X-ray Photoelectron Spectroscopy (XPS) and Diffusion- Reflection Infrared Fourier Transformed Spectroscopy (DRIFTS) data obtained for an Rh/La_2O_3 catalyst [6] indicate that the only visible surface species in this case are lanthanum oxycarbonate and mainly Rh^o, whereas both linear and bridge-bonded CO adsorbed on metallic rhodium are present during MDR.

The metal–support interface may also contain active sites for subsequent CHO formation and decomposition. Hence, the support may affect the catalyst activity by altering the stability of any intermediate species at the metal–support interface. To gain insights into the specificity of MDR over Me-supported fluorite-like doped ceria–zirconia catalysts, a combination of transient kinetic methods including TAP and SSITKA with pulse microcalorimetry and spectral studies such as in situ FTIRS techniques has been used [10]. One of the most significant findings of this study is that both CH_4 and CO_2 dissociate independently of each other on metal and support sites, respectively. However, metal alloy nanoparticles (Pt, Ru, Ni + Ru) are not only involved in CH_4 activation, but may mediate CO_2 activation as well. Thus, in the case of Ru + Ni/PrSmCeZrO catalyst, for CO_2 molecules adsorbed at the metal–support interface, the C–O bond rupture is facilitated by Ni + Ru surface species.

4.3 TAP studies

Generally, Ni + Ru-supported catalysts in their oxidized state are most active for the complete oxidation (combustion) of methane. The catalysts pretreated by reduction with hydrogen are effective for dry reforming [11–13]. The TAP studies were performed with the reduced catalyst [10]. Successive pulses of CO_2 and CH_4 are efficiently transformed into CO and CO + H_2, respectively. Furthermore, the amount of CO generated during a CO_2 pulse is similar to that produced during a CH_4 pulse, indicating that no carbon is left on the catalyst surface. In the pump-probe experiments with heavy isotope oxygen (^{18}O) labeled carbon dioxide, only formation of $C^{16}O$ was observed when the reduced Ru + Ni/ PrSmCeZrO catalyst was exposed to pulse of $C^{18}O_2$. Since the only possible source of ^{16}O in this experiment was the support, it was suggested that $C^{18}O_2$ was able to exchange oxygen with the surface very fast.

According to Sadykov et al. [11, 14, 15] and Galdikas et al. [16], a complex exchange mechanism is involved with two oxygen atoms participating in the heteroexchange. For $Ce_xZr_{(1-x)}O_2$ solid solutions [16], the diffusion coefficient was found to depend on Ce percentage and was the highest, that is, ~1.6×10^{-18} m^2/s at 850 °C, for $Ce_{0.15}Z_{0.85}O_2$. Oxygen diffusion coefficients along interfaces (i.e., surface as well as domain boundaries) for both catalysts (i.e., Pt/PrCeZrO and LaNiPt/PrSmCeZrO) were found to be much higher (see Table 4.1). Application of $C^{18}O_2$ SSITKA allowed us to estimate the oxygen mobility in catalysts under steady-state MDR conditions when the catalysts are in the reduced state. The oxygen diffusion coefficients along interfaces remained high, thus demonstrating that, indeed, the surface diffusion can provide the required fast transfer of reactive oxygen species from support sites to the metal–support interface under reaction conditions.

Table 4.1: Oxygen diffusion coefficients in Pt/PrCeZr and LaNiPt/PrSmCeZr catalysts as measured by $^{18}O_2$ and $C^{18}O_2$ SSITKA [10].

Sample	D_{eff}*, s^{-1}	D_{bulk}**, 10^{-18} m^2/s	$D_{interfaces}$***, 10^{-16} m^2/s
	Pt/PrCeZrO		
$^{18}O_2$ [31]	0.04	4	>33
$C^{18}O_2$	0.003	–	>2
	LaNiPt/PrSmCeZrO		
$^{18}O_2$ [31]	>0.03	3	>25
$C^{18}O_2$	0.008	–	>5

*D_{eff} – effective average oxygen diffusion coefficient for catalyst estimated by solving the system of differential equations describing SSITKA data.
**D_{bulk} – oxygen diffusion coefficient within oxide domains estimated by using D_{eff} and domain size.
***$D_{interfaces}$ – oxygen diffusion coefficient along domain boundaries estimated with a regard for the relative amount of oxygen located within domain boundaries [10, 15].

During subsequent CH_4 pulses only $C^{16}O$ was produced, indicating that oxygen was able to be supplied from the support during 0.6 s time lag between CO_2 and CH_4 pulses.

After a prolonged supply of $C^{18}O_2$ pulses without any CH_4 present, the response spectra for $C^{18}O$ appeared. A total of 8,000 pulses were required for this, indicating that the support oxygen was partially replenished during the previous CO_2 pulse series. Therefore, either $C^{18}O_2$ dissociation takes place near the metal/support interface or it occurs on oxygen vacancies on the support, and the oxygen isotope is in equilibrium with the surface oxygen at the metal/support interface. This may be related to the fact that a mobile oxygen pool is present in the fluorite-like doped ceria–zirconia.

In the subsequent series of $^{13}CO_2$ and $^{12}CH_4$ pulses upon pump-probe experiments with heavy isotope (^{13}C) labeled carbon dioxide (Figs. 4.2 and 4.3), only ^{13}CO was observed during the first pulse, while a $^{12}CH_4$ pulse resulted in ^{12}CO production. Hence, any exchange of carbon atoms between the reactants is absent, which excludes the existence of any common intermediate.

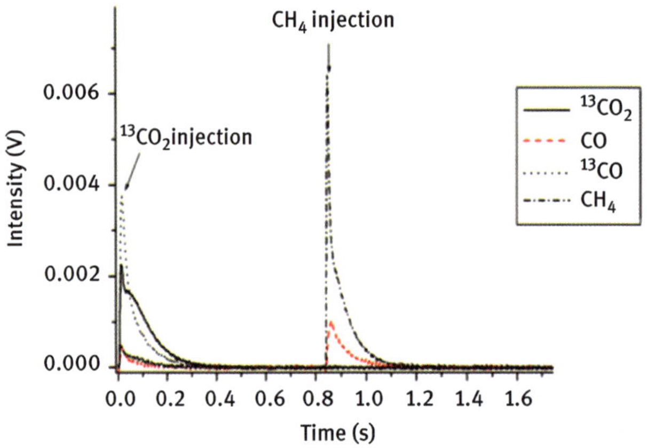

Fig. 4.2: $^{13}CO_2$, ^{12}CO, ^{13}CO and CH_4 responses corresponding to $^{13}CO_2$ and $^{12}CH_4$ pump-probe experiment over reduced Ru + Ni/ PrSmCeZrO catalyst at 750 °C. Injection times were 0 s for $^{13}CO_2$ and 0.8 s for CH_4.

Isotopic labeling studies have shown that the rate of the catalyst reoxidation by CO_2 as an oxygen supplier greatly exceeds that of its reduction by methane. CH_4 dissociation is the rate-limiting step. Hence, the rate coefficients for the elementary steps involved in CH_4 and CO_2 dissociation were estimated at ~1–10 s^{-1} and >10^2 s^{-1}, respectively. Moreover, the removal of carbon remaining on the metal sites occurs during CO_2 pulses due to the reactivity of the oxygen species and the mobility of the surface carbon species on the catalyst. Rapid oxygen redistribution between adsorption sites located on metal and oxide surface occurs during catalysis. Being strongly adsorbed on basic supports carbon dioxide can be activated through the formation of the surface carbonate species.

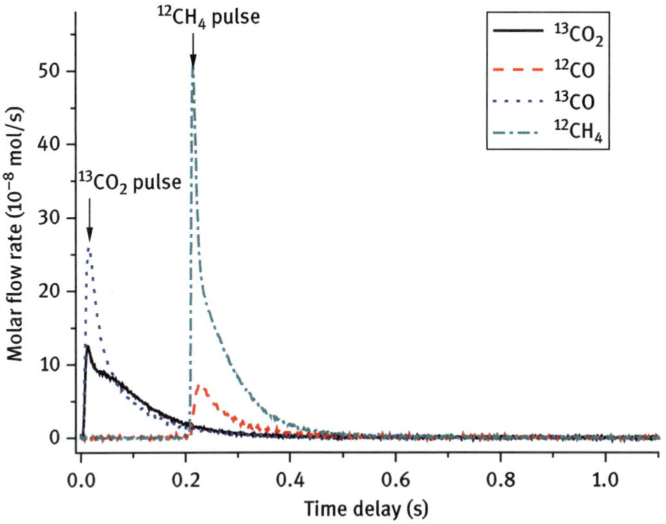

Fig. 4.3: $^{13}CO_2$, ^{12}CO, ^{13}CO and CH_4 outlet molar flow rates corresponding to $^{13}CO_2/CH_4$ alternating pulse experiment over reduced Ru + Ni/ PrSmCeZrO catalyst at 1,023 K. Time lag for CH_4 was 0.2 s.

4.4 Pulse microcalorimetry studies

In general, the changes in CH_4 conversion and CO/CO_2 selectivity with the CH_4 pulse number and, hence, the reduction degree of the catalyst sample (Fig. 4.4) reasonably agree with trends observed for TAP experiments (vide supra). The CO formation, which already occurs during the first CH_4 pulse admitted onto the oxidized sample surface, supports the hypothesis about a primary route of syngas formation via a CH_4 pyrolysis-partial oxidation route. The observation of rather high degrees of CH_4 conversion after the removal of about one monolayer of oxygen from the

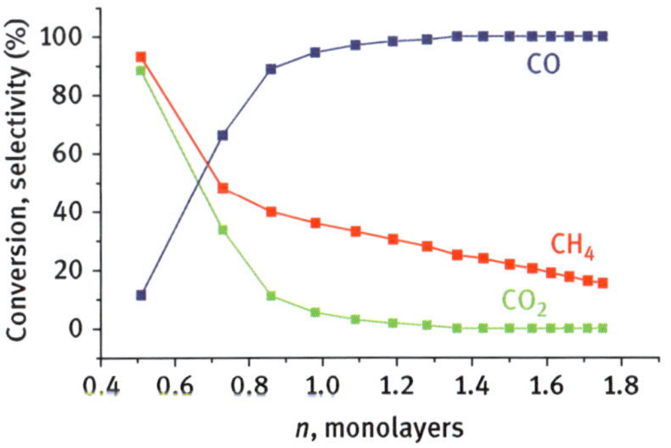

Fig. 4.4: Dependence of CH_4 conversion, CO and CO_2 selectivity on the degree of Pt/PrCeZrO sample reduction (*n*) by pulses of 7% CH_4 in He at 600 °C.

sample again underlines a high rate of oxygen diffusion from the bulk of oxide particle to the surface, in order to compensate the used surface oxygen during the preceding pulses. The average heat of oxygen adsorption on a partially reduced surface, that is, ~600 kJ/mol O_2 (Fig. 4.5), is close to values corresponding to bonding strength of bridging (M_2O) oxygen forms located at Ce cations [10].

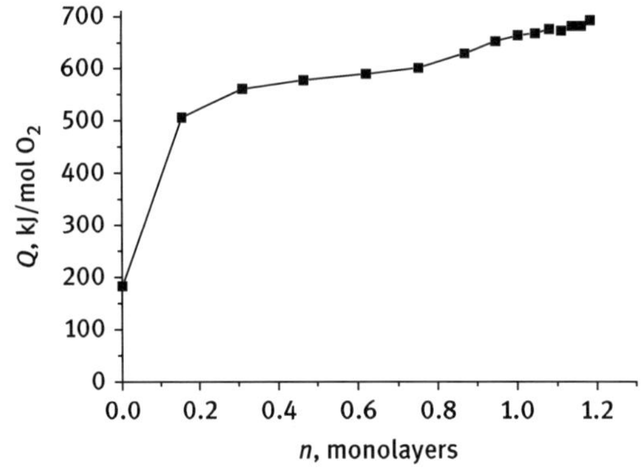

Fig. 4.5: Enthalpy of oxygen adsorption (*Q*) versus reduction degree (*n*) estimated from the heats of 1.4% Pt/$Pr_{0.3}Ce_{0.35}Zr_{0.35}O_2$ sample reduction by CO pulses at 600 °C.

The average oxygen bonding strength increases with the catalyst reduction degree (Fig. 4.5), which results in the increase of the enthalpy of CH_4 interaction with the catalyst (Fig. 4.6) corresponding to the enthalpy of methane transformation into deep and partial oxidation products with a due regard for syngas selectivity. Some decline of the heat of reduction at a reduction degree exceeding 1.5 monolayer can be explained by the increasing contribution of CH_4 cracking in agreement with CH_4 TPR results []. A practically linear (i.e., rather weak) dependence of the CH_4 conversion on the reduction degree agrees with this weak variation of the heat of CH_4 transformation.

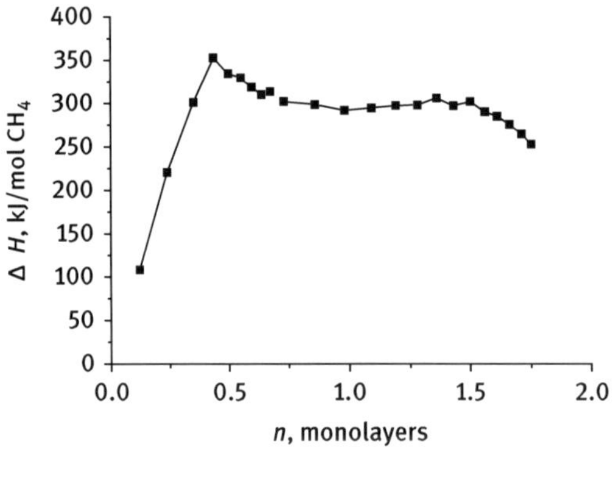

Fig. 4.6: Enthalpy of CH_4 interaction with 1.4% Pt/$Pr_{0.3}Ce_{0.35}Zr_{0.35}O_2$ sample as a function of reduction degree (*n*) at 600 °C.

Since the surface diffusion rate, that is, reverse oxygen spillover from the support to Pt, is high, this provides some coverage of Pt by adsorbed oxygen, hence favoring CH_4 activation.

For all investigated catalysts in the steady-state of MDR, reactant conversions in mixed pulses and in pulses containing only pure components were practically identical (Figs. 4.7 and 4.8). Moreover, the product selectivity was the same, that is, methane producing $CO + H_2$, $CO_2 - CO$, in several pulses as long as less than ~ 30% of oxygen is removed/replenished. In agreement with the isotope transient studies and TAP, this underlines the independent activation of the two reactants on different active sites with rapid oxygen migration between them.

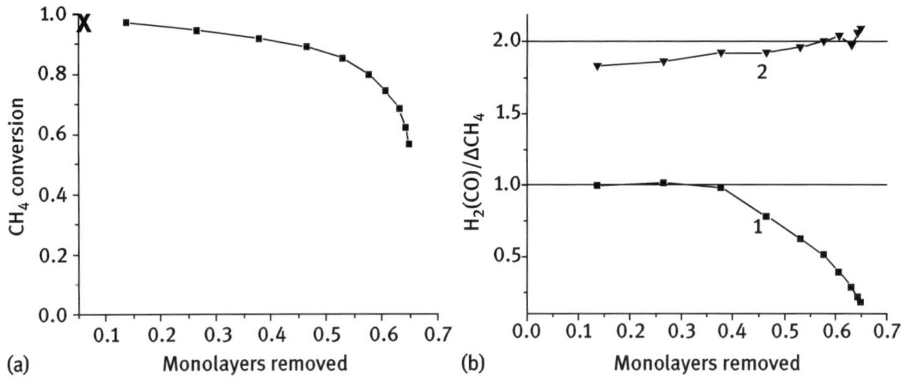

Fig. 4.7: Variation of CH_4 conversion (a, *x* on the ordinate axes marks CH_4 conversion in the mixed $CH_4 + CO_2$ pulse) and products selectivity (b, 1 – $CO/\Delta CH_4$; 2 – $H_2/\Delta CH_4$) in the course of steady-state Ru + Ni/PrSmCeZr catalyst reduction by pulses of 7% CH_4 in He at 700 °C.

The measured heat values correspond to removal/replenishing of the bridging surface/interface oxygen forms with a desorption heat ~600–650 kJ/mol O_2 (Table 4.2).

Note that CO_2 conversion rapidly declines with the pulse number as the catalyst is reoxidized (Fig. 4.8). In agreement with calorimetric data demonstrating that the heat of surface reoxidation by CO_2 is constant, such kinetics reasonably fitted by a first-order equation, suggests that the surface sites are occupied following a uniform adsorption energy. The rate coefficients for CO_2 consumption are very close for Pt and $LaNiO_3$-supported samples (i.e., ~10^2 s^{-1}) while being an order of magnitude higher for Ni + Ru-supported sample (i.e., ~10^3 s^{-1}). In agreement with SSITKA results (see above), this suggests that Ni–Ru alloy nanoparticles participate in CO_2 activation, perhaps, by favoring the C–O bond rupture in CO_2 molecules adsorbed at metal–support interface.

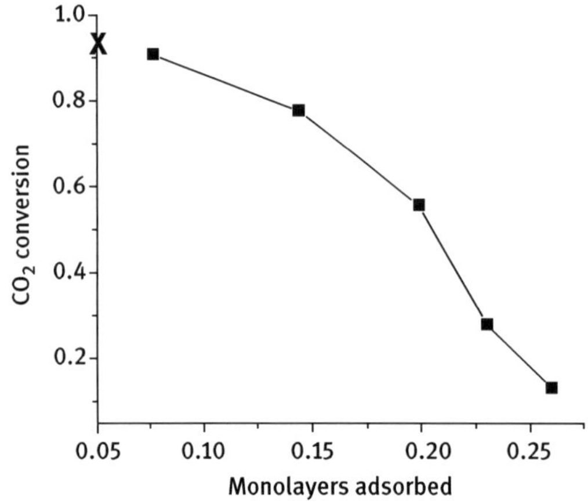

Fig. 4.8: Variation of CO_2 conversion in the course of steady-state Ru + Ni/PrSmCeZr catalyst oxidation by pulses of 7% CO_2 in He at 700 °C; *x* on the ordinate axes marks CO_2 conversion in the mixed CH_4 + CO_2 pulse.

Table 4.2: Characteristics of bonding strength of reactive bridging oxygen forms for catalysts in the steady-state by reduction of CH_4 or CO pulses and reoxidation by CO_2 or O_2 pulses at 700 °C.

Catalyst composition	Heat of oxygen desorption, kJ/mol O_2			
	CO_2^*	CH_4^*	CO^*	O_2^*
Pt/PrCeZr	640	650	640	640
LaNi/PrSmCeZr	660	630	650	560
Ru + Ni/PrSmCeZr	630	670	635	550
Ru + Ni/PrSmCeZr/YSZ	640	740	645	550

*Estimated by using pulses of respective component.

4.5 SSITKA studies

Figure 4.9 illustrates typical SSITKA responses corresponding to feed switches from ^{12}C normal isotope composition to that containing labeled $^{13}CH_4$. As can be seen, the transients are fast indicating that the steady-state surface coverage by carbon-containing species is quite low, that is, not exceeding 10% of a monolayer, calculated on the basis of the metallic surface. After switching the feed stream from $^{12}CH_4 + {}^{12}CO_2 + He$ to $^{12}CH_4 + {}^{13}CO_2 + He + 1\%Ar$ for the Ru + Ni/PrSmCeZr catalyst at 650 °C, the fractions of ^{13}C in CO and CO_2 in the effluent increased without any delay relative to the Ar tracer concentration, and at each moment the total number of ^{13}C atoms in CO and CO_2 was equal to that in the inlet $^{13}CO_2$, so there was no carbon isotope accumulation on the surface, in addition to the amount of carbon adsorbed on the catalyst surface under steady-state conditions. This demonstrates that the concentration of C-containing intermediates, that

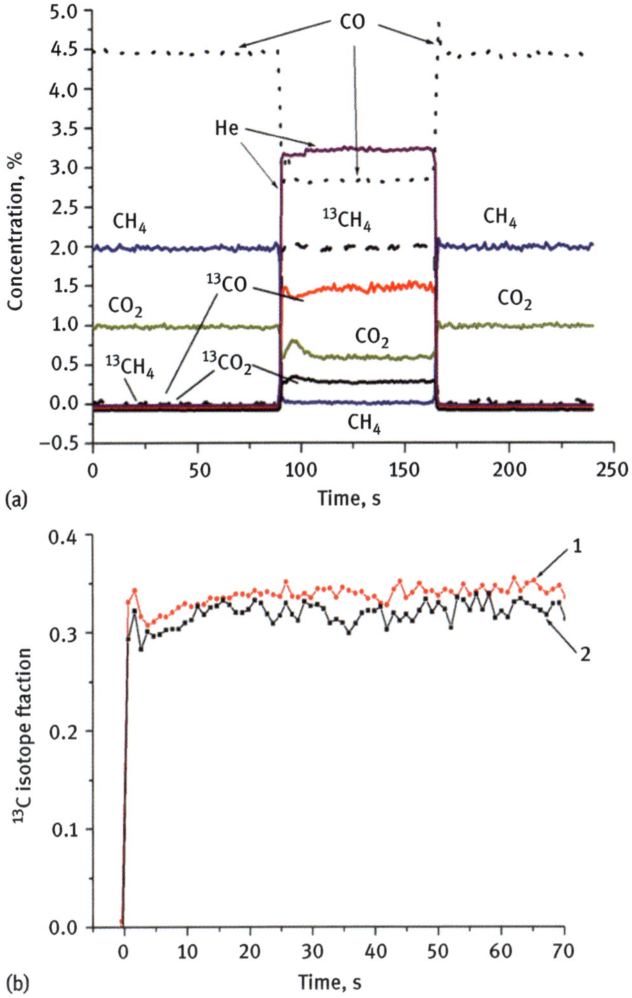

Fig. 4.9: (a) Steady-state isotopic transients after the feed switches $CH_4 + CO_2 + He \rightarrow {}^{13}CH_4 + CO_2 + He \rightarrow CH_4 + CO_2 + He$ for Pt/PrCeZr sample at 830 °C, 15 ms contact time and inlet concentration of CH_4 and CO_2 4%. (b) Time dependence of ^{13}C isotope fraction in CO (1) and CO_2 (2) after switching $CH_4 + CO_2 + He \rightarrow {}^{13}CH_4 + CO_2 + He$ for Pt/PrCeZr sample at 830 °C, 15 ms contact time and inlet concentration of CH_4 and CO_2 4%.

is, carbonates, carbides, etc. on the steady-state surface remains small but stable. This also suggests that the fraction of ^{13}C in CO should be equal to a half a sum of ^{13}C fractions in CO_2 and CH_4:

$$\alpha_{CO} = \frac{\alpha_{CO_2} + \alpha_{CH_4}}{2} .$$

In the present experiments, the ^{13}C fractions in the reactants were equal to $\alpha_{CO_2} = 0.7$ and $\alpha_{CH_4} = 0$, respectively. In this case, the estimated ^{13}C fraction in CO should be equal to 0.35, while the experimental value was found to be equal to 0.4. At the same time, the ^{13}C fraction of CO_2 in the effluent is much lower than that in the inlet feed being equal to 0.46. This means that in the course of reaction some exchange of carbon atoms between CO and CO_2 proceeds, while there is no transfer of ^{13}C into CH_4. This proves that the interaction of the catalyst with CH_4 and CO_2 occurs independently, the first one being irreversible and the second one reversible.

The simplest mechanism corresponding to this statement can be presented as follows:

1) $[ZO] + CH_4 \rightarrow [Z] + CO + 2H_2,$
2) $[Z] + CO_2 \leftrightarrow [ZO] + CO.$

Here methane irreversibly interacts with oxidized sites leading to CO and H_2 formation. Most likely metallic sites are involved in these steps. The methane transformation is followed by a reversible reoxidation of the reduced sites by carbon dioxide.

The total CO formation rate according to this scheme is equal to

$$w_{\Sigma_{CO}} = w_1 + w_2 - w_{-2}.$$

In the case of labeled CH_4, the redistribution of ^{13}C among all C-containing feed compounds is observed. The kinetic parameters characterizing the ^{13}C exchange rate and specific rates of CH_4 and CO_2 consumption are shown in Table 4.3. The rate coefficients of CO_2 transformation as estimated by SSITKA exceed those of CH_4 consumption by approximately an order of magnitude. As already mentioned, these results are in line with the resistance to coking of the studied catalysts.

Hence, under steady-state conditions, the MDR rate is limited by methane interaction with the catalyst. Note that both w_1 (i.e., the rate of CH_4 transformation) and w_2 (i.e., the rate of catalyst reoxidation by CO_2) at a lower temperature (550 °C) are apparently higher for catalysts containing Ru + Ni as compared with those for Pt/PrCeZr at a higher temperature (735 °C). Since the specific activity of Pt/PrSmCeZr is even lower than that of Pt/PrCeZr (Table 4.3), this difference could not be explained by the effect of support composition. Hence, this result suggests that Ru + Ni alloy nanoparticles are not only involved in CH_4 activation but also help to activate carbon dioxide. This confirms the key role of interface sites for this bi-functional reaction mechanism.

4.6 FTIRS in situ studies

The question whether the surface carbonate species participate in the catalytic cycle by providing oxygen which reacts with the carbon produced from methane activated

Table 4.3: Elementary step rate assessment in MDR using SSITKA.

T,°C	α_{CO_2}	α_{CO}	w_2/w_1	w_1, mkmol/g min	w_2, mkmol/g min	w_{-2}, mkmol/g min
			Pt/PrCeZr			
735	0.12	0.14	42	0.6	25	24
785	0.21	0.23	34	1.0	34	33
830	0.32	0.34	30	1.6	48	46
			Ru + Ni/PrSmCeZr			
550	0.52	0.46	7.7	4.2	32	28
600	0.49	0.42	6.0	6.2	36	30
650	0.46	0.39	5.5	7.6	42	34
			Ru + Ni/PrSmCeZr/YSZ			
550	0.49	0.46	11.5	1.5	17	15
600	0.44	0.42	24	2.0	47	45
650	0.40	0.39	43	2.5	106	103

on the metal sites is still open. To elucidate whether the main route of reductive transformation of CO_2 at reaction conditions proceeds via carbonates as intermediates, the reactivity of carbonate species on La-containing catalysts was studied, because carbonates bound to La cations were expected to be very stable. Interaction of CO_2 with the catalysts could yield a variety of carbonate-like species due to the surface basicity.

Results of in situ FTIRS study are shown in Fig. 4.10. In this study, thin pellets, that is, 2.8–3.6 cm^2 in area and 60–90 mg in weight, of the LaNi/PrSmCeZrO_2 catalyst in the IR cell adjusted to experimental temperature (600, 650 °C) were firstly exposed to CO_2. After recording spectra carbon dioxide was removed from the gas phase by freezing in a trap cooled by liquid nitrogen, and again the IR spectrum was registered. Then, CH_4 was injected and the IR spectra were registered.

No matter which is the active metal, lanthanum oxycarbonates are formed by the interaction of La_2O_3 with CO_2. The lanthanum oxycarbonate bands are observed in the range 1,300–1,500 cm^{-1}. As follows from Fig. 4.10, bands at 1,385 and 1,460 cm^{-1} are observed corresponding to bridging surface carbonates. Though their intensity strongly declines after removing gas-phase CO_2, they are still retained on the surface. The existence of a pool of carbonate species, mostly dioxomonocarbonate $La_2O_2CO_3$, at 700–800 °C under a partial pressure of CO_2 which corresponds to the reforming conditions was demonstrated by Slagtern et al. [17].

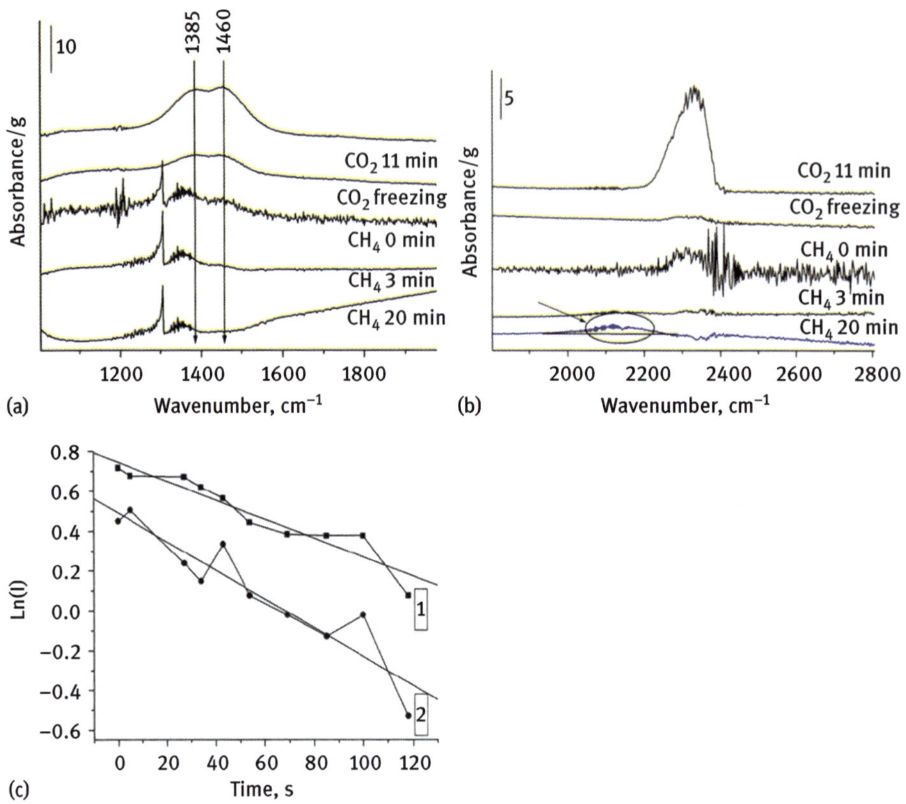

Fig. 4.10: FTIR spectra of LaNi/PrSmCeZrO_2 sample in the range of surface carbonates (a) and CO and CO_2 gas-phase molecules' (b) vibrations under contact with CO_2 followed by admission of CH_4 at 600 °C; variation of the intensity of carbonate bands with time under contact with CH_4 in coordinates of the first-order rate equation (c) 1 – band 1,385 cm^{-1}, 2 – 1,460 cm^{-1}.

The type of carbonate species seems to be different for various supports. Fast equilibrium was found to be achieved between the carbonates and the carbon dioxide in the gas phase, but essential irreversibility was observed under steady-state reaction conditions [17]. The lanthanum carbonate species readily react with CH_4 with formation of gas-phase CO, since respective bands disappear (Fig. 4.10a). Indeed, as it was studied by Krylov et al. [18], surface carbonate species may react with carbon formed via methane decomposition. Again, admission of CH_4 into the cell causes the formation of new carbonates (Fig. 4.10b). Linearization of the intensity variation in coordinates of the first-order equation revealed their identical slopes for both bands, which confirms their assignment to the same bridging carbonate (Fig. 4.10c).

The rate coefficients of a first-order reaction of gas-phase CH_4 with the surface carbonates over the nanocomposite fluorite-like doped ceria–zirconia catalysts were found to be very small compared to that of the rate-limiting step of CH_4 dissociation (Table 4.4). Therefore, the surface carbonate may assist with continuous supply of activated oxygen to the metal sites preserving them from deactivation. However, they can only be considered as spectator species in the catalytic cycle.

Table 4.4: Rate constants of the interaction of the surface carbonates with gas-phase CH_4.

Sample	Temperature, °C	Constant, s^{-1}
Ru/PrSmCeZr	600	0.003
LaNi/PrSmCeZr	600	0.007
LaNi/PrSmCeZr	650	0.009

4.7 Summary of mechanism

The investigation of the Me/PrSmCeZrO catalysts has revealed the most important features of nanocomposite catalysts for MDR:
- reducibility of the support
- oxygen mobility on the surface and in the bulk of the specific catalyst
- reoxidation of the support by CO_2

The mechanistic scheme of syngas formation during methane dry reforming on Me/PrSmCeZrO catalyst and sequence of elementary steps over Ni–Ru/$Sm_{0.15}Pr_{0.15}Ce_{0.35}Zr_{0.35}O_2$ catalyst can be proposed as it is presented in Fig. 4.11.

The mechanistic approach suggests that during reduction, methane is activated and decomposed to carbon and hydrogen on the metal sites. Hence, methane dissociation on Ru may result in the formation of different hydrocarbonaceous species, namely methylidyne (CH), vinylidene (CCH_2), and ethylidyne (CCH_3) species which, at temperatures exceeding 700 K, may transform into the graphitic phase [19]. Hydrogen species might be present on the surface in the form of either H_s or CH_{xs}. When the metal particle is preoxidized, H_2O is formed as a result of the reduction of the metal sites. The cationic metal particles cause labilization of lattice and surface oxygen. The lattice oxygen from oxide support is consumed for CO or H_2O formation and oxygen vacancies are generated again. Similar to the literature findings [4, 20], it was suggested that CO produced from methane originates from the removal of carbon that is deposited on the metal particle during CH_4 decomposition. The carbon originating from methane partially reduces the oxide support near the perimeter of the metal particle, thus creating oxygen vacancies in the support. The oxygen

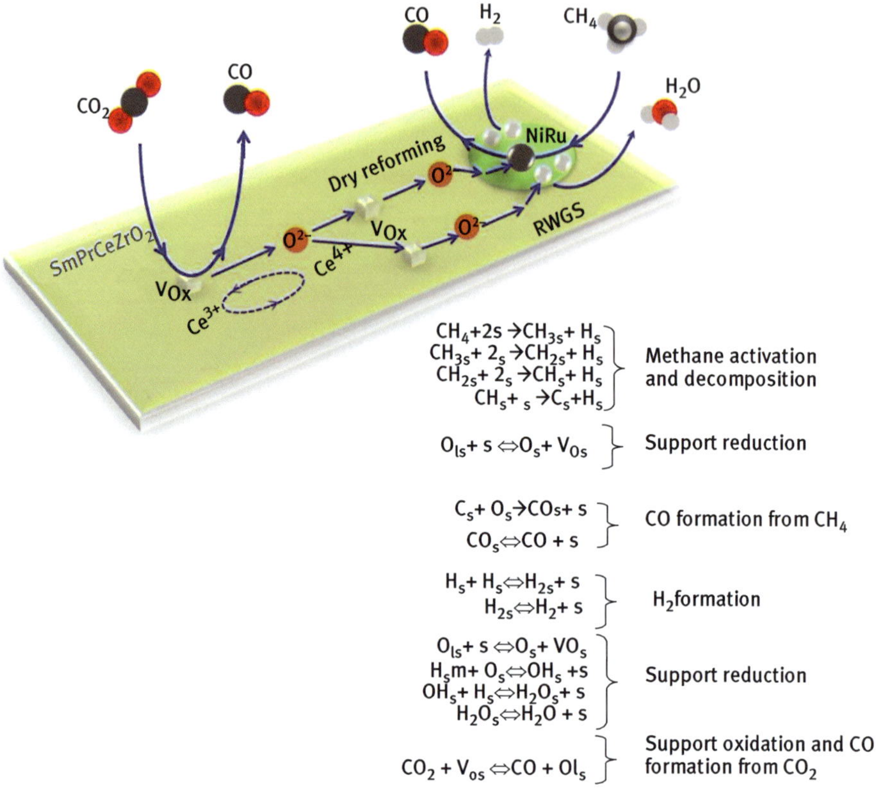

Fig. 4.11: Simplified scheme of methane dry reforming and sequence of elementary steps over Ni–Ru-$Sm_{0.15}Pr_{0.15}Ce_{0.35}Zr_{0.35}O_2$ catalyst (→ – irreversible step; ⇔ – reversible steps; s – surface sites; O_{ls} – lattice oxygen atoms; V_{OS} – lattice oxygen vacancy).

formed during CO_2 dissociation replenishes the oxygen in the lattice, thus providing a redox mechanism. The long-term activity of the composite catalyst would depend upon the balance between the ability for CO_2 dissociation, oxygen exchange and removal of carbon from the metal surface.

References

[1] Bradford, M.C.J., & Vannice, M.A. CO_2 reforming of CH_4. *Catalysis Reviews* 1999, 41(1), 1–42.

[2] Wei, J., & Iglesia, E. Structural requirements and reaction pathways in methane activation and chemical conversion catalyzed by rhodium. *Journal of Catalysis* 2004, 225, 116–127.

[3] Tauster, S.J. Strong metal–support interactions. *Accounts of Chemical Research* 1987, 20, 389–394.

[4] Tomishige, K., Yamazaki, O., Chen, Y., Yokoyama, K., Li, X., & Fujimoto, K. Development of ultra-stable Ni catalysts for CO_2 reforming of methane. *Catalysis Today* 1998, 45, 35–39.

[5] Erdöhelyi, A., Cserényi, J., Papp, E., & Solymosi, F. Catalytic reaction of methane with carbon dioxide over supported palladium. *Applied Catalysis A: General* 1994, 108, 205–219.

[6] Múnera, J.F., Irusta, S., Cornaglia, L.M., Lombardo, E.A., Cesar, D.V., & Schmal, M. Kinetics and reaction pathway of the CO_2 reforming of methane on Rh supported on lanthanum-based solid. *Journal of Catalysis* 2007, 245, 25–34.

[7] Bradford, M.C.J., & Vannice, M.A. Catalytic reforming of methane with carbon dioxide over nickel catalysts II. Reaction kinetics. *Applied Catalysis A: General* 1996, 142, 97–122.

[8] O'Connor, A.M., Schuurman, Y., Ross, J.R.H., & Mirodatos, C. Transient studies of carbon dioxide reforming of methane over Pt/ZrO_2 and Pt/Al_2O_3. *Catalysis Today* 2006, 115, 191–198.

[9] Tsipouriari, V.A., & Verykios, X.E. Kinetic study of the catalytic reforming of methane with carbon dioxide to synthesis gas over Ni/La_2O_3 catalyst. *Catalysis Today* 2001, 64, 83–90.

[10] Bobin, A.S., Sadykov, V.A., Rogov, V.A., Mezentseva, N.V., Alikina, G.M., Sadovskaya, E.M., Glazneva, T.S., Sazonova, N.N., Smirnova, M.Yu., Veniaminov, S.A., Mirodatos, C., Galvita, V., & Marin, G.B. Mechanism of CH_4 dry reforming on nanocrystalline doped ceria–zirconia with supported Pt, Ru, Ni, and Ni–Ru. *Topics in Catalysis* 2013, 56, 958–968.

[11] Sadykov, V., Mezentseva, N., Muzykantov, V., Efremov, D., Gubanova, E., Sazonova, N., Bobin, A., Paukshtis, E., Ishchenko, A., Voronin, V., Ross, J., Mirodatos, C., & van Veen, A. Real structure–oxygen mobility relationship in nanocrystalline doped ceria–zirconia fluorite-like solid solutions promoted by Pt, *Materials Research Society Symposium Proceedings*, 1122, 1122-O05-03.1-6, 2009.

[12] Sadykov, V., Sobyanin, V., Mezentseva, N., Alikina, G., Vostrikov, Z., Fedorova, Y., Pelipenko, V., Usoltsev, V., Tikhov, S., Salanov, A., Bobrova, L., Beloshapkin, S., Ross, J.R.H., Smorygo, O., Ulyanitskii, V., & Rudnev, V. Transformation of CH_4 and liquid fuels into syngas on monolithic catalysts. *Fuel* 2010, 89, 1230–1240.

[13] Sadykov, V., Mezentseva, N., Alikina, G., Bunina, R., Pelipenko, V., Lukashevich, A., Vostrikov, Z., Rogov, V., Krieger, T., Ishchenko, A., Zaikovsky, V., Bobrova, L., Ross, J., Smorygo, O., Smirnova, A., Rietveld, B., & van Berkel, F. Nanocomposite catalysts for steam reforming of methane and biofuels: design and performance, in *Nanocomposite materials, theory and applications*, INTECH, Vienna, 2011, 909–946.

[14] Sadykov, V.A., Mezentseva, N.V., Alikina, G.M., Lukashevich, A.I., Borchert, Yu.V., Kuznetsova, T.G., Ivanov, V.P., Trukhan, S.N., Paukshtis, E.A., Muzykantov, V.S., Kuznetsov, V.L., Rogov, V.A., Ross, J., Kemnitz, E., & Mirodatos, C. Pt-supported nanocrystalline ceria–zirconia doped with La, Pr or Gd: factors controlling syngas generation in partial oxidation/autothermal reforming of methane or oxygenates. *Solid State Phenomena* 2007, 128, 239–248.

[15] Sadykov, V., Muzykantov, V., Bobin, A., Mezentseva, N., Alikina, G., Sazonova, N., Sadovskaya, E., Gubanova, L., Lukashevich, A., & Mirodatos, C. Oxygen mobility of Pt-promoted doped CeO_2–ZrO_2 solid solutions: Characterization and effect on catalytic performance in syngas generation by fuels oxidation/reforming. *Catalysis Today* 2010, 157, 55–60.

[16] Galdikas, A., Bion, N., Duprez, D., Virbickas, V., & Maželis, D. Modeling of diffusion process in the isotopic oxygen exchange experiments of $Ce_xZr_{1-x}O_2$ catalysts, Materials science = Medžiagotyra. *Kaunas University of Technology, Academy of Sciences of Lithuania, Kaunas*, 19(1), 83–88, 2013.

[17] Slagtern, A., Schuurman, Y., Leclercq, C., Verykios, X., & Mirodatos, C. Specific features concerning the mechanism of methane reforming by carbon dioxide over Ni/La_2O_3 catalyst. *Journal of Catalysis* 1997, 172, 118–126.

[18] Krylov, O.V., Mamedov, A.K., & Mirzabekova, S.R. Interaction of carbon dioxide with methane on oxide catalysts. *Catalysis Today* 1998, 42, 211–215.
[19] Choudhary, T.V., Aksoylu, E., & Goodman, D.W. Non-oxidative activation of methane. *Catalysis Reviews. Science and Engineering* 2003, 45, 151–203.
[20] Stagg-Williams, S.M., Noronha, F.B., Fendley, G., & Resasco, D.E. CO_2 reforming of CH_4 over Pt/ZrO_2 catalysts promoted with La and Ce oxides. *Journal of Catalysis* 2000, 194, 240–249.

5 Kinetics and mechanism of high-temperature N_2O decomposition

Catalytic decomposition of N_2O is generally described by the following scheme [1, 2]:

1) $N_2O + S \rightarrow N_2O - S$
2) $N_2O - S \rightarrow N_2 + O - S$
3) $O - S + N_2O \rightarrow N_2 + O_2 + S$ (the Eley–Rideal mechanism)
4) $2O - S \leftrightarrow O_2 + 2S$ (the Langmuir–Hinshelwood mechanism)

where S is the surface active site, N_2O–S is the adsorbed or chemisorbed N_2O, and O–S is the adsorbed or chemisorbed O.

For many oxides, it is known that in the temperature range 300–600 °C, N_2O conversion decreases in the presence of oxygen in the reaction mixture, which can be due to the competitive adsorption of N_2O and oxygen on the same active sites [3, 4] being oxygen vacancies or metal cations. Having studied the activity of manganites in N_2O decomposition at 400–600 °C vs lattice parameters, Raj et al. [5] came to the conclusion that oxygen desorption (stage 4) appeared to be the rate-controlling step of the reaction. The isotopic oxygen exchange carried out by Sazonov et al. on perovskites [6] and by Winter on a number of oxides [7] at 400–600 °C confirmed these results, because a linear correlation was observed between the rate of oxygen homoexchange and that of N_2O decomposition. It means that at medium temperatures (300–600 °C), the catalytic activity of manganites is determined by the binding strength of adsorbed or chemisorbed oxygen. However, at higher temperatures (700–900 °C), the correlation between the rate of N_2O decomposition and the rate of bulk oxygen diffusion found for $La_{1-x}Sr_xMnO_{3\pm\delta}$ (x = 0, 0.3, 0.5) samples in our previous work [8] indicated that reactivity of perovskites can depend on the lattice oxygen mobility. To understand what this correlation means in respect to the mechanism of N_2O decomposition, we studied the dynamics of oxygen transfer from N_2O to O_2 during the reaction of N_2O decomposition on $LaMnO_{3+\delta}$. The question was how can the lattice oxygen mobility influence the catalytic properties at the steady state.

If oxygen desorption is the rate-limiting step of the reaction, then the correlation should exist between the rate of N_2O decomposition and the rate of the surface oxygen exchange. It is known that heat treatment of $LaMnO_{3+\delta}$ at 900 °C in an air atmosphere can result in creation of the cation vacancies, V_{La} and V_{Mn}, which causes some Mn^{3+} to convert into Mn^{4+} state [9, 10] . In such mixed-valence manganites the holes are the charge carriers, thus determining their ferromagnetic

https://doi.org/10.1515/9783110587777-005

metallic behavior and good electrical conductivity at high temperatures [11]. At the same time, it is exactly a small amount of oxygen vacancies that is responsible for a lower bulk oxygen mobility of $LaMnO_{3+\delta}$ in comparison to $LaFeO_{3-\theta}$ and $LaCoO_{3\pm\gamma}$.Earlier, De Souza and Kilner [12] demonstrated a strong correlation between the coefficient of bulk oxygen diffusion *D* and the constant of surface oxygen exchange *R* for $(La,Sr)(Mn,Co)O_{3-\gamma}$ systems. It means that the surface concentration of oxygen vacancies, which are scarce on $LaMnO_{3+\delta}$, rather than the energetic barrier of the $O_s \leftrightarrow O_s^-$ charge transfer, can be of critical importance for dissociative adsorption of O_2 molecule. Using N_2O instead of molecular O_2 increases the rate of labeled oxygen transfer, because dissociative adsorption of N_2O can take place on coordinatively unsaturated surface cations Mn^{3+} too [13], whereas surface oxygen vacancies are the only sites for oxygen adsorption. Thus, the observed correlation between the tracer oxygen diffusion coefficient and the rate of N_2O decomposition is likely to be due to the correlation between *D* and *R*.

We failed to properly estimate the surface tracer exchange constant in $La_{1-x}Sr_xMnO_{3\pm\delta}$ ($x = 0, 0.3, 0.5$) because of strong limitation of the kinetics by bulk oxygen diffusion [8]. At the same time, we found a strong correlation between the rate of N_2O decomposition on the $La_{1-x}Sr_xFeO_{3-\theta}$ system and the surface exchange rate constant *R*, but there was no correlation between catalytic activity and the coefficient of bulk oxygen diffusion [14]. As shown in [12], the correlation between bulk oxygen diffusion and surface exchange depends on the ratio of the ionic to electronic conductivity. Indeed, Adler et al. and ten Elshof et al. [15] supposed that the rate-limiting step of the surface oxygen exchange in the $La_{1-x}Sr_xFeO_{3-\theta}$ system with high concentration of oxygen vacancies can be the O_2 chemisorption. In this case, there is no correlation between *D* and *R*, because at high availability of surface oxygen vacancies the rate of surface rate exchange will be determined most likely by charge transfer.

The revealed correlations between high-temperature catalytic activity and oxygen exchange properties indicate that oxygen desorption from the surface is likely to be the rate-determining step of N_2O decomposition for both the $La_{1-x}Sr_xMnO_{3\pm\delta}$ and $La_{1-x}Sr_xFeO_{3-\theta}$ systems. In consequence, when methane, as a reductant increasing the rate of oxygen desorption from the surface, is introduced in the reaction mixture, N_2O conversion rises (Fig. 5.1)[16].

Hence, the isotopic transient experiment with $^{18}O_2/N_2{}^{16}O$ on $LaMnO_{3+\delta}$ at 900 °C revealed that labeled oxygen transfer from N_2O to O_2 proceeds faster as compared to oxygen exchange $^{18}O_2/^{16}O_2$ and includes the interaction of N_2O with two lattice oxygen atoms of perovskite. As a result, a strong correspondence should exist between the tracer coefficient of bulk oxygen diffusion and the rate constant of surface tracer oxygen exchange. It explains the earlier observed correlation between the rate of N_2O decomposition (900 °C) and the rate of lattice oxygen diffusion (900 °C) for $La_{1-x}Sr_xMnO_{3\pm\delta}$ and its absence for the $La_{1-x}Sr_xFeO_{3-\delta}$ system.

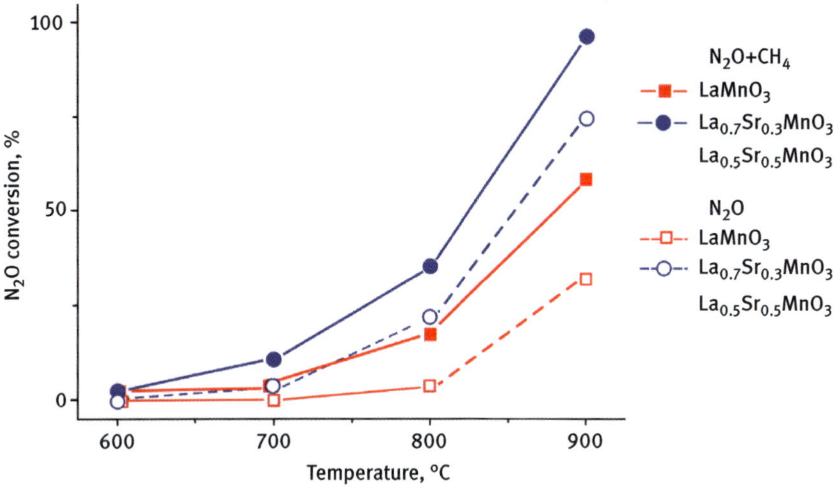

Fig. 5.1: The increase in N_2O conversion on $LaMnO_{3+\delta}$ and $La_{0.7}Sr_{0.3}MnO_{3\pm\delta}$ in the presence of methane in the reaction mixture feed.

References

[1] Kondratenko, EV., Kondratenko, VA., & Santiago, M. Perez-Ramirez. Mechanism and microkinetics of direct N_2O decomposition over $BaFeAl_{11}O_{19}$ hexaaluminate and comparison with Fe-MFI zeolites. J. Appl. Catal.: B-Environ. 2010, 99, 66–73.

[2] Li, ML., Yang, XL., Tang, L., Xiong, XM., Ren, SL., & Hu, B. Catalysts for catalytic decomposition of nitrous oxide. Prog. Chem. 2012, 24(9), 1801–1817.

[3] Kapteijn, F., Rodriguez-Mirasol, J., & Moulijn, JA. Heterogeneous catalytic decomposition of nitrous oxide. Appl. Catal.: B-Environ. 1996, 9, 25–64.

[4] Zabilskiy, M., Djinovic, P., Erjavec, B., Drazic, G., & Pintar, A. Small CuO clusters on CeO_2 nanospheres as active species for catalytic N_2O decomposition. Appl. Catal.: B-Environ. 2015, 163, 113–122.

[5] Raj, SL., & Srinivasan, V. Decomposition of nitrous oxide on rare earth manganites. J. Cat. 1980, 65(1), 121–126.

[6] Sazonov, LA., Mosvina, ZV., & Artamonov, EV. Investigation of catalytic properties of $LnMeO_3$ oxides in oxygen homomolecular exchange reaction. Kinet. Katal. 1970, 15, 120–126, in Russian.

[7] Winter ERS. Exchange reactions of oxides. Part IX. J. Chem. Soc. A 1968, 2889-2902.

[8] Ivanov, DV., Sadovskaya, EM., Pinaeva, LG., & Isupova, LA. Influence of oxygen mobility on catalytic activity of La–Sr–Mn–O composites in the reaction of high temperature N_2O decomposition. J. Catal. 2009, 267, 5–13.

[9] Van Vanroosmalen, JAM., & Cordfunke, EHP. The defect chemistry of $LaMnO_{3\pm\delta}$: 4. Defect model for $LaMnO_{3+\delta}$. J. Solid State Chem. 1994, 110, 109–112.

[10] Kilner, JA., Berenov, A., & Rossiny, J. Diffusivity of the Oxide Ion in Perovskite Oxides, T. Ishihara, Ed., Perovskite Oxide for Solid Oxide Fuel Cells, US, Springer, 2009, 95–116.

[11] Bouwmeester HJM, Burggraaf AJ. Dense ceramic membranes for oxygen separation. In: A.J. Burggraaf, L. Cot (Eds.) Membrane Science and Technology, Elsevier Science (Netherlands) 1996, 435-528.
[12] De Souza, RA., & Kilner, JA. Oxygen transport in $La_{1-x}SrxMn_{1-y}CoyO_{3\pm\delta}$ perovskites: Part II. Oxygen surface exchange. Solid State Ionics 1999, 126, 153–161.
[13] Li, GN., Pidko, EA., Filot, IAW., van Santen, RA., Li, C., & Hensen, EJM. Catalytic properties of extra framework iron-containing species in ZSM-5 for N_2O decomposition. J. Catal. 2013, 308, 386–397.
[14] Ivanov, DV., Pinaeva, LG., Isupova, LA., Sadovskaya, EM., Prosvirin, IP., Gerasimov, EY., & Yakovleva, IS. Effect of surface decoration with $LaSrFeO_4$ on oxygen mobility and catalytic activity of $La_{0.4}Sr_{0.6}FeO_{3-\delta}$ in high-temperature N_2O decomposition, methane combustion and ammonia oxidation. Appl Catal.: A-General 2013, 457, 42–51.
[15] Ten_Elshof, JE., Bouwmeester, HJM., & Verweij, H. Oxygen transport through $La_{1-x}Sr_xFeO_{3-\delta}$ membranes II. Permeation in air/CO, CO2 gradients. Solid State Ionics 1996, 89, 81–92.
[16] Ivanov, DV., Pinaeva, LG., Sadovskaya, EM., & Isupova, LA. Isotopic transient kinetic study of N_2O decomposition on $LaMnO_{3+\delta}$. J. Mol. Catal. A: Chem. 2016, 412, 34–38.

6 Application of mechanochemical methods in catalysis

Mechanochemistry is a field that deals with the effect of mechanical factors on chemical and physicochemical properties and transformations of substances in various fields of science and technology. The literature data demonstrate that mechanochemical methods can widely be applied in catalysis, too. Thus, various applications of mechanochemical factors are to:

- investigate the features of active sites and the catalytic action of oxides;
- prepare mixed oxide catalytic systems without precipitation steps associated with high water consumption;
- control rheological characteristics of pastes and grain strength during the synthesis of granulated catalysts; and
- control the catalytic process.

The least studied are the catalytic reactions with mechanochemical treatment (MT, the so-called mechanocatalytic reactions), probably due to technical difficulties associated with such reactions.

Let us use oxide catalysts for deep oxidation reactions as an example to consider the above-listed application aspects of mechanochemical activation in catalysis.

6.1 Effect of mechanochemical treatment on the catalytic activity of oxides

A review of studies concerning the effect of MT on a solid demonstrates that reactivity of oxides increases in the reactions of decomposition, dissolution and solid-phase synthesis, and changes in the catalytic activity are also observed. Thermochemical interpretation of the enhanced reactivity is based on the growth of free energy of the system due to increased concentration of defects [1]. Reactivity of a substance in each reaction can often be changed only by a certain type of defects rather than by the total defectness. So, one of the goals of studies on the effect exerted by MT on the reactivity of oxides is to reveal not only the total defectness but also the particular defects affecting the given reaction [2–5]. This task is complicated by various phenomena that occur during MT; such phenomena are divided into four groups [6].

The first group is related to brittle failure, which increases the dispersion of particles. This facilitates the growth of specific surface area of the oxide catalyst particles. However, it was found that the specific surface area of oxides not only increases upon MT but can also decrease owing to aggregation; as a result, the

https://doi.org/10.1515/9783110587777-006

steady surface area is established. Along with the surface area, the reactivity should also change during MT because for heterogeneous processes the reaction rate referred to unit mass is proportional to the interface area. Examples of symbate changes in specific surface area and reactivity for Fe_2O_3 and CuO are reported in [1].

The second group of phenomena is connected with the plastic shear flow. Plastic flows are based on a dislocation mechanism [7–9]. Exactly this group of phenomena leads to the formation of various points and extended defects. Similar to other factors that increase defectness of a solid, such as radiation, phenomena of this group activate reactivity of a solid by increasing the reactivity of unit surface area of the oxide catalyst.

The third group of phenomena is caused by heating of a substance upon mechanochemical treatment. For example, treatment of $CaCO_3$ or $Ag_2C_2O_4$ results in their thermal decomposition [10, 11].

The fourth group of phenomena implies structural changes of a substance during MT; in this case, the notion of defect becomes meaningless. Here, an increase in reactivity is related to amorphization degree of a solid [8, 12–16].

The accumulated data on MT of oxides allow revealing the regularities that are pertinent to certain groups of oxides.

Avvakumov and coworkers [7, 17–21] investigated the effect of MT on simple mixed-valence oxides having the framework (loose) structure (TiO_2, SnO_2, WO_3, PbO_2). Since MT of oxides is accompanied by intense plastic shear deformation, the elementary acts of which include generation, motion and interaction of dislocations, the authors supposed that oxygen abstraction from the oxide, that is, the mechanochemical dissociation, can occur during such treatment. The body of data acquired for the oxides of elements belonging to groups 5, 4, 6 and 8 of the Mendeleev's periodic table indicates that MT in vacuum actually leads to mechanochemical dissociation of such oxides [18–21].

Pavlyukhin and coworkers [8, 12–16] studied the effect of MT on double oxides (zinc, nickel, cobalt, copper and magnesium ferrites), which have a closely packed anionic sublattice, and determined structural features of a substance during its deep treatment. It was found that when a substance becomes X-ray amorphous and the long-range order is destroyed, the short-range order is preserved. This is accompanied by an increase in the number of cations in octahedral positions as compared to the initial spinel structure, which testifies to the formation of a substance having a new crystal structure at a constant chemical composition. Physicochemical properties of such substance, particularly the dissolution rate, differ from the initial substance. The increase in the number of occupied octahedral positions as compared to the spinel structure and the disturbance of the long-range order were attributed by the authors to the shear deformations along the [111] plane. A study on MT of perovskites (lanthanum ferrite) showed that the substance rapidly becomes X-ray amorphous without structural changes. The appearance of new properties of the substance is related primarily to the fraction of X-ray amorphous substance.

Mechanochemical reduction was not observed by the authors during MT of oxides with the closely packed structure.

When MT of oxides is carried out in a medium, the emerging structural distortions promote the interaction of the oxides with the atmosphere to form new chemical compounds. Thus, after MT of WO_2 in the ethanol (96%) atmosphere, water, hydrogen and carbon oxide were detected in gaseous products releasing from the surface of the activated oxide [22]. MT of magnesium oxide in a planetary mill in the air atmosphere leads not only to structural distortions (an increase in the lattice parameter of MgO) but also to the interaction of the freshly formed surface with water vapor and air CO_2, resulting in the formation of hydroxide and basic magnesium carbonate [23]. The formation of hydrogen during MT of copper oxide in water was observed [24]. It is accepted that mechanochemical reactions are caused by an increase in the concentration of defects; therewith, the places of termination of dislocations and the adjacent regions are supposed to serve as the active surface sites.

Butyagin et al. [25–39] proposed a "radical" approach to describe the mechanochemical phenomena. They think that MT is accompanied by bond cleavage or deformation producing reactive radicals on the surface of oxides. This approach adequately explains the interaction of activated surface with gaseous atmosphere and allows characterizing the active surface sites. Thus, activation of GeO_2 (covalent binding) leads to stabilization of the cleft ($\equiv Ge^{\cdot}$ and $\equiv GeO^{\cdot}$) and deformed ($\equiv$Ge-O-Ge$\equiv$) bonds on the surface. Upon MT of MgO (ionic binding), surface active sites are represented by the deformed Mg–O bonds that enhance donor properties of the anion (the charge transfer band in diffuse reflectance spectra shifts by 3–4 eV to the long-wave region). Activation of SnO_2 (intermediate binding) results in a partial reduction of the oxide and appearance of various electronic states in the band gap up to the conduction band (an increase in electrical conductivity). Therewith, the authors assume the existence of at least four regions in the activated particle: 1 – the surface layer containing chemisorbed gases; 2 – the strongly disordered surface layer; 3 – the plastically deformed subsurface layer; and 4 – the weakly distorted core of the particle.

The effect of MT on the catalytic properties of oxides and metals was discovered and investigated by Shrader and coworkers [40–47], Heiniche and coworkers [48–51], Japanese researchers [52, 53] and later by Buyanov and coworkers [54–62]. It was found that activity in different catalytic reactions changes with extending the MT time: in some cases, the activity increases or goes over a maximum, while in other cases it decreases.

Changes in the catalytic activity may be caused by the formation of new chemical compounds that are more active than the initial compound. For example, according to Heiniche and coworkers [48–51], the formation of nickel carbonyls from carbon oxide and nickel in the presence of sulfur or sulfide compounds is accelerated by the formation of nickel sulfide, which serves as the actual catalyst. It seems interesting that the highest activity of nickel sulfide is observed during mechanical action and decreases upon its termination.

Changes in the catalytic activity are often compared with changes in the crystal lattice parameter of oxides (iron, calcium, magnesium, copper and aluminum oxides) [40–47]. In most cases, the activity changes symbatically with the lattice parameter. For instance, the catalytic activity of copper oxide in decomposition of nitrous oxide [40] and ortho–para conversion of hydrogen [41] was considered by Shrader from these positions. Pure copper oxide milled in the oxygen atmosphere stimulates the ortho–para conversion of hydrogen. The mechanism of such conversion depends on magnetic properties of the oxide. Magnetic susceptibility is a structure-sensitive property of a solid; so it depends on distortions in a solid (on the crystallite sizes and mostly on the presence of dislocations). A substantially elevated hole conduction was observed even in the activated copper oxide samples that were calcined at 600 °C for 3 h. It was noted that the catalytic activity of copper oxide in the ortho–para conversion of hydrogen changes symbately with magnetic susceptibility and destruction degree of the CuO lattice. A complicated dependence of the nitrous oxide decomposition rate on accumulation of structural distortions in copper oxide was revealed in [41]. The extreme dependence of rate constant was attributed to changes in the reaction mechanism after accumulating a certain concentration of defects.

The effect of MT in a vibrating ball mill on powdered α-Fe_2O_3 was studied in the oxidation of carbon oxide with an excess of oxygen and in its absence [42]. A similarity was found in the dependences of CO conversion on the milling time for the reactions performed in an excess of oxygen ($2\,CO + O_2 + \alpha\text{-}Fe_2O_3 \rightarrow CO_2 + \alpha\text{-}Fe_2O_3$) and in its absence ($CO + 3\,Fe_2O_3 \rightarrow Fe_3O_4 + CO_2$) – both dependences are extremal and have a maximum at 16 h of milling. However, X-ray diffraction (XRD) differences between mechanically activated α-Fe_2O_3 samples were not found in this case. It was supposed that MT enhances the activity of the oxide by facilitating the reduction step. The authors noted that at temperatures above 300 °C the activities were irreproducible.

The enhancement of CaO activity toward nitrous oxide decomposition during MT was attributed to the increase not only in specific surface area but also in defectness [45]. A prolonged grinding of a nickel oxide powder destroys its lattice and increases activity of the powder toward hydrogen peroxide decomposition [52]. Heating of the activated sample above 300 °C for 3 h restores the properties of the initial nickel oxide.

A review [63] considers changes in the structure, chemical composition and properties of alumina and aluminosilicates under grinding. An ambiguous effect of MT on powders is noted. For alumina, the activity toward dehydration of isopropanol changes symbately with accumulation of structural distortions. For titanium pyrophosphate, structural distortions disappear under grinding, and activity toward dehydrogenation of isopropanol increases. According to Takashashi and Tsutsumi [53], the catalytic activity of zinc oxide in hydrogen peroxide decomposition and photooxidation of isopropyl alcohol decreases with increasing the amount of structural distortions in the oxide under the action of MT.

Thus, in the early studies devoted to the effect of MT on the catalytic properties of oxides, such an effect was revealed and its sufficient stability in the reaction medium was demonstrated. The improved reactivity of activated oxides was attributed either to the total increase in disorder of a solid or to the appearance of reactive radicals or new compounds on the surface. However, this could not explain the observed processes of mechanochemical deactivation related to an increase in the total defectness. So, in both the chemistry of solids and the heterogeneous catalysis, it became necessary to reveal the effect exerted on catalytic activity and reactivity not by the total defectness but by a particular type of defects responsible for a certain reaction.

The indicated approach was implemented for the first time at the Institute of Catalysis SB AS USSR. Thus, Molchanov and coworkers [64, 65] have developed a technique for inducing crystallographic shear defects in titania (using mechanochemical activation and subsequent annealing in an inert medium) and established a correlation between catalytic activity in CO oxidation and concentration of such extended defects. According to them, this correlation unambiguously indicates that active sites are the crystallographic shear planes terminating on the oxide surface. For zinc oxide, a correlation was also found between activity in CO oxidation and density of extended defects (dislocations and packing defects). The activity is enhanced by accumulation of such defects and deteriorated when density of defects decreases upon structural relaxation caused by milling of crystallites; this reflects the extremal dependence of activity on the milling time.

A cycle of studies considering the effect of MT of dispersed 3d oxides [copper, cobalt, iron and manganese (II–IV) oxides] on their catalytic activity in the deep oxidation of CO and butane, despite various phenomena accompanying the MT (decomposition of aggregates, changes in the density of extended defects, reduction of oxides, hydration and carbonization of the surface), has established a relation between changes in the catalytic activity of oxides observed during MT and changes in the density of extended defects in the oxides, which were determined by infrared (IR) spectroscopy of probe molecules (termination of extended defects on the surface leads to the formation of reduced cation clusters serving as the active sites), high-resolution electron microscopy and small-angle X-ray scattering (SAXS) [66–73]. It was found that MT of dispersed oxides, which ultimately decreases the density of extended defects in the oxides despite a simultaneous increase in the number of vacancies, deteriorates their activity in the deep oxidation reactions. The data obtained have reliably elucidated the role of extended defects in the formation of active sites in the reactions of deep catalytic oxidation.

Thus, Fig. 6.1a and 6.1b shows the catalytic activity of copper oxide (specific surface area 25 m^2/g) in butane oxidation versus MT time and displays the spectra of CO

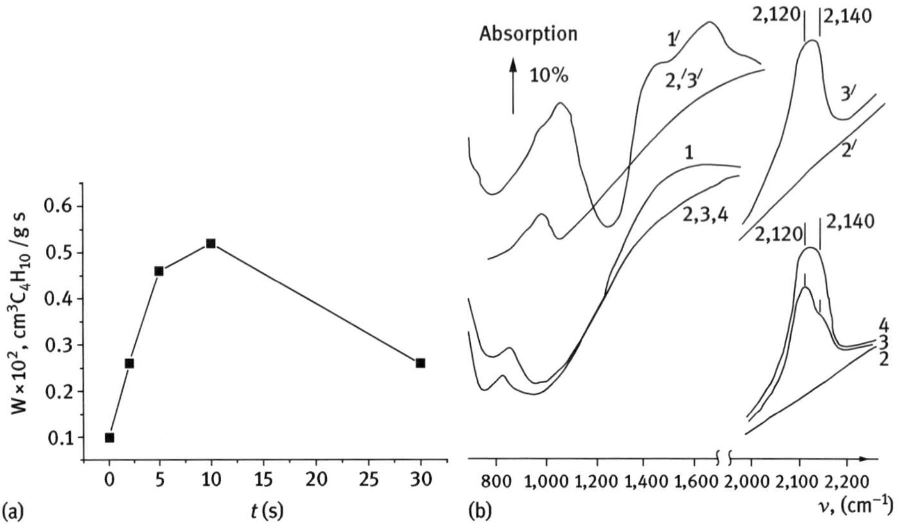

Fig. 6.1: (a) Catalytic activity of CuO (1–2 mm particles) samples in butane oxidation at 300 °C (W*10^2 cm^3 C_4H_{10}/g s) in dependence on MT time (t, s). (b) IR spectra of adsorbed CO: 1 and 1′ – spectra of the samples without high-temperature treatment: initial and after MT for 10 s, respectively; 2 and 2′ – spectra of the samples after high-temperature treatment; 3 and 3′ – spectra of carbon oxide adsorbed at 25 °C (30 Torr) on the samples after their oxidation treatment; 4 – the spectrum of CO adsorbed on the initial copper oxide after its reduction treatment.

adsorbed at room temperature (30 torr) on the initial and mechanically activated for 10 s samples after their oxidation and reduction treatments [66, 67]. MT first enhances the activity by a factor of 3–5 but later, as MT time exceeds 10 s, deteriorates it. At a constant phase composition of the samples and similar specific surface area, pronounced changes are caused primarily by changes in the state of surface active sites. Indeed, according to IR spectroscopy of CO probe molecules, the adsorption of carbon oxide at room temperature on the initial copper oxide sample after oxidation treatment leads to appearance of the absorption band at 2,120 cm^{-1} (Fig. 6.1b). After holding the sample in the carbon oxide atmosphere, the intensity of the 2,120 cm^{-1} band increases and a band appears at 2,140 cm^{-1}. The observed bands correspond to Cu^+–CO complexes [66, 73–81] that are formed on coordinatively unsaturated Cu^+ cations on the copper oxide surface. Sites with a greater effective charge $Cu^{1+\delta}$ are formed after the reduction (the band at 2,140 cm^{-1}). The causes of this phenomenon are discussed in Chapter 3. The reduction treatment expectedly increases the intensity of the 2,140 cm^{-1} band, which testifies to an increase in the number of such sites. The adsorption of carbon oxide on the CuO sample activated for 10 s after its oxidation treatment immediately results in the appearance of a broad intense band at 2,120–2,140 cm^{-1}

(Fig. 6.1b), which indicates an increase in the concentration of reduced Cu^{+} sites, particularly at the points where dislocations terminate.

Quantitative evaluation of adsorption sites from the optical density showed that the number of sites is 0.6% of a monolayer for the initial copper oxide after oxidation treatment, and increases 2.5-fold for activated samples, reaching 1.5% of a monolayer [66].

Figure 6.2 shows the catalytic activity of cobalt oxide (specific surface area 100 m^2/g) in butane oxidation versus MT time (Fig. 6.2a) and displays the spectra of CO adsorbed at room temperature (Fig. 6.2b). One can see that the catalytic activity of cobalt oxide decreases and the number of active adsorption sites also decreases according to IR spectroscopy of a probe molecule; however, no changes are observed in the phase composition and specific surface area of the oxides. Taking into account the data of Sadykov et al. [82, 83] indicating that the active site on the cobalt oxide surface includes a pair of cobalt cations located in adjacent octahedra at the points where dislocations terminate at the surface, it can be concluded that MT destroys the dislocation network, thus decreasing the number of active sites capable of adsorbing weakly bound oxygen. This is accompanied by reduction of the oxide and hence an increase in the density of point defects, which lead only to a decrease in coordination of the initial Co^{3+} cations. The data obtained by IR spectroscopy of adsorbed CO are quantitatively consistent with the kinetic data. Thus, assuming a constant extinction coefficient, the optical density ratio of 2,075 and 2,085 cm^{-1} bands gives a 1.8-fold difference for uncalcined initial and mechanically treated for 10 s Co_3O_4 samples, and a 1.27-fold difference for the calcined samples, which agree with the observed decrease in catalytic activity of these samples by a factor of 2 and 1.25, respectively [67, 68].

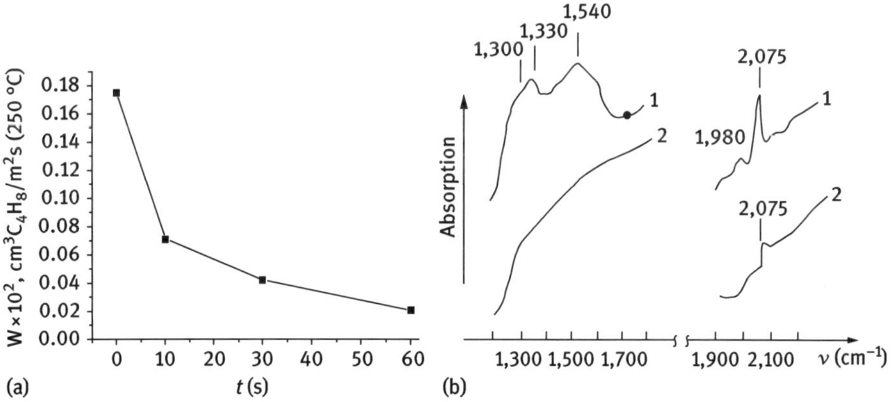

Fig. 6.2: (a) Effect of MT of Co_3O_4 on the catalytic activity toward butane oxidation at 250 °C ($W * 10^2$ $cm^3C_4H_{10}$ $/m^2s$, 1–2 mm particles). (b) IR spectra of CO adsorbed on cobalt oxide samples: 1 – initial Co_3O_4; 2 – Co_3O_4 after MT during 10 s.

Figure 6.3 shows the catalytic activity of iron oxide (specific surface area 14 m^2/g) in CO oxidation versus MT time (Fig. 6.3a) and displays the spectra of CO adsorbed at room temperature (Fig. 6.3b). As follows from the data obtained, MT of iron oxide decreases the number of clustered Fe^{2+} cations on the surface with the formation of isolated ions or their pairs (the band shift from 2,170 to 2,180 cm^{-1}). It is seen that the specific catalytic activity (SCA) of iron oxide upon mechanochemical activation decreases symbately with the number of active adsorption sites (Fe^{2+} clusters) [71].

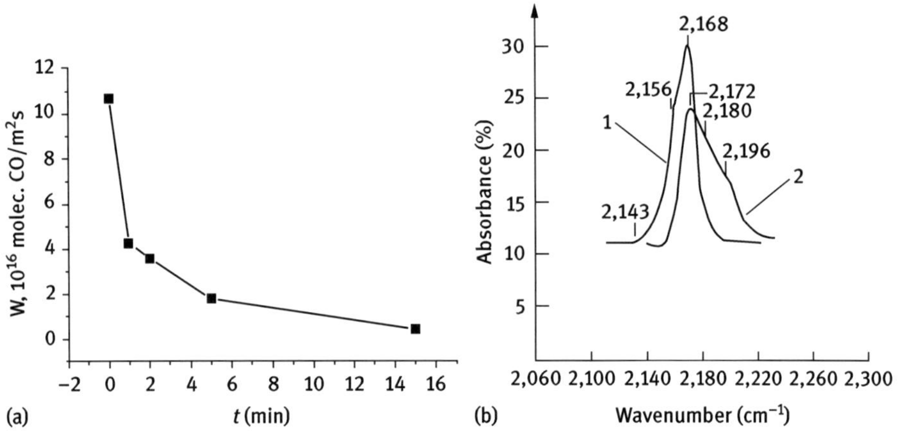

Fig. 6.3: (a) Effect of MT on the catalytic activity of iron oxide in CO oxidation ($W * 10^{16}$ mol CO/m^2 s, treatment at 400 °C). (b) IR spectra of CO adsorbed on iron oxide samples: 1 – the initial sample, 2 – the sample after 15 min MT.

MT of manganese oxides in different oxidation states, which have low specific surface areas, is accompanied by the dispersion and defect formation processes, which are testified by the broadened lines in diffraction patterns and the increased specific surface area (Table 6.1). Therewith, the phase composition of β-MnO_2 and β-Mn_2O_3 does not change even at large MT times, and the γ-Mn_2O_3 phase appears in β-Mn_3O_4 and MnO oxides after 20 min MT [72, 73]. Table 6.1 lists data on the stoichiometry of initial and mechanochemically treated oxides that were calculated from thermal analysis data [72, 73]. Changes in the stoichiometry of manganese oxides after MT correspond to their thermal stability upon heating in air. The data obtained show an increase in the concentration of point defects (oxygen vacancies) for MnO_2. For Mn_2O_3, changes are of extremal nature. Quantitative evaluation of the density of extended defects by SAXS revealed that MT of coarsely dispersed manganese oxides may increase the density of extended defects with MO time (Table 6.1). It was observed that for MnO_2 extremal change of relative integrated intensity with activation time, which is shown for the samples after testing, may be caused by

Table 6.1: Effect of MT on the properties of manganese oxides.

Sample	MT time (s)	S_{sp} (m^2/g)	I (SAXS) (a.u.)	W (140 °C) * 10^{-17} mol CO/m^2 s initial	steady
$MnO_{2-0.027}$	0	0.3	10.4	5.8	0.85
$MnO_{2-0.039}$	30	6.2	70.5*	28	4.2
$MnO_{2-0.050}$	300	19	181*	17	3.2
$MnO_{2-0.077}$	900	29	130*	8.8	2.1
$Mn_2O_{3-0.09}$	0	0.15	28.2	1.5	0.62
$Mn_2O_{3-0.11}$	30	1.2	42.7	14	2.3
$Mn_2O_{3-0.037}$	300	7.6	98.3	6.8	1.2
$Mn_2O_{3-0.008}$	900	14	113	2	0.5
Mn_3O_4	0	2.8	33.3	1.1	0.18
Mn_3O_4	30	4.8	62.2	1.6	0.3
$Mn_3O_{4+0.26}$	900	11	107*	5.9	1.9
$Mn_3O_{4+0.37}$	1,200	14	139*	3.4	1.2
MnO	0	0.5	11.2	0.64	0.16
	30	0.9	38.7	1.4	0.32
	900	3.7	–	1.6	0.45
	1,200	6.1	94.5	1.6	0.35

The composition of manganese oxides was found from thermal analysis data obtained in a He flow.
*Data were obtained on the samples after measuring their catalytic activity.

a faster relaxation of a part of extended defects in the reaction medium, because relative integrated intensity usually decreases by 10–20 units for the samples after testing in comparison with the initial samples.

As follows from Table 6.1, for the oxides reducible during MT, the activity (both initial and steady) goes over a maximum corresponding to a short treatment time (30 s), whereas for the oxidizable oxides the activity increases to a certain level.

After MT, changes were observed not only in the initial SCA of the oxides, but also in its steady level; this shows that MT produces stable changes in the state of active sites on the surface of manganese oxides, which may be caused by rearrangement of the defect structure and by changes in the phase composition of oxides. Since differences in the point defectness of samples are commonly leveled in the reaction medium, the observed differences can be attributed to different density of extended defects.

Thus, even a short MT of coarsely dispersed manganese oxides increases their dispersion, specific surface area and density of point and extended defects. It was shown that in the absence of phase changes and amorphization, which are observed during a prolonged MT, and in spite of fast relaxation of the surface in the reaction medium, which is typical of manganese oxides, a correlation can be established between changes in catalytic activity and density of extended defects, thus providing a better understanding of the features of active sites for these oxides. For example, it was accepted that manganese oxides have regular surface sites: Mn^{4+} in MnO_2 and the cationic pair Mn^{2+}–Mn^{3+} in Mn_2O_3 and Mn_3O_4 [74]. However, this conclusion was made without investigating a large series of samples having different origin and hence different defectness. Data on the effect of MT on the catalytic properties of manganese oxides obtained in this work suggest that active sites of MnO_2 may include the termination sites of extended defects. Exactly the increase in their concentration can be responsible for the enhanced activity of the sample mechanochemically treated for 30 s. The subsequent decrease in activity may be caused by a decrease in the oxygen content of the samples because it is known that the catalytic activity of the oxide decreases upon its reduction [75, 76]. In the case of Mn_2O_3, active sites may be related to the existence of steps on the (120) face because an increase in the number of exactly such faces upon activation improves the catalytic activity of the sample treated for 30 s, and the subsequent decrease in activity is caused by the surface amorphization, that is, the destruction of the stepwise shape of the surface. The enhancement of Mn_3O_4 activity correlates also with an increase in the density of extended defects, an increase in the oxygen content of the sample, and the formation of the Mn_2O_3 phase on the surface, whereas the deterioration of activity of the sample treated for 20 min may be attributed to amorphization of the surface. The improved activity of MnO may be related also to the formation of the surface Mn_2O_3 phase, the amount of which increases only due to the growing layer of the product. A prolonged MT, leading to amorphization of the surface, decreases the specific surface area; therewith, a similar level of activity is established in the reaction medium for all the oxides subjected to MT.

Thus, studies on the effect exerted by MT on the catalytic activity of 3d oxides in oxidation reactions have revealed the role of extended defects in the formation of active sites of deep oxidation catalysts. It was shown for the first time that even a short (less than a minute) MT of dispersed 3d oxides, along with the emerging high density of point defects, decreases SCA of the oxides in the catalytic oxidation of hydrocarbons and CO by lowering the density of extended defects and changing the state of adsorption sites connected with the points where such extended defects terminate on the surface.

MT of coarsely dispersed oxides, which increases the density of extended defects, first enhances the SCA; however, as the treatment time is extended, the activity deteriorates owing to amorphization (disordering) of the surface, which changes the state of active sites. Thus, a prolonged MT decreases the specific surface area. Thermal annealing of the activated oxides does not restore their catalytic properties.

It has been demonstrated that MT of 3d oxides in air, which is accompanied by milling at different levels (aggregates, crystallites and micromonoliths) and changes in the density of point and extended defects in the bulk and on the surface, alters the chemical features of the oxides. Thus, oxides with a low charge state of cations after MT contain the oxide phase in a higher oxidation state of cation. Oxides with a high charge state of cations are reduced without any change in their phase composition. Hydration and carbonization of the surface are observed and changes occur in the state of adsorption sites.

6.2 Mechanochemical synthesis of mixed oxides

For solid-phase synthesis reactions, many studies have demonstrated that highly reactive mixed oxides can be obtained by MT of initial reagents with subsequent calcination at moderate temperatures for a short time [84, 85, 86, 1, 2, 9, 25, 56, 86–131], To reveal the effect of MT on the interaction between oxides, thermodynamically allowed reactions are commonly considered [84, 9, 88, 107, 96]. The beneficial effect of MT is attributed to the increase in diffusion mobility of ions caused by elevation of temperature upon impact and by the formation of defects. Of great importance is also the transition to kinetic mode of reaction because a layer of the reaction product is removed, thus exposing the fresh surface. For example, an electron paramagnetic resonance study showed the intercalation of Cu^{2+} ions into the lattice of titania or silica to occur during MT of CuO and TiO_2 or CuO and Al_2O_3 for 5–10 min (the powders were not subjected to additional thermal treatment) [96].

Among studies on the mechanism of mechanochemical synthesis of mixed oxides, the work performed under the guidance of E.G. Avvakumov took into account the acid–base properties of initial reagents and developed the theory of mild mechanochemical synthesis upon interaction of hydroxides or hydrated oxides with the proton transfer step under nearly hydrothermal conditions [90–94, 131]. The authors synthesized mullite and other aluminosilicates as well as aluminates, titanates, zirconates, tungstates and vanadates of alkali metals to demonstrate that the proposed approach is promising for the acid–base interaction reactions proceeding with a release of water.

Zyryanov [9, 86–89] noted that the chemical composition of the mechanochemical reaction product is determined not only by thermodynamics of the process but also by the difference in hardness between mechanically activated oxides. It was shown that the stoichiometric product is formed from the oxides having equal hardness. If the initial oxides are of different hardness, the resulting mixed oxide will have a deficit of a harder reagent. Assuming that mixing of components upon spalling of molecular layers in the MT step is based on the roller mechanism, the author concluded also that the activation step results in the formation of an amorphous product, which crystallizes in the second step.

Butyagin and Boldyrev emphasized that the interaction proceeds under the action of both the mechanical and chemical forces. They assumed that deformation mixing is performed by mechanical forces, but stability of the emerging structures is provided by the action of chemical forces [96, 132].

In a cycle of studies [73, 97–99, 133–137] on mechanochemical synthesis of mixed perovskite-like oxides such as $Me^1Me^2O_3$ (where Me^1 = La, Sr, Ca, Ba,. . ., and Me^2 = Mn, Co, Fe), the effects exerted by the composition of oxides, features of the initial reagents, time of mechanical treatment and subsequent calcination on their physicochemical and catalytic properties were investigated.

Mechanochemical synthesis of lanthanum manganite and cobaltite was used to demonstrate the effect of the transition metal oxidation state in oxides used for the synthesis [97, 138]. It was found that the perovskite phase is formed by calcination of activated oxide powders at a temperature above 600 °C. Therewith, the interaction of reagents after activation and calcination is not complete if oxide in the lowest oxidation state (MeO) is used; on the contrary, the interaction proceeds to a completion in the case of transition metal oxide in a higher oxidation state (MeO_x, III or IV) (Table 6.2). Without mechanochemical activation the perovskite phase is not formed under the given thermal treatment conditions. Noteworthy is the fact that the use of $Co(OH)_2CoCO_3$ instead of CoO leads to the complete interaction, probably due to the formation of Co_3O_4 and hydrothermal conditions in the mill drum during decomposition of basic cobalt carbonate upon activation [97, 138]. Thus, other conditions (activation and thermal treatment) being equal, the use of transition metal oxides in the highest oxidation state provides a more complete interaction between initial reagents.

Table 6.2: Effect of initial reagents on the formation of perovskites (according to XRD data) by mechanochemical synthesis (activation for 5 min, calcination at 700 °C, 4 h) [97, 138].

Initial reagents	Formation of $LaMeO_3$ according to XRD (%)
La_2O_3, CoO	50
La_2O_3, Co_3O_4	100
$La_2(CO_3)_3$, $Co_2(OH)_2CO_3$	100
La_2O_3, MnO_2	100
La_2O_3, Mn_2O_3	60
La_2O_3, Mn_3O_4	40

Effect of the features of anions was studied in the synthesis of calcium ferrite [98–99]. It was shown that MT of oxides leads to the formation of the $Ca_2Fe_2O_5$ phase even without subsequent calcination, whereas MT of hydroxides or calcium carbonate in a mixture with iron hydroxide or oxide produces calcium ferrite only after thermal treatment of the activated mixture. Thus, a short-term (2–3 min) activation of various initial compounds followed by thermal treatment at moderate (400–700 °C) temperatures makes it possible to obtain crystalline calcium ferrite with the specific surface area of 10–15 m^{-2} g^{-1}. Without activation (the ceramic method), calcium ferrite can be formed from the same initial reagents only after calcination at 1,100 °C for 100 h.

The data obtained can be described in terms of differences in the acid–base properties of initial reagents using the Usanovich general theory of acid–base properties. Thus, an increase in the charge of a transition metal cation increases its acidity, whereas basicity of the cation, which is determined by the chemical features of cation and anion in a compound, decreases in the order CaO > $Ca(OH)_2$ > $CaCO_3$. From this point of view, the interaction degree in a mixture of activated oxides can be raised by increasing the difference in their acid–base properties. These results reflect the chemical aspect of interaction upon mechanical activation and allow extending the notion of the acid–base interaction. Thus, for the synthesis of perovskites with mechanical activation, more suitable are hydroxides and oxides of transition metals with the highest oxidation state of metal [73].

The effect of mechanical treatment and subsequent calcination on microstructural features and catalytic activity was studied in the mechanochemical synthesis of lanthanum ferrite from oxides [73]. According to XRD data (Fig. 6.4), all the samples subjected to MT (both calcined and noncalcined) were single-phase ones.

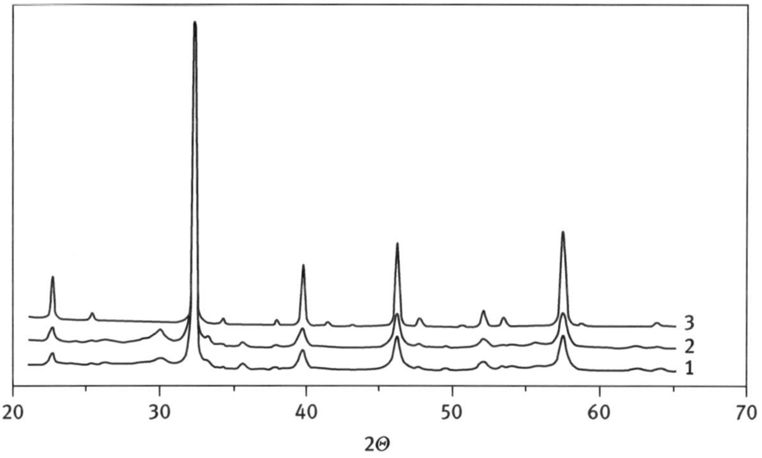

Fig. 6.4: Diffraction patterns of the MT mixture of lanthanum and iron oxides in dependence on calcination temperature: 1 – without thermal treatment; 2 – 500 °C; 3 – 900 °C.

Based on high-resolution transmission electron microscopy (HRTEM) data, particles of the noncalcined sample are of micron size and have regions with the ordered crystal structure that are surrounded by regions with the disordered structure (Fig. 6.5a). The ordered regions are ~100 Å in size, whereas the disordered regions are 80–300 Å. The size of the ordered regions coincides with that of the coherent scattering domain (CSD)

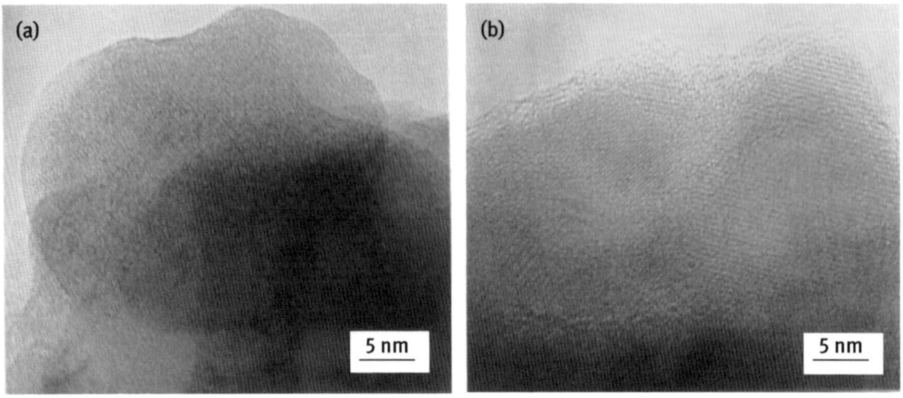

Fig. 6.5: (a) The $LaFeO_3$ sample synthesized by mechanochemical activation without thermal treatment. (b) The $LaFeO_3$ sample calcined at 500 °C.

estimated by XRD. Thermal treatment at 500 °C leads to crystallization of the regions with the disordered structure, increases the size of crystalline regions and results in the development of interphase boundaries (Fig. 6.5b). After calcination at 900 °C, the particles have a classical micromonolithic structure (the size of micromonoliths is 500–600 Å). Point diffraction from such a particle (Fig. 6.6) looks diffuse, which corresponds to disoriented junction of micromonoliths. Calcination at 1,100 °C leads to annealing of the micromonolithic structure and formation of single-crystal perovskite particles with the typical point microdiffraction (Fig. 6.7). Therefore, the crystal structure of perovskite also improves (Table 6.3).

Fig. 6.6: Microdiffraction from the $LaFeO_3$ sample calcined at 900 °C.

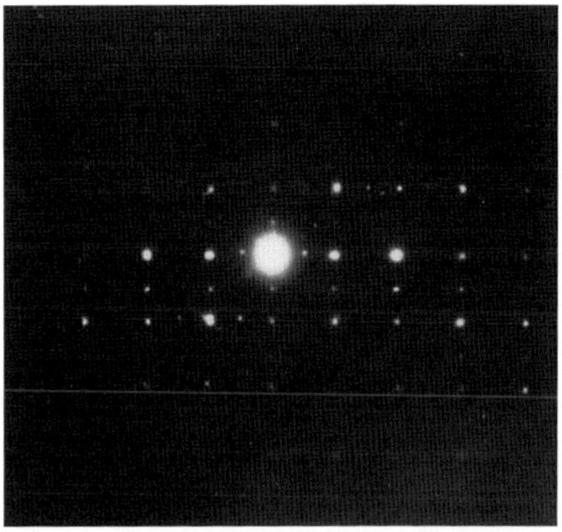

Fig. 6.7: Microdiffraction from the $LaFeO_3$ sample calcined at 1,100 °C.

Samples that are not subject to thermal treatment had the specific surface area of ca. 5 m^2/g and the lowest SCA. Raising the calcination temperature to 900 °C decreased the specific surface area to 1 m^2/g and lowered the relative content of iron cations on the surface (Fig. 6.8); therefore, SCA increased first and then decreased (Fig. 6.9) symbately with the formation of micromonolithic structure and its annealing. The discrepancy between catalytic activity maxima (900 °C) and the maximum density of interphase boundaries estimated by SAXS (700 °C) may be caused by the presence of disordered regions on the surface of samples calcined at 700 °C.

Similar changes were observed in the real structure of calcium ferrite obtained by joint MT of hydroxides with subsequent calcination at 400–1,100 °C [99]. According to HRTEM data, the sample calcined at 600 °C contains the micron-size particles consisting of chaotically joined crystallites with the size of ~ 250 Å because microdiffraction has the ring shape, which is typical of polycrystalline powders. Regions with disordered structure are observed on the surface of crystallites. Raising the calcination temperature to 900 °C increases the size of crystallites in particles up to ~500 Å and improves their surface structure; microdiffraction from the particles is typical of mosaic crystal (the extended reflections). Further rising of calcination temperature results in the formation of single-crystal particles with point microdiffraction (1,100 °C). In this case, SCA also increases with the calcination temperature symbately with improvement of the structure and formation of micromonoliths, and then decreases upon annealing of the micromonolithic structure [99].

For lanthanum cobaltite and manganite synthesized using MT, specific surface area decreases from 15 to 20 m^2/g (600 °C) to less than 1 m^2/g (1,100 °C) when calcination temperature is raised. On the surface of low-temperature samples, as in the case of lanthanum ferrite, the ion current ratio Me/La estimated by SIMS is higher than in the bulk. So, high activity of the low-temperature samples may be caused here by segregation of highly active MeO_x on the surface of perovskites,

Table 6.3: Crystal structure parameters of lanthanum ferrite.

T_{calc} (°C); R_{fact} (%)	Atom	Parameters of positions			B (Å^2)	Occupancy of position	Parameters of orthorhombic cell (Å)		
		x	y	z			a	b	c
Without calcination	La	−0.0003	0.0005	¼	0.1	0.75	5.556	5.565	7.862
R = 5%	Fe	0	1/2	0	0.1	0.75			
	O_1	0.0558	0.4746	¼	2.5	1.0			
	O_2	−0.2934	0.2697	0.0036	3.0	1.0			
500 °C,	La	−0.0032	0.0014	¼	0.6	0.75	5.556	5.565	7.862
R = 4.6%	Fe	0	1/2	0	0.4	0.75			
	O_1	0.0663	0.4409	¼	2.5	1.0			
	O_2	−0.2340	0.2964	0.01	1.0	1.0			
	Fe_2	0	0	0	1.0	0.07			
900 °C,	La	−0.0028	0.0265	¼	0.5	0.9	5.555	5.561	7.867
R = 4.9%	Fe	0	1/2	0	0.3	0.9			
	O_1	0.0675	0.4899	¼	0.3	1.0			
	O_2	−0.2653	0.2917	0.0363	0.6	1.0			
	Fe_2	–	–	–	–	–			

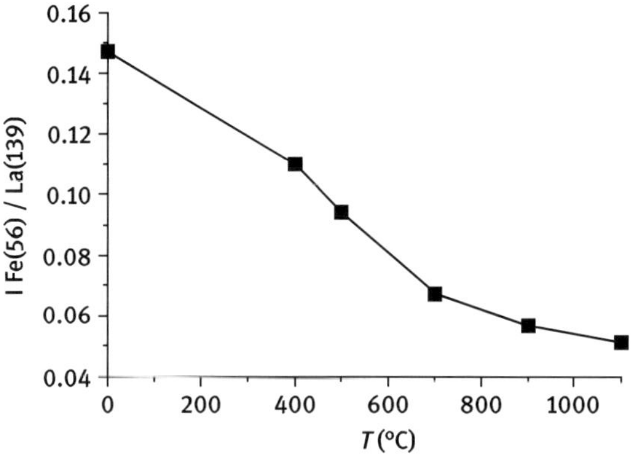

Fig. 6.8: Secondary ion mass spectrometry (SIMS) data on the ion current ratio I_{Fe}/I_{La} for $LaFeO_3$ samples in dependence on calcination temperature.

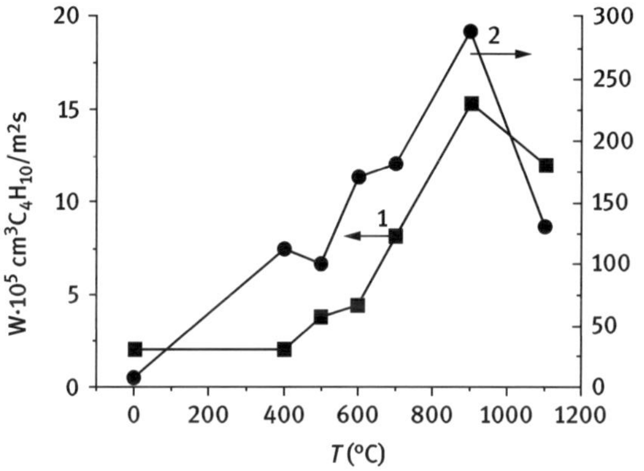

Fig. 6.9: Steady activity of lanthanum ferrite samples versus calcination temperature in butane oxidation at the testing temperatures: (1) 300 °C and (2) 400 °C.

and the observed decrease in activity of the samples after high-temperature treatment is related to intercalation of cations from the segregated oxides into the perovskite lattice and hence to a lower, in comparison with simple oxides, activity of perovskites in the low-temperature oxidation reactions [97, 138]. Thus, although the $LaBO_3$ (B = Co, Mn, Fe) perovskites synthesized by mechanochemical method have virtually the single-phase composition based on XRD, their structure is characterized by the presence of micromonoliths, and 3d oxides are segregated on the surface, which determine their high activity.

In substituted perovskites $La_{1-x}Me^1_xMe^2O_3$ (Me^1 = Ca, Sr, Ba; Me^2 = Fe, Co, Mn), the formation of micromonoliths in the particles may be caused also by the occurrence of morphotropic phase transitions in these systems; alkali-earth cations can segregate on the surface, which will also affect the catalytic activity of oxides in deep oxidation reactions.

$La_{1-x}Ca_xFeO_{3-0.5x}$ system. The $La_{1-x}Ca_xFeO_{3-0.5x}$ system, as was reported by J. Greier, V. Alario-Franco et al., may contain the vacancy-ordered phases of a homologous series with the general formula $(A^1A^2)_nB_nO_{3n-1}$, where $n = 2, 3,\ldots,\infty$, which have different alternations of octahedral (O) and tetrahedral (T) layers [139–143].

According to our data, when the synthesis was performed by ceramic method, only three phases of the homologous series were detected in the system: lanthanum ferrite with the perovskite structure ($n = \infty$), calcium ferrite with the brownmillerite structure ($n = 2$) and the Grenier phase ($n = 3$ or $x = 0.67$), which is characterized by the presence of twins and the superlattice parameter 11.3 Å appearing in the sequence OOTOOTOOT.... The formation of a limited ($x = 0$–0.17) homogeneous solid solution of Ca in the perovskite structure was also observed. For the remaining x values, the system consists of phases with close compositions that are coherently intergrown within a particle (Fig. 6.10) [133].

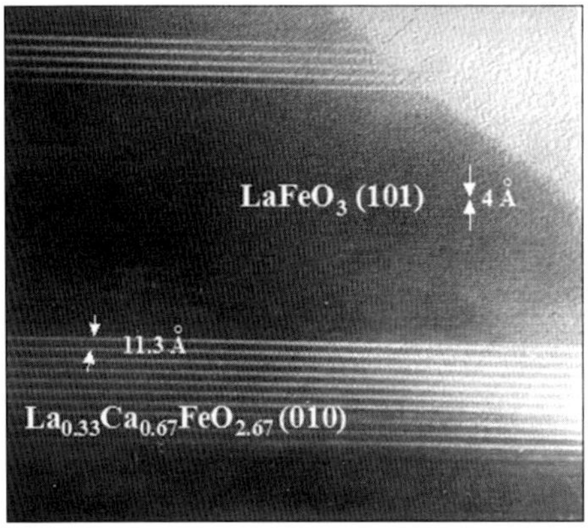

Fig. 6.10: HRTEM image of the $La_{0.4}Ca_{0.6}FeO_{2.7}$ sample from ceramic series.

In the case of mechanochemical synthesis, the vacancy-ordered phase with $n = 3$ was not found in the system. Calcination of the mechanically activated mixture at 900 °C for 4 h leads to the formation of a metastable solid solution of calcium in the perovskite structure with the maximum content $x = 0.34$ and the brownmillerite phase. Metastability of the solid solution is proved by the fact that after a longer calcination (15 h) at 900 °C or after calcination at a higher temperature (1,100 °C, 4 h), the content of calcium in the solution decreases to $x = 0.17$. For the samples calcined at 900 °C, these phases exist, according to HRTEM data, as the regions chaotically joined within a particle (Fig. 6.11), and after calcination at 1,100 °C – as individual particles [134–135].

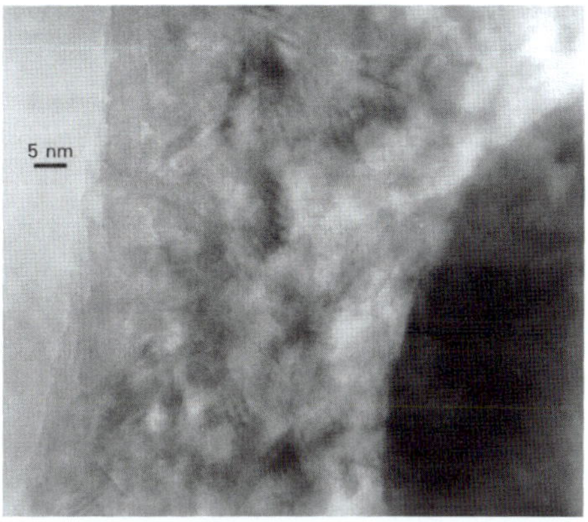

Fig. 6.11: HRTEM image of the $La_{0.6}Ca_{0.4}FeO_{2.8}$ sample from mechanochemical series calcined at 900 °C.

In both series, SCA of the samples toward CO oxidation depends on x and goes over a maximum for x = 0.4–0.6. Therefore, SCA of the substituted samples from mechanochemical series, irrespective of calcination temperature, is lower or comparable with the samples from ceramic series (Fig. 6.12). As per SIMS data (Fig. 6.13), the relative content of iron cations on the surface in dependence on x changes monotonically; hence, the presence of a maximum on the activity curve cannot be related to the concentration factor. Since changes in SCA correlate well with changes in the density of intermonolith and interphase boundaries (according to SAXS data), active sites may be connected with coordinatively unsaturated clusters of Fe^{2+} cations that are formed on the surface at the termination sites of such boundaries [136, 137, 144–147].

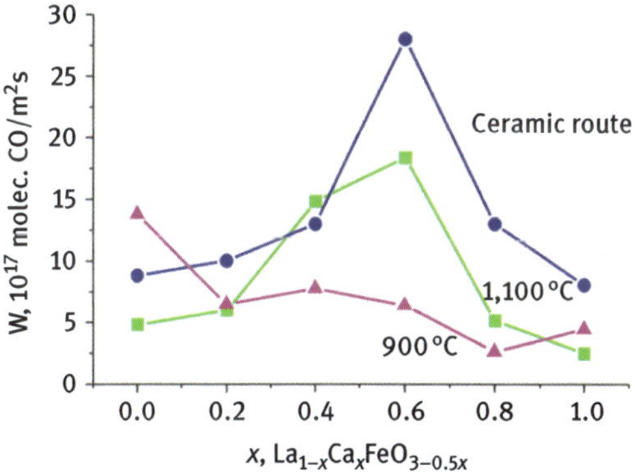

Fig. 6.12: Catalytic activity of $La_{1-x}Ca_xFeO_{3-0.5x}$ perovskites synthesized by ceramic and mechanochemical methods (calcination temperatures are indicated) toward CO oxidation at 450 °C.

$La_{1-x}Sr_xFeO_{3-0.5x}$ system. In the $La_{1-x}Sr_xFeO_{3-\delta}$ system synthesized by mechanochemical method with subsequent thermal treatment at 900 or 1,100 °C, virtually the

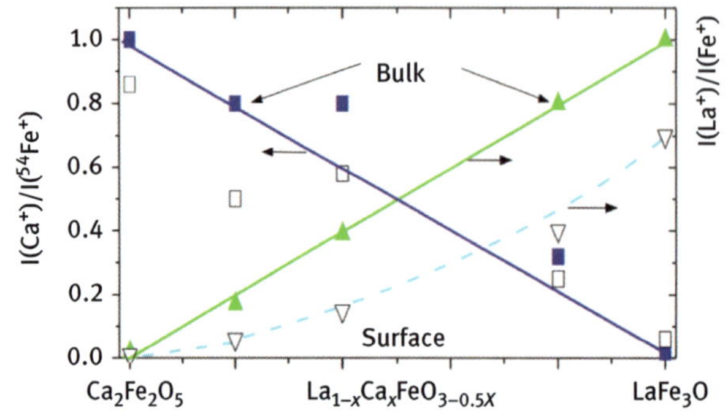

Fig. 6.13: SIMS data for samples from mechanochemical series, T_{calc} 900 °C.

single-phase perovskites are formed and the morphotropic phase transition in the regions with $x = 0.3–0.4$ is observed [148–154]. Thus, according to XRD data obtained using synchrotron radiation, *the* rhombic modification of perovskite is observed *for* compositions *with* $x < 0.3$ (Table 6.4), *while at* $x > 0.4$ the diffraction patterns *correspond to the* perovskite phase *with cubic modification*. Diffraction patterns of the samples having the composition with $x = 0.3$ and 0.4 show a substantial broadening of peaks due to the overlap of diffraction peaks corresponding to two phases of solid solutions with close lattice parameters. This is indicated by the experiment carried out at a high-resolution station: all the diffraction peaks are split, as it is shown for the diffraction peak 200 (indices of the cubic phase) (Fig. 6.14a) for the sample

Table 6.4: Lattice parameters of $La_{1-x}Sr_xFeO_{3-d}$ (1,100 °C) samples.

x	Rhombic phase Symmetry group Pnma ($Z = 4$)					Cubic phase Pm3m ($Z = 1$)	
	a, Å	*b*, Å	*c*, Å	*V*, Å^3	(*V*/*Z*), Å^3	*a*, Å	*V*/*Z*, Å^3
0	5.563(1)	7.858(1)	5.562(1)	243.14	60.78	–	–
0.2	5.551(4)	7.865(8)	5.544(3)	242.03	60.51	–	–
0.3	5.545(7)	7.860(1)	5.540(5)	241.38	60.35	3.892(2)	58.95
0.4	5.604(4)	7.991(8)	5.374(6)	240.66	60.17	3.889(2)	58.82
0.6	–	–	–	–	–	3.880(1)	58.41
0.7	–	–	–	–	–	3.874(1)	58.14
0.75	–	–	–	–	–	3.871(1)	58.01
1.0	–	–	–	–	–	3.868(1)	57.87

with $x = 0.4$ [151]. This sample possesses a disordered micromonolithic structure; perovskite phases of rhombic and cubic modifications exist supposedly as individual crystal monoliths within a particle (Fig. 6.14b).

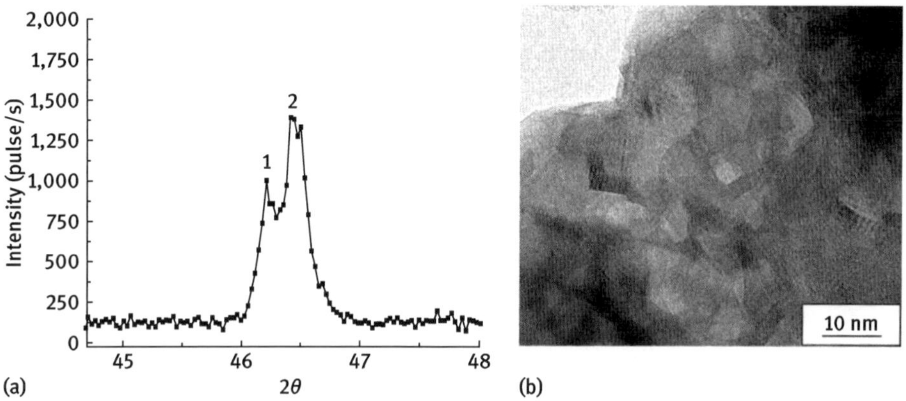

Fig. 6.14: A region of high-resolution diffraction pattern for the $La_{0.6}Sr_{0.4}FeO_{3-\delta}$ sample (1 – rhombic phase, 2 – cubic phase) obtained using synchrotron radiation ($\lambda = 1.5406$ Å) (a) and microstructure of the $La_{0.6}Sr_{0.4}FeO_{3-\delta}$ sample (b).

SCA of strontium-substituted ferrites toward CO oxidation changes nonmonotonically with increasing the fraction of strontium; two activity maxima are observed for compositions with $x = 0.3$ and 0.8 (Fig. 6.15a), the distinctive feature of which is the microheterogeneous structure of particles (HRTEM data). An increase in the integrated intensity of small-angle scattering on inhomogeneities in these samples was revealed also by SAXS.

In methane oxidation, SCA first decreases upon introduction of strontium and then remains virtually constant as x is increased (Fig. 6.15b). The intermediate activity maxima were not observed in this case [149]. Taking into account the polymorphous phase transition occurring in the samples with $x = 0.3–0.4$ at temperatures ~450 °C and the formation of a high-temperature cubic homogeneous vacancy solid solution, the absence of the intermediate maximum may be caused by microstructural changes of particles under the testing conditions (>500 °C) (Fig. 6.16).

Based on the thermal analysis data, the samples are oxidized (contain virtually no vacancies) at room temperature, whereas upon heating they lose oxygen (Fig. 6.17); thus, the microheterogeneous solid solutions observed in the morphotropic phase transition region are formed during cooling and oxidation of single-phase cubic perovskites produced at the synthesis temperature. One can see that the samples demonstrate an enhanced activity exactly in the temperature region of existence of microheterogeneous solid solutions [152–154].

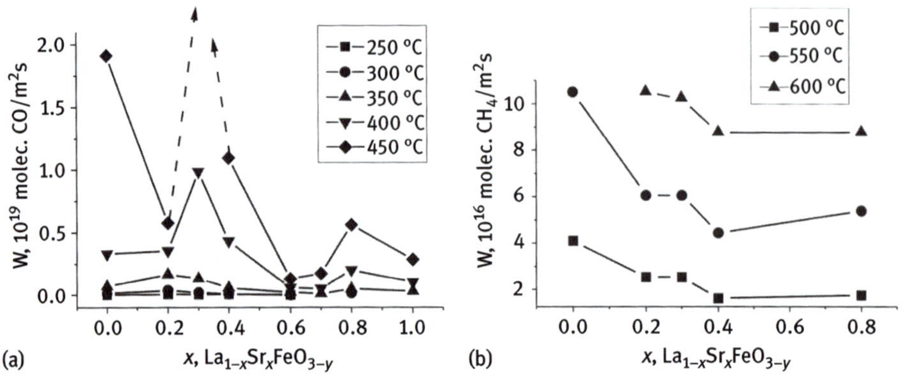

Fig. 6.15: Catalytic properties of the $La_{1-x}Sr_xFeO_{3-y}$ series samples in the oxidation of CO (a) and CH_4 (b).

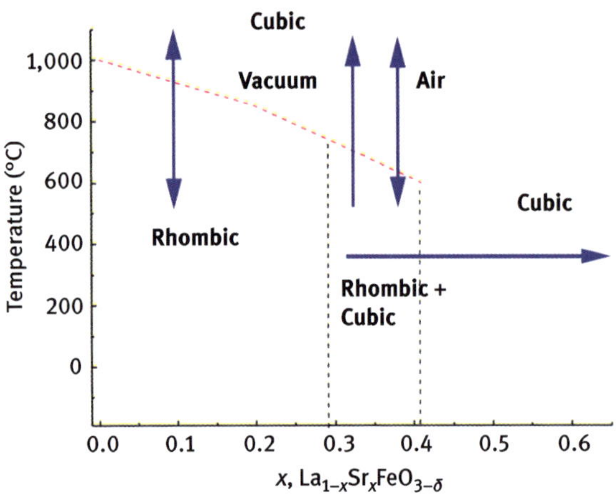

Fig. 6.16: A region of the phase diagram for the series of $La_{1-x}Sr_xFeO_{3-\delta}$ samples.

$La_{1-x}Ba_xFeO_{3-\delta}$ system. Similar results were obtained for samples of the $La_{1-x}Ba_xFeO_{3-y}$ (x = 0–1) series that were obtained by calcination at 1,100 °C for 4 h of a mixture of initial simple oxides mechanically preactivated for 3 min in a centrifugal-planetary mill [155–156]. Based on XRD data, all the synthesized samples up to the compositions with x = 0.8 are single-phase oxides with the perovskite structure (Fig. 6.18). The sample with x = 1 is a mixture of phases belonging to the perovskite and brownmillerite structural types. In this series, the morphotropic phase transition from rhombic modification to cubic one is observed in the region of compositions with x = 0.3.

In this series, the linear dependence of the catalytic activity toward CO oxidation on the composition (barium content in the samples) is also absent, and high

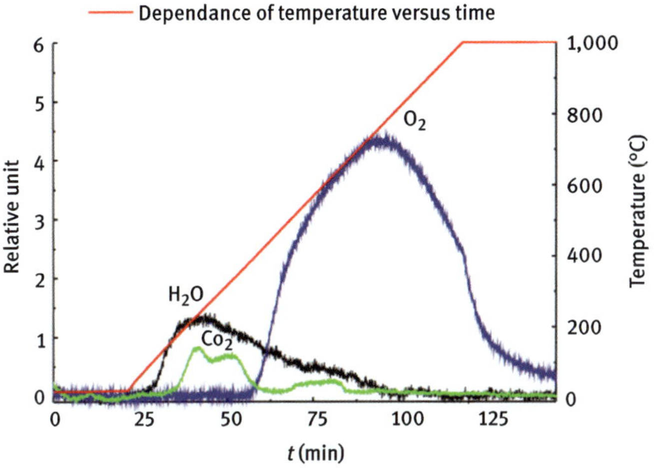

Fig. 6.17: Mass spectrometric analysis of the products that are released upon heating of the $La_{0.4}Sr_{0.6}FeO_{3-\delta}$ sample in a helium flow.

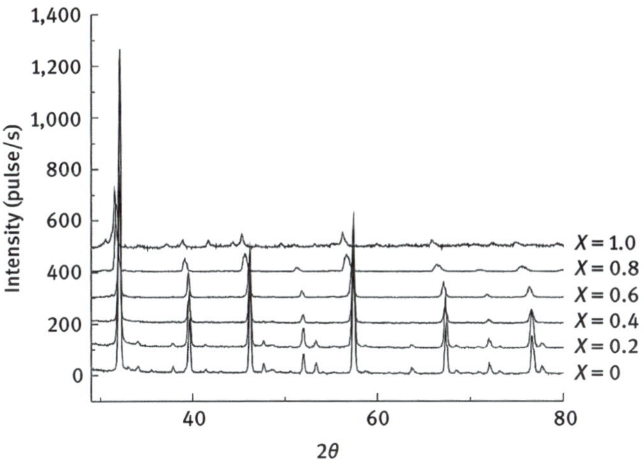

Fig. 6.18: X-ray diffraction pattern of $La_{1-x}Ba_xFeO_{3-y}$ (x = 0–1) samples.

activity is observed for compositions with x = 0.3 and x = 0.8 (Fig. 6.19) having the microheterogeneous structbiure [155, 156].

$La_{1-x}Ca_xCoO_{3-\delta}$ system. The mechanochemical method, against the ceramic method, allows really the substituted solutions with larger x to be prepared during shorter time of the calcination at 1,100 °C [157–161]. From XRD data (Fig. 6.20), the prepared samples (after mechanochemical activation the mixtures of oxides were calcined at 1,100 °C for 5 h) with $x \leq 0.5$ are virtually single-phase perovskites with

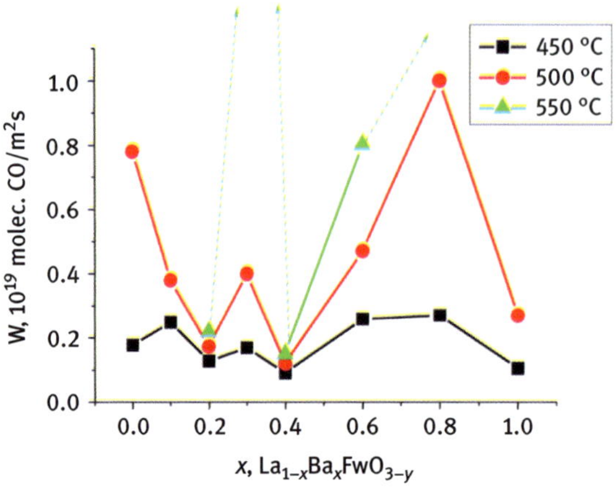

Fig. 6.19: The rate of CO oxidation at 450–550 °C versus the composition of $La_{1-x}Ba_xFeO_{3-y}$ samples.

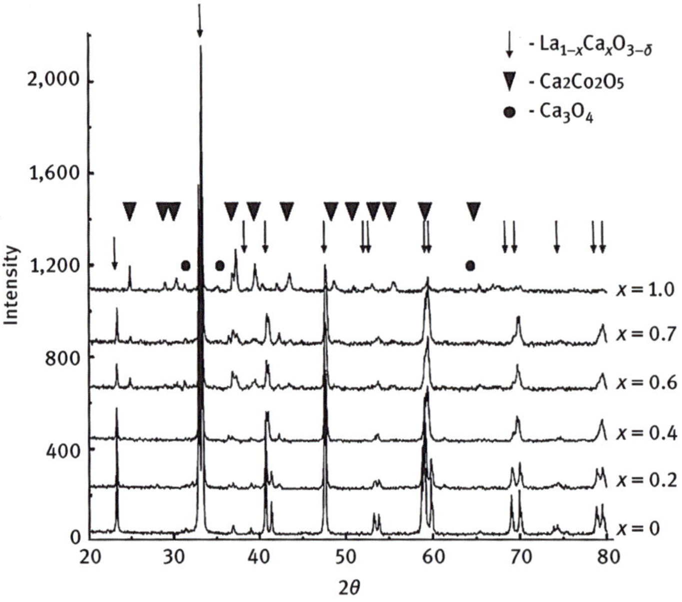

Fig. 6.20: X-ray diffraction data for mechanochemically prepared $La_{1-x}Ca_xCoO_{3-\delta}$ (x = 0–1).

a small admixture of cobalt oxides. The lattice parameter and unit cell volume decreased with increasing the x value to 0.4, which indicated the intercalation of Ca cations into lanthanum sublattice (Table 6.5). In samples with $x \geq 0.6$, the phase of calcium cobaltite with the brownmillerite structure is detected along with the

Table 6.5: Specific surface area (S_{sp}, m^2/g) of mechanochemically prepared $La_{1-x}Ca_xCoO_{3-\delta}$ oxides and unit cell parameters for the perovskite phases.

Sample	*a*, Å	*c*, Å	*V*, Å³	S_{sp}, m^2/g
$LaCoO_{3-\delta}$	5.444	13.097	56.027	0.24
$La_{0.9}Ca_{0.1}CoO_{3-\delta}$	5.443	13.104	56.018	0.16
$La_{0.8}Ca_{0.2}CoO_{3-\delta}$	5.437	13.109	55.938	0.14
$La_{0.7}Ca_{0.3}CoO_{3-\delta}$	5.429	13.120	55.825	0.26
$La_{0.6}Ca_{0.4}CoO_{3-\delta}$	3.818		55.652	0.2
$La_{0.5}Ca_{0.5}CoO_{3-\delta}$	3.816		55.564	0.4
$La_{0.4}Ca_{0.6}CoO_{3-\delta}$	3.818		55.669	0.34
$La_{0.3}Ca_{0.7}CoO_{3-\delta}$	3.819		55.702	1.0
$La_{0.1}Ca_{0.9}CoO_{3-\delta}$	3.819		55.700	0.36
$CaCoO_{3-\delta}$				0.35

perovskite phase, the former phase increasing in proportion with x to become the main phase with cobalt and calcium oxides as impurities at $x = 1$. The XRD data indicate the morphotropic phase transition from hexagonal ($x < 0.4$) to cubic ($x > 0.4$) modification of the perovskite structure.

From HRTEM data, $La_{1-x}Ca_xCoO_{3-\delta}$ samples at $x = 0$–0.3 are mainly built up by the particles of rhombohedral perovskite phase with the size from 100 nm to micron (Fig. 6.21). There are ensembles (10–50 nm in size) of fine cobalt oxide particles ($x = 0$), as well as lanthanum and calcium oxides ($x = 0.2$–0.4 and $x > 0.5$, respectively) on the perovskite surface. With an increase in the amount of calcium, the amount of admixtures in the samples (Fig. 6.21) increases. The size of such particles is ca. 10 nm (Fig. 6.21). The sample with $x = 0.4$ is cubic perovskite. In the samples with $x \geq 0.5$, an additional brownmillerite phase occurs along with the cubic perovskite modification. From HRTEM and EDX data, the composition of the rhombohedral phase changes according to the Vegard rule up to $x < 0.4$. The HRTEM data allow the compositions with $x = 0.3$–0.4 to be assigned to the region of morphotropic phase transition, too.

The substitution of calcium for lanthanum results in a nonmonotonic (with an intermediate maximum at $x = 0.2$–0.5) decrease in the SCA of $La_{1-x}Ca_xCoO_{3-\delta}$ oxides (Fig. 6.22). The observed variations in the catalyst activity toward methane oxidation do not correlate with variations in the quantity of the most weakly bound surface oxygen species, probably due to the absence (desorption) of these species under conditions of the catalytic studies, or with calcium content (or with Co^{4+} content determined from the total hydrogen consumption during H_2-TPR) [161]. It is not

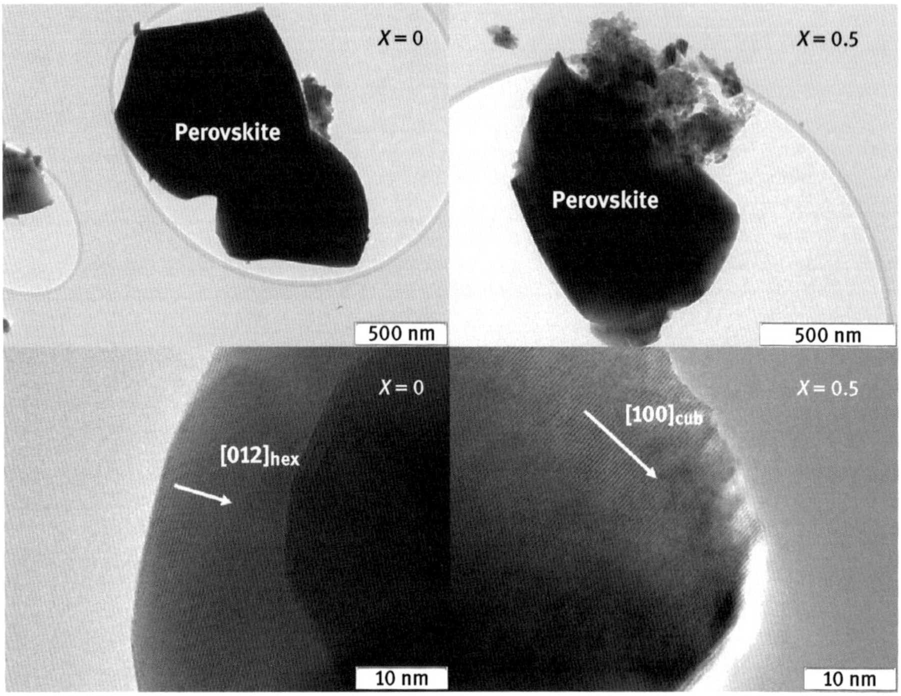

Fig. 6.21: Microstructure of two kinds of particles of mechanochemically prepared oxides: the $LaCoO_{3-\delta}$ particle with R3m perovskite structural modification x=0; the $La_{0.5}Ca_{0.5}CoO_{3-\delta}$ particle with Pm3m perovskite structural modification x=0.5.

impossible that the higher catalytic activity of lanthanum cobaltite prepared by the mechanochemical method results from the presence of disperse cobalt oxide on the surface, while other samples of the series are free of this oxide. The increasing activity in the row from $x = 0.1$ up to the intermediate maximum at $x = 0.3$–0.5 may be accounted for by an increase in the content of Co^{4+} cations and by microheterogeneity of the $x = 0.3$–0.4 samples (due to phase transformation), while the activity lowering at $x > 0.5$ – by the emergence of the less active brownmillerite phase in the samples and calcium oxide on the surface.

$La_{1-x}Sr_xCoO_3$ system. According to XRD data for the ceramic series of samples (Table 6.6), in the regions with $x \leq 0.3$ and $x \geq 0.5$ the system is single phase and has the hexagonal or cubic structure, respectively, which is consistent with the literature data (Table 6.6) [162–164]. In diffraction pattern of the $La_{0.6}Sr_{0.4}CoO_{3-\delta}$ sample, along with the unsplit diffraction maxima that can be assigned to both the cubic and hexagonal phases, there are peaks corresponding only to the hexagonal phase. However, their intensity is much lower than the typical intensity of single-phase samples with the hexagonal structure. In addition, a weak splitting of the diffraction peak 2.2.2 (cubic indices) is observed, although for the hexagonal phase

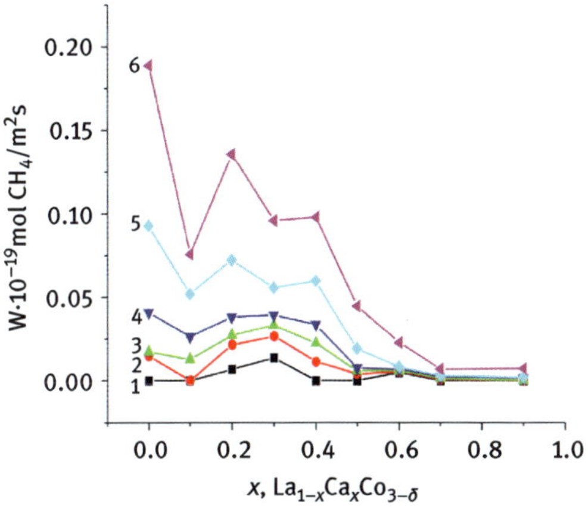

Fig. 6.22: Catalytic activity (reaction rate) in methane oxidation of mechanochemically prepared $La_{1-x}Ca_xCoO_{3-\delta}$ versus x at different temperatures: 350 °C (1), 400 °C (2), 450 °C (3), 500 °C (4), 550 °C (5), 600 °C (6).

Table 6.6: X-ray diffraction analysis of $La_{1-x}Sr_xCoO_{3-\delta}$ samples from ceramic series.

Sample	Hexagonal cell		Rhombohedral/cubic cell			CSD	Microdistortion
	a (nm)	*c* (nm)	*a* (nm)	*α* (deg.)	*V* (cm³)	*D* (nm)	ε
$LaCoO_3$	0.5445(4)	1.3092(1)	0.3827	90.7	0.0560	>200	0.0001
$La_{0.9}Sr_{0.1}CoO_{3-\delta}$	0.5451(1)	1.3129(4)	0.3833	90.6	0.0563	>200	0.0014
$La_{0.8}Sr_{0.2}CoO_{3-\delta}$	0.5452(1)	1.314(3)	0.3835	90.6	0.0564	>200	0.0013
$La_{0.7}Sr_{0.3}CoO_{3-\delta}$	0.5437(1)	1.3193(3)	0.3832	90.4	0.0563	100	0.0006
$La_{0.6}Sr_{0.4}CoO_{3-\delta}$	0.5433(2)	1.316(1)	0.3829	90.4	0.0562	–	–
	–	–	0.38324(6)	90.0	0.0563		
$La_{0.5}Sr_{0.5}CoO_{3-\delta}$			0.38312(9)	90.0	0.0562	>200	0.0013
$La_{0.2}Sr_{0.8}CoO_{3-\delta}$			0.38325(5)	90.0	0.0563	>200	0.0004

this peak should also remain unsplit. These features of the diffraction pattern made it possible to conclude that two phases – cubic and hexagonal – are present in the sample. As the strontium content is increased, the hexagonal distortion of the perovskite unit cell becomes less pronounced [164].

As shown by HRTEM data, particles in all the samples of ceramic series are quite large (1.5–2 μm). The initial sample has a virtually perfect structure with point microdiffraction. As the x value is increased, microstrains appear in the samples, the pointwise nature of the diffraction pattern is retained, but reflections become diffuse. At x = 0.3, microdiffraction is virtually ring-shaped, which testifies to the polycrystalline (micromonolithic) structure of the particles. At x – 0.5, point diffraction is observed again, which corresponds to the cubic lattice. For the two-phase sample ($x = 0.4$), the particles were detected for which both the ring and point diffraction was observed simultaneously at different sides. It means that regions with the cubic phase (point microdiffraction) and regions with the hexagonal micromonolithic structure (ring microdiffraction) exist within a single particle. In the cubic phase ($x \geq 0.5$), microdistortions become more pronounced again with an increase in x (Table 6.6).

Samples of mechanochemical series calcined at 900 or 1,100 °C for 4 h are virtually single phase according to XRD and DD data [165]. Only for the samples calcined at 900 °C, the presence of a minor amount of initial oxides is observed. As in the case of ceramic synthesis, $LaCoO_3$ is a hexagonal perovskite. The introduction of Sr changes the structural modification of the perovskite solid solution from hexagonal to cubic at $x = 0.4$. However, in distinction to ceramic sample, peaks in diffraction patterns remain narrow (the size of CSD is greater than 1,000 Å). The DD data, which confirmed the single-phase nature and stoichiometric composition of the synthesized samples, revealed also the presence of excess oxygen in the low-temperature samples; its amount decreases when calcination temperature is raised and strontium is introduced. Specific surface area changes from 1.4 to 5.4 m^2/g for the samples calcined at 900 °C, and from 0.4 to 1.3 m^2/g for those calcined at 1,100 °C. Samples with $0 < x < 1$ possess a higher specific surface area as compared to the extreme terms of the series.

As shown by HRTEM, subsurface layers of the particles with 200 Å thickness are disordered. As in the case of lanthanum ferrite, these layers have crystalline and disordered regions (in the sample with $x = 0.3$ calcined at 900 °C) or disordered crystalline regions (in the sample with $x = 0.8$ calcined at 1,100 °C). Evidently, the product formed in this system also consists of disordered and crystalline regions, as demonstrated for lanthanum ferrite. The subsequent thermal treatment leads to crystallization of disordered regions and increases the size of crystalline regions up to ~1,000 Å. However, the subsurface layer and the surface of such perfect crystallites in the case of lanthanum cobaltite remain disordered even after thermal treatment at 1,100 °C [165].

For the samples of both series, a nonmonotonic dependence of SCA on the chemical composition was found (Figs. 6.23 and 6.25). For the samples of ceramic series, the initial SCA correlates with the concentration of cobalt cations on the surface (according to SIMS, Fig. 6.24), whereas the steady SCA has a distinct maximum for the composition with $x = 0.4$, which is characterized by intermonolith and interphase boundaries. There is a good correlation between steady catalytic activity and density of extended defects according to SAXS data (Figs. 6.23 and 6.24). A similar dependence was obtained for the La-Sr-Co-Fe-O system [166], which has the activity maximum for samples with micromonolithic structure. The activity maxima in samples of mechanochemical series (Fig. 6.25) at $x = 0.3$ and $x = 0.8$ may be caused by microstructural features of subsurface layers in their particles. The increased content of the most weakly bound surface oxygen species and the correlation between

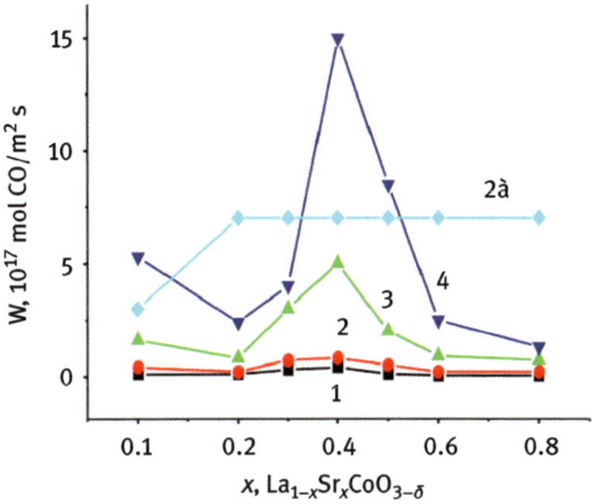

Fig. 6.23: Dependence of the steady (1–4) and initial (2a) SCA on the composition of $La_{1-x}Sr_xCoO_{3-\delta}$ samples from ceramic series and the testing temperature: 1 – 100 °C; 2 and 2a – 140 °C; 3 – 185 °C; 4 – 227 °C.

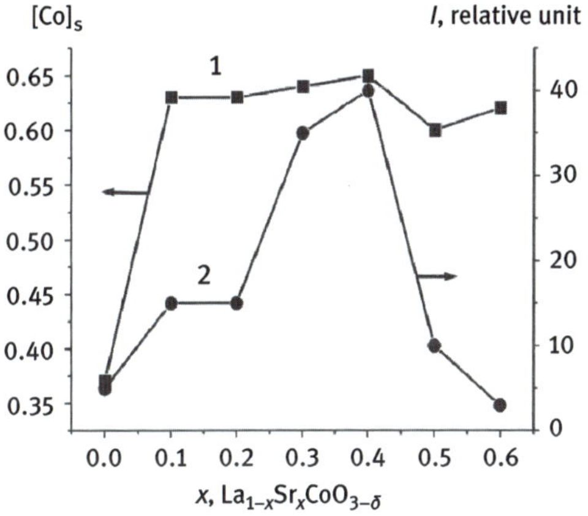

Fig. 6.24: Dependence of the surface concentration of Co cations (SIMS) and intensity of small-angle X-ray scattering (SAXS) on the chemical composition of $La_{1-x}Sr_xCoO_{3-\delta}$ samples from ceramic series.

activity toward CO oxidation and amount of such oxygen species are observed exactly for such samples [167].

$La_{1-x}Me_xMnO_{3\pm\delta}$ system with Me = Ca, Sr. According to authors [168–177], the extreme terms of the series can have different crystallographic modifications based on δ. The introduction of the alkali-earth cation in lanthanum manganite results in the formation of homogeneous solid solutions based on orthorhombic perovskite. In this system, electroneutrality is maintained by the formation of Mn^{4+}. According

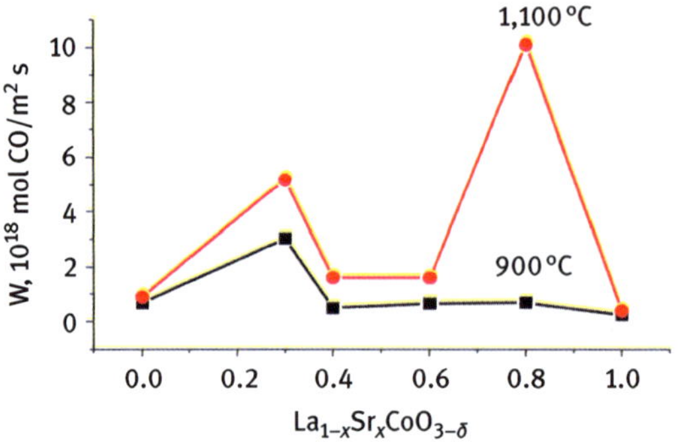

Fig. 6.25: SCA of $La_{1-x}Sr_xCoO_{3-\delta}$ samples from mechanochemical series versus the composition (x) and calcination temperature (900 and 1,100 °C). Composition of the gas mixture: 1% CO + 1% O_2 in He; testing temperature 220 °C.

to XRD data, the initial lanthanum manganite from ceramic series is a two-phase sample comprising two hexagonal phases that may differ in the content of excess oxygen [178]. Calcium manganite from ceramic series is a mixture of two phases ($CaMnO_3$ and $Ca_2Mn_2O_5$). In the region of $0.1 < x \leq 0.4$, homogeneous orthorhombic solid solutions are formed; therefore, the lattice parameter decreases monotonically, which agrees with the earlier obtained data. For the compositions with $0.4 < x \leq 0.8$, not only the shift is observed but also the broadening and splitting of diffraction maxima, which may indicate the phase inhomogeneity, although according to DD data these are also the single-phase samples. According to HRTEM, a sample with the composition $x = 0.8$ has a modulated structure (the long-range order is preserved, whereas the short-range order is destroyed), which may lead to broadening of the diffraction maxima [178].

Lanthanum manganite from mechanochemical series, in distinction to the sample from ceramic series, is a single-phase hexagonal perovskite, and calcium manganite contains not only the cubic $CaMnO_3$ phase but also the brownmillerite $Ca_2Mn_2O_5$ phase. The amount of the latter phase is much smaller as compared to ceramic series. In the samples of mechanochemical series, a homogeneous solid solution forms only in the region of $0 < x \leq 0.4$. For compositions with $1 > x \geq 0.6$, the samples consist of two phases, $La_{0.6}Ca_{0.4}MnO_3$ and $CaMnO_{3-\delta}$. The observed narrowing of the region where homogeneous solid solutions are formed during mechanochemical synthesis may testify that solutions with $x > 0.4$ formed upon ceramic synthesis are unstable at lower temperatures [176, 179].

SCA toward CO oxidation of all calcium-substituted perovskites was lower as compared to the activity of the extreme terms of the series (Fig. 6.26). Therefore, the SIMS study revealed that for all intermediate samples the relative content of calcium on the surface was much higher than in the bulk. So, the decreased catalytic activity of intermediate samples in this series may be caused by the presence of Ca

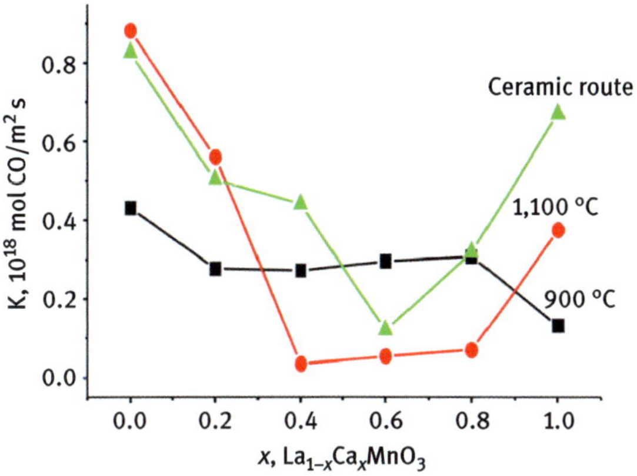

Fig. 6.26: Activity of $La_{1-x}Ca_xMnO_3$ samples in CO oxidation at 300 °C versus the composition and calcination temperature. Composition of the gas mixture: 1% CO + 1% O_2 in He.

cations on the surface, which serve as a catalytic poison for deep oxidation reactions [179].

In the La-Sr-Mn-O samples of ceramic series, surface enrichment with Sr and changes in the steady catalytic activity were not observed in the region of homogeneous solid solutions ($x \leq 0.4$). A maximum was noted only for the initial SCA at $x = 0.2$, which testifies to the fast relaxation of point defectness in the reaction medium [178–179].

Thus, the reported studies demonstrate that the mechanochemical method is promising for the synthesis of mixed oxide catalytic systems because it allows mixed oxides to be produced by a nonwaste technology and with a lower energy consumption. It should be noted that the synthesis of unsubstituted single-phase perovskites proceeds under milder conditions as compared to substituted ones.

It was shown that calcination temperatures of substituted perovskites provide the formation of cubic vacancy solid solutions that add oxygen upon cooling, which may lead to polymorphous phase transitions. Temperature of the polymorphous transition decreases with increasing the content of the substituting alkali-earth cation; at 300–600 °C this results in the formation of micromonolithic structure in the perovskite particles. The presence of such extended defects in oxides enhances their catalytic activity toward low-temperature deep oxidation reactions. In the case of surface segregation of simple oxides (alkali-earth or 3d), this decreases or increases the activity, respectively. Segregation of 3d oxides is typical of $LaMeO_3$ double oxides. The surface of calcium-containing perovskite-like oxides is commonly enriched with calcium, which is represented by the oxide or by the vacancy-ordered phase with the brownmillerite structure, $CaMeO_{2.5}$.

6.3 Control of rheological characteristics of pastes

For practical application, the catalyst should be formed as grains, whereas for the processes occurring at high linear velocities, preferable are the monolithic catalysts or monoliths. Depending on the content of active component and method of synthesis, supported and bulk oxide catalysts are distinguished. The most widely used and simple granulation method is the extrusion molding of plastic pastes obtained by mixing of powders, electrolytes and structure-forming additives. The fundamentals of this method were developed by Rebinder and coworkers [180–184].

Within the approaches developed by Rebinder's school and his followers, moldability of pastes and their grain strength can be enhanced by optimizing the moisture content, milling of powders, changing the surface reactivity of particles toward a binder to improve wettability of the surface and increase the strength of hydrated shells, amorphization of the surface, introduction of surfactants and electrolytes, and mixing of materials having different structural-mechanical properties [180–189].

Hence, MT is an efficient method for controlling the rheological characteristics of pastes and structural-mechanical properties of grains; it can affect some rheologically important parameters [73, 109, 190–203]. Milling of oxides as well as hydration and carbonization of their surface, which are observed during MT, improve moldability of oxides due to a better wetting of their particles with electrolyte and the complete development of hydrated shells at a lower moisture content of the paste. The complete development of hydrated shells facilitates mutual slipping of the particles during extrusion molding. Thus, separation of the pastes containing mechanically activated copper, cobalt and iron oxides can be prevented even by the addition of water, which makes it possible to synthesize the granulated bulk (without reinforcing additives) catalysts (Tables 6.7–6.9) [73, 109, 190–203].

The catalyst grain strength can be improved by extending the MT time. The analysis of factors underlying this phenomenon suggests that at short MT times the grain strength is improved mostly due to increasing the strength of unit contact, because there are no changes in the mean pore volume or radius and hence, within the Shchukin–Rebinder model, the number of contacts in the grain does not change. An increase in the strength of unit contact in the grain is caused by the extended area of contact between particles, which may occur due to acceleration of surface diffusion of cations to the contact sites of particles owing to structural distortions in their surface layers during MT. At longer MT times, the number of contacts can also increase due to decreasing the size of particle aggregates, because a decrease in the total pore volume is observed. Therefore, the extension of MT time not only increases the grain strength but can also decrease the specific surface area and hence the activity of catalysts owing to intensive transfer processes during thermal treatment. Thus, the MT time should be optimized to obtain a durable and active catalyst. Tables 6.7–6.9 illustrate the effect of mechanical treatment on the properties of granulated bulk oxide catalysts.

Table 6.7: Effect of MT time of CuO powder in an RPM-3 mill on the properties of granulated catalyst (T_{calc} 500 °C, moisture content of the paste 20%).

No.	MT time (s)	S_{sp} (m²/s)	P_o (kg/cm²)	W*10² cm³ butane/g s 350 °C for 1–2 mm grains	400 °C
1	0	25	Not molded	1.6	4.36
2	2	25	3	1.44	4.36
3	5	23	13	1.59	5.25
4	10	21	23	1.66	5.62
5	20	20	24	1.63	5.60
6	30	17	25	1.60	5.30
7	60	9	27	0.80	3.60
8	120	8	29	0.83	2.99
9	180	7	30	0.81	3.03

Table 6.8: Characteristics of cobalt oxide catalysts.

No.	MT time (s)	L (mL): T (100 g)	S_{sp} (m²/g)	P_o (kg/s²)	W*10² m³ C_4H_{10}/g*s for 1–2 mm grains 250 °C	400 °C
1	0	100	23	0.84	2.34	21.9
2	10	50	23	23	1.86	19.0
3	30	40	22	35	1.60	18.2
4	60	30	21	40	0.83	13.5

Calcination temperature 500 °C.
*MT was performed in an RPM-3 mill.

Table 6.9: Effect of MT in a VCM-25 mill on the properties of granulated iron oxide (T_{calc} 500 °C).

No.	Number of runs through VCM-25	Moisture content of paste (%)	S_{sp} (m²/s)	P_o (kg/cm²)	V (cm³/g)
1	0	37.5	12	10	0.53
2	1	31.5	11	25	–
3	3	20.6	9.4	34	–
4	5	19.4	9.8	37	0.21

The beneficial effect of MT on rheological properties (elasticity, viscosity and plastic strength) of the pastes intended for extrusion molding was revealed in [109, 198–203]. It was shown, for example, that treatment in a bead mill increases wettability of the oxide particles with chromic acid, which leads to a complete development of hydrated shells upon mixing of reagents, enhances the molding process, increases the grain strength and improves the catalyst vendibility.

References

[1] Boldyrev VV, Avvakumov EG. Mechanochemistry of inorganic materials. Uspekhi khimii 1971, 40, 1835–1856 (In Russian).
[2] Boldyrev VV. Experimental methods in mechanochemistry of inorganic materials. Novosibirsk, Russia, Nauka, 1983 (In Russian).
[3] Boldyrev VV. Mechanochemistry and mechanical activation of solids. Solid State Ionics 1993, 63–65, 537–543.
[4] Boldyrev VV. Mechanochemistry and Mechanical Activation. Mater. Sci. Forum. 1996, 225–227, 511–520.
[5] Boldyrev VV. Mechanochemistry and mechanical activation of solids. Russ. Chem. Rev. 2006, 75 (3), 177–189.
[6] Lyakhov NZ, Boldyrev VV. Mechanochemistry of inorganic materials. Analysis of initiative chemical process factors. Izvestiya SB AN USSR. Seria khimicheskih nauk 1983, 2(5), 3–8.
[7] Avvakumov EG. Mechanical methods for chemical process activation, Novosibirsk, Nauka, 1986 (In Russian).
[8] Pavlyukhin YT. Structural changes during mechanical activation of complex oxides with close-packed constitution motive. Dr.Sci. Thesis. Novosibirsk, Institut khimii tverdogo tela SB RAS, 2000.
[9] Zyryanov VV. Mechanochemical phenomenon in oxide systems. Dr.Sci. Thesis. Novosivirsk, Institute of Solid State Chemistry SB RAS, 1999.
[10] Molchanov VI, Gordeeva VI, Korneeva TA, et al. Dissociation of carbonates during comminuting. In: Mechanochemical phenomenon during comminuting. Novosibirsk, Nauka, 1971, 155–161 (In Russian).
[11] Khairetdinov EF, Galitsyn YG, Iost G. Influence of mechanochemical treatment on future thermal decomposition of $Ag_2C_2O_4$. Izvestiya Sibirskogo Otdeleniya AN USSR. Seriya khimicheskih nauk 1979, 14(6), 50–55 (In Russian).
[12] Pavlyukhin YT, Medikov YY, Boldyrev VV. Investigation of magnetic properties of amorphous ferrite magnetic by Mossbauer diffraction. Physics of Solid-State 1983, 5(3), 630–638.
[13] Pavlyukhin YT, Medikov YY, Boldyrev VV. Change in cationic distribution in ferrites- spinels during mechanical activation. Doklady AN USSR 1982, 266(6), 1420–1423 (In Russian).
[14] Pavljukhin YT, Medikov YY., Boldyrev VV. Magnetic and chemical properties of mechanically activated zinc and nickel ferrites. Mat. Res. Bull. 1983, 18, 1317–1327.
[15] Rykov AI, Pavlyukhin YT, Sirotinas NI, Boldyrev VV. Specifics of amorphous lanthanum, dysprosium and bismuth ferrites composition. Izvestiya SO AN USSR. Seriya khimicheskih nauk 1988, 14 (4), 15–24 (In Russian).
[16] Pavlukhin Yu. T., Rykov A. I., Boldyrev V. V., Medikov Ya.Ya. Structural transformation in oxides with close-packed sublattice during mechanical activation. In: Proceedings. of 2nd Japan-Soviet Symposium on mechanochemistry. Tokyo, 1988, 119–129.
[17] Avvakumov EG, Kosova NV, Alexandrov VV. Formation of defects during mechanical activation of titanium, tin and tungsten oxides. Izvestiya AN USSR. Neorganicheskie materialy 1983, 19, 1118–1121 (In Russian).

[18] Avvakumov EG, Kosova NV, Alexandrov VV. Influence of mechanical activation on lead dioxide decomposition. Izvestiya SO AN USSR. Seriya khimicheskih nauk 1983, 7(3), 25–30 (In Russian).
[19] Avvakumov EG, Anufrienko BF, Vossel SV, et al. Investigation of structural changes in mechanically activated titanium oxide by EPR. Izvestiya SO AN USSR. Seriya khimicheskih nauk 1986, 6(17), 16–21 (In Russian).
[20] Avvakumov EG, Varenek VA, Mazalov LN. Investigation of defects annealing in mechanically activated tin dioxide powder by Mossbauer diffraction. Izvestiya SO AN USSR. Seriya khimicheskih nauk 1980, 2 (1), 119–123 (In Russian).
[21] Avvakumov EG, Krechman AF, Marksi TY at al. Mossbauer effect in fine-grain ferric oxide powders. Izvestiya SO AN USSR. Seriya khimicheskih nauk 1977, 4(2), 3–8 (In Russian).
[22] Kosova NV, Petrov ES, Alexandrov VV, Avvakumov EG. About gas products escaped during heating of mechanically activated tungsten dioxide. Izvestiya SO AN USSR. Seriya khimicheskih nauk 1982, 2, 84–88 (In Russian).
[23] Steinike U, Kretzshmar U, Tolochko B. Structure Change in MgO by activation in Planetary Mill. Crystall. Res. Technol. 1983, 18(6), 793–798.
[24] Gusev GM, Novgorodtseva SV. Behavior of copper oxides during hyperfine disintegration and mechanical activation. In: Fiziko-khimicheskie issledovaniya mehanicheski aktivirovannyh mineral'nyh veshestv. Novosibirsk, IGiG, 1975, 46–52 (In Russian).
[25] Butyagin YP. Disordered structures and mechanochemical reactions in solid state. Uspekhi khimii 1984, 53(2), 1769–1780 (In Russian).
[26] Berestetskaya IV, Butyagin PY, Kolbanev IV. Reactivity of friction surface. Kinetika i kataliz 1983, 24(2), 441–448.
[27] Bystrikov AV, Berestetskaya IV, Srteletsky AN, Butyagin PY. Mechanochemistry of silica surface. 1. Reaction products with hydrogen. Kinetika I kataliz 1980, 21(3), 765–769.
[28] Srteletsky AN, Butyagin PY. Mechanochemistry of silica surface. 2. Friction role. Kinetika I kataliz 1980, 21(3), 770–775.
[29] Bystrikov AV, Srteletsky AN, Butyagin PY. Mechanochemistry of silica surface. 3. Active sites in reaction with hydrogen. Kinetika I kataliz 1980, 21(4), 1013–1018.
[30] Berestetskaya IV, Bystrikov AV, Srteletsky AN, Butyagin PY. Mechanochemistry of silica surface. 4. Interaction with oxygen. Kinetika I kataliz 1980, 21(4), 1019–1022.
[31] Bystrikov AV, Srteletsky AN, Butyagin PY. Mechanochemistry of silica surface. 5. CO oxidation. Kinetika I kataliz 1980, 21(5), 1148–1153.
[32] Kolbanev IV, Butyagin YP. Mechanochemical reaction of silica with water. Kinetika I kataliz 1982, 23(2), 326–333.
[33] Kolbanev IV, Butyagin YP. Investigation of silica disintegration process by EPR. In: Mechanoemission and mechanochemistry of solids. Frunze, Ilim, 1971, 215–218 (In Russian).
[34] Radtsig VA, Bystrikov AV. Investigation of reactive sites on silica surface by EPR. Kinetika I kataliz 1978, 19(3), 713–718.
[35] Radtsig VA. Reactive sites on milled silica surface. In materials: All-USSR symposium on mechanoemission and mechanochemistry, Tashkent, 1981, 1, 24–28 (In Russian).
[36] Radtsig VA. Paramagnetic sites on silica fault surface. Interaction with CO and H_2O molecules. Kinetika I kataliz 1979, 20(1), 248–255.
[37] Radtsig VA. Paramagnetic sites on silica fault surface. Interaction with H_2 and D_2 molecules. Kinetika I kataliz 1979, 20(2), 456–464.
[38] Vlasova MV, Cossack NT. Investigation of mechanical activation process by EPR. Izvestiya SB AN USSR. Seriya khimicheskih nauk 1983, 12(5), 40–45 (In Russian).

[39] Bobushev AA. Structure and surface reactivity of activated germanium, tin and magnesium oxides. Ph.D. Thesis. Moscow, 1983 (In Russian).
[40] Shrader R, Deren J, Fritsche B, Ziolkovski J. Kupfer(II)-oxide als Kontakt fur den N_2O Zerfall. Z. Fur Anorganische und Allgemeine Chemie 1970, 379(1), 25–34.
[41] Shrader R, Frietsche B. Kupfer(II)-oxide als Kontakt fur die orto-para wasserstoffumwandlung. Z. Fur Anorganische und Allgemeine Chemie 1970, 379(1), 17–24.
[42] Shrader R, Jacob G. Untersuchung von mechanich activierten die Kohlenmonoxidoxydation an -Fe_2O_3. Chemische Technik 1966, 18(17) 414–416.
[43] Shrader R, Tetzner G, Grund H. Der Aktive Zustand eines mechanisch aktivierten Kontaktes aus reduzierten Kobaltpulwer. Z. Fur Anorganische und Allgemeine Chemie 1966, 343, 308–314.
[44] Shrader R, Stedter W. Catalytic hydrogenation of Phenol and higher Alcohols with mechanically activated commercial Nickel powder. Acta Chimica Academial Scientiarum Hungarical 1968, 55(1), 39–47.
[45] Shrader R, Thien E. Mechanical activation CaO for use as contact catalyst. Z. Phys. Chem 1968, 238(1–2), 41–50.
[46] Shrader R, Nobst P, Tetzner G, Petzold D. Mechanical activation of nickel-silica carrier catalysts. Z. Fur Anorganische und Allgemeine Chemie 1969, 365(5–6), 225–261.
[47] Shrader R, Vogelsberger W. Mechanically activated Fe_2O_3. Z. Fur Anorganische und Allgemeine Chemie 1969, 368(3–4), 187–195.
[48] Heiniche G, Bock N, Harenz H. Chemical modification of mechanically activated reactions. III. Mechanism of tribomechanically activated metal carbonyl formation under the influence of sulfur-containing substances. Z. Fur Anorganische und Allgemeine Chemie 1970, 372(2), 162–170.
[49] Heinicke G, Harenz H. Chemische Aktivierung der mechanisch angeregten Nickel und Eisencarbonylbildung. Z. Fur Anorganische und Allgemeine Chemie 1963, 324(1–6), 185–196.
[50] Heinicke G, Harenz H, Sigrist K. Zur Kinetic der Reaction Ni+4CONi$(CO)_4$ bei tribomechanischer Bearbeitung des Nickels. Z. Fur Anorganische und Allgemeine Chemie 1967, B352, 168–183.
[51] Heinike G. Tribochemistry. Berlin: Akademie-Verlag, 1984.
[52] Sadahiro J, Shimazu K. Effect of dry grinding on physico-chemical properties of NiO powder. J. Chem. Soc. Japan 1968, 71(9), 1874–1878.
[53] Takashashi H, Tsutsumi K. Correlation between the structural disorder and catalytic activity of ZnO. J. Chem. Soc. Japan 1968, 71(9), 1345–1349.
[54] Buyanov RA, Molchanov VV, Boldyrev VV. Mechanochemical Activation as a Tool of Increasing Catalytic Activity. Catal. Today 2009, 144 (3–4), 212–218.
[55] Molchanov VV. Influence of mechanochemical activation on catalytic properties of iron-chrome-potassium dehydration catalyst. Khimicheskay promyshlennost' 1992, 7, 386–388 (In Russian).
[56] Molchanov VV, Plyasova LM, Goidin VV, Lapina OV, Zaikovsky VI. New compound in MoO_3-V_2O_5 system. Neorganicheskie materialy 1995, 31(9), 1225–1229 (In Russian).
[57] Molchanov VV, Goidin VV. Application of mechanochemical activation to increasing the phosphate dehydration catalyst strength. Khimicheskaya promyshlennost' 1993, 12, 613–615 (In Russian).
[58] Moroz EM, Bogdanov SV, Zaikovsky VI, Molchanov VV, Buyanov RA. Investigation of mechanochemically activated catalysts. Change of structural and morphological characteristics of polycrystalline zinc oxide. Kinetika I kataliz 1989, 30(4), 993–996.

[59] Chesnokov VV, Molchanov VV, Paukshtis EA, Konovalova TA. The influence of mechanochemical activation on coke formation on aluminum oxide. Kinet. Catal. 1995, 36(5), 698–700.
[60] Molchanov VV, Maksimov GM, Plyasova LM. Mechanochemical synthesis of alkali metals vanadates. Neorganicheskie materialy 1993, 29(4), 555–558 (In Russian).
[61] Molchanov VV, Buyanov RA, Goiudin VV. Possibilities of mechanochemical method application for preparation of supported catalysts. Kinetika i kataliz 1998, 39(3), 465–471 (In Russian).
[62] Molchanov VV, Buyanov RA. Scientific bases for application of mechanochemical method for catalysts preparation. In: Thesis 4th Russian conference: Scientific bases of catalysts preparation and technology. Sterlitamak, 2000, 48 (In Russian).
[63] D. Restrepo. Mechanochemistry for solid-state chemistry and catalysts. Dissertation D.Sci, Universitet of Centrasl Florida, 2013.
[64] Molchanov VV, Buyanov RA, Avvakumov EG, Boldyrev VV. In: Thesis II all-USSR conference on catalysts deactivation. Ufa, 1989, 2, 3-4 (In Russian).
[65] Molchanov VV, Buyanov RA. Mechanochemistry of catalysts. Russ. Chem. Rev. 2000, 69(5), 435–450.
[66] Isupova LA, Aleksandrov VY, Popovsky VV, Balashov VA, Davydov AA, Budneva AA, Kryukova GN. Effect of mechanical activation on the catalytic, physical and chemical properties of copper oxide. React. Kinet. Catal. Lett 1986, 31(1),195–202.
[67] Isupova LA, Aleksandrov VY, Popovsky VV, Moroz EM, Litvak GS, Kryukova GN. Influence of mechanochemical activation on the physico-chemical properties of cobalt oxide (11) and (111). Izvestiya SO AS USSR. Seriya khimicheskih nauk 1989, 1, 39–44 (In Russian).
[68] Chyong TG, Isupova LA, Davydov AA. A study of surface active centers of cobalt (II) and (III) oxide by IR-probe molecule (CO). Izvestiya SO AS USSR. Seriya khimicheskih nauk 1989, 3, 26–29 (In Russian).
[69] Isupova LA, Aleksandrov VY, Popovsky VV, Litvak GS, Davydov AA. Study of influence of mechanical activation on the catalytic and physico-chemical properties of Co_3O_4.In: Thesis of 6 all-USSR conference on oxidative heterogeneous catalysis. Baku, 1988, 153–154 (In Russian).
[70] Isupova LA, Sadykov VA, Tsybulya SV, G. Litvak GS, Burgina EB, Kustova GN, Kryukova GN, Kolomiychuk VN, Ivanov VP, Paukštys EA, Golovin, AV, Avvakumov EG. Effect of Mechanochemical activation on the physico-chemical properties (-Fe_2O_3). In: Thesis XI all USSR conference on the kinetics and mechanism of chemical reactions in a solid. Minsk, 1992, 176–178 (In Russian).
[71] Sadykov VA, Isupova LA, Tsybulya SV, Cherepanova SV, Litvak GS, Burgina EB, Kustova GN, Kolomiichuk VN, Ivanov VP, Paukshtis EA, Golovin AV, Avvakumov EG. Effect of mechanical activation on the real structure and reactivity of iron (III) oxide with corundum-type structure. J. Solid State Chem. 1996, 123(2), 191–202.
[72] Isupova LA, Sadykov VA, Pauli IA, Andryushkova OV, Poluboyarov VA, Litvak GS, Kryukova GS, Burgina EB, Solovieva LP, Kolomiichuk VN. Influence of mechanical activation on physical and chemical properties of MnO_x. In: Mehanokhimiya I mehanicheskaya aktivatsya, Petersburg, 1995, 6 (In Russian).
[73] Isupova LA, Sadykov VV. Physical and chemical bases of bulk oxide catalysts for deep oxidation process preparation by using mechanochemical method. 1. Specificity of mechanochemical method using for preparation of bulk oxide catalyst. Kataliz v promyshlennosti 2003, 4, 3–16 (In Russian).
[74] Sadykov VA, Tsyrul'nikov PG, Popovskii VV, Tikhov SF. Interaction of manganese oxides with reaction medium in heterogeneous catalytic oxidation of carbon monoxide by molecular

oxygen. 3. Study of relationships in formation of stationary surface layer. Kinetics and Catalysis 1981, 22(5), 951–957.
[75] Sadykov VA, Tsyrulnikov PG, Popovskii VV, Tikhov SF. Interaction of manganese oxides with the reaction medium in the heterogeneous catalytic oxidation of carbon monoxide by molecular oxygen. 4. Effect of the reaction mixture and catalyst on the steady-state kinetic parameters. Kinetics and Catalysis 1982, 23(2), 339–344.
[76] Sadykov VA, Tikhov SF, Popovskii VV, Bulgakov NN. Interaction of copper oxides with the reaction medium in heterogeneous catalytic-oxidation of carbon-monoxide by molecular-oxygen. 2. Investigation of the kinetics and thermodynamics of topochemical transformations in reduction of cuprous-oxide and a copper oxygen solid-solution. Kinetics and Catalysis 1983, 24(4), 665–671.
[77] Little LH, Infrared Spectra of Adsorbed Species. Pergamon Press, 1966.
[78] Davydov AA. Infrared spectroscopy of adsorbed species on the surface of transition metal oxides. Rochester CH. ed.. Chichester; New York: Wiley, 1990.
[79] Lokhov YA, Sadykov VA, Tikhov SF, Popovskii VV. Interaction of copper oxides with reaction medium in heterogeneous catalytic oxidation of carbon monoxide by molecular oxygen. IV. IR spectroscopic study of carbon monoxide interaction with copper oxide surface. Kinetics and Catalysis 1985, 26(1), 152–158.
[80] Lokhov YA., Davydov AA. Investigation of the state of transition-metal cations on catalyst surfaces by IR spectroscopy of adsorbed indicator molecules (CO, NO). 2. Reduced centers on the surface of catalysts containing copper. Kinetics and Catalysis 1979, 20(6), 1239–1245.
[81] Davydov AA, Budneva AA. IR-spectra of CO and NO adsorbed on CuO. React. Kinet. Catall. Lett.1984, 25(12), 121–124.
[82] Sadykov VA, Lokhov YA, Tikhov SF, Kryukova GN, Razdobarov VA, Popovskii VV, Bulgakov NN, Bredihin MN, Solovieva LP, Olenkova IP, Golovin AV. The Nature of the Surface active Centers and Mechanism of CO Catalytic Oxidation of Cobalt Oxide Catalysts. In: Proc. VI Int. Symp. Heterogeneous Catalysis. Sofia, 1987, 359.
[83] Sadykov VA, Razdobarov VA, Veniaminov SA, Bulgakov NN, Kovalenko ON, Pankratiev YD, Popovskii VV, Kryukova GN, Tikhov SF. Nature of the active oxygen of Co_3O_4. React. Kinet. Catal. Lett. 1988, 37(7), 109–114.
[84] Avvakumov EG. Mechanical methods for activation of chemical processes. Novosibirsk, Nauka, 1986 (In Russian).
[85] Petrov AN, Cherepanov VA, Zuev AY, Zhukovsky VM. Thermodynamic stability of ternary oxides in Ln-M-O (Ln=La, Pr, Nd; M=Co, Ni, Cu). J. Solid State Chem. 1988, 77, 1–14.
[86] V V Zyryanov VV. Mechanochemical synthesis of complex oxides. Russ Chem Rev 2008, 77(2),105–135.
[87] Zyryanov VV. Mechanochemical ceramic technology: opportunities and prospects. In: Avvakumov E ed, Mechanochemical synthesis of inorganic chemistry. Novosibirsk, Nauka, 1991, 102–1125 (In Russian).
[88] Zyryanov VV, Sysoev VF, Boldyrev VV. Mechanochemical ceramic technology. Doklady chemistry 1988, 298–99, 51–53.
[89] Zyryanov VV. Reaction zone model during mechanical treatment of powders in planetary mill. Neorganicheskie materialy 1998, 34(12), 1525–1534 (In Russian).
[90] Kosova NV, Deviatkina ET, Denisova TA, Zhuravlev NA, Avvakumov EG. Proton transfer in the mechanochemical reactions of hydrated oxides. Zurnal neorganicheskoi khimii 1999, 44(6), 912–916 (In Russian).
[91] Kosova NV, Deviatkina ET, Avvakumov EG. Surface basic and acidic centers and mechanochemical reactions in mixtures of hydrated oxides. Doklady akademii nauk 1996, 347 (4), 489–492 (In Russian).

[92] Deviatkina ET, Avvakumov EG, Kosova NV, Lyakhov NZ. Mechanical activation in the cordierite synthesis. Neorganicheskie materialy 1994, 30(2), 237–240 (In Russian).

[93] Boldyrev VV, Khabibulin FH, Kosova NV, Avvakumov EG. Hydrothermal reaction during mechanochemical activation. Neorganicheskie materialy 1997, 33(11), 1350–1353 (in Russian)

[94] Avvakumov E. G. Soft Mechanochemical Synthesis as Basis for New Chemical Processes. //Chemistry for Sustainable development.-1983.-V.2.-P.1–15.

[95] Avvakumov EG, Karakchiev LB. Mechanochemical synthesis as a method for preparation of nanodisperse particles and oxide materials. Chemistry for sustainable development 2004, 12, 287–291.

[96] Butyagin PY. Physical and chemical ways of relaxation of elastic energy in solids. Mechanochemical reactions in the two components systems.//In: Avvakumov EG ed. Mechanochemical synthesis in inorganic chemistry. Novosibirsk, Nauka, 1991 (In Russian).

[97] Pauli IA, Avvakumov EG, Isupova LA, Sadykov VA, Poluboyarov VA. Influence of mechanical activation on the synthesis and catalytic activity of lanthanum cobaltite. In: Proc. 1st Int. Conf. on Mechanochem, InCoMe -93 Kosice. Cambridge Intersci. Pub, 1993, 140–144.

[98] Kosova NV, Devyatkina ET, Avvakumov EG, Gainutdinov II, Rogachev AY, Pavlyukhin YT, Isupova LA, Sadykov VA. Mechanochemical synthesis of dicalcium ferrite with the perovskite structure. Inorganic materials 1998, 34(4), 385–390.

[99] Isupova LA, Tsybulya SV, Kryukova GN,. Budneva AA, Paukshtis EA, Litvak GS, Ivanov VP, Kolomiichuk VN, Pavlyukhin YT, Sadykov VA. Mechanochemical Synthesis and Catalytic Properties of the Calcium Ferrite $Ca_2Fe_2O_5$. Kinet. Catal. 2002, 43(1), 122–128.

[100] Do J-L, Frisčič T. Mechanochemistry: A Force of Synthesis. ACS Cent. Sci. 2017, 3(1), 13–19.

[101] Pal P, Saha S, Banik A, Sarkar A. Biswas K. All-Solid-State Mechanochemical Synthesis and Post-Synthetic Transformation of Inorganic Perovskite-type Halides. Chemistry 2018, 24(8), 1811–1815.

[102] James SL, Adams CJ, Bolm C, Braga D, Collier P, Friščić T, Fabrizia G, Harris KDM, Hyett G, Jones W, Krebs A, Mack J, Maini L, Orpen AG, Parkin IP., Shearouse WC, Steed JW, Waddell, DC. Mechanochemistry: opportunities for new and cleaner synthesis. Chem. Soc. Rev. 2012, 41 (1), 413–447.

[103] Zyryanov VV, Uvarov NF, Kostrovskii VG, Sadykov VA, Kuznetsova TG, Rogov VA, Zaikovskii VI, Frolova YV, Alikina GM, Litvak GS, Burgina EB, Moroz EM, Neophytides S. Design of New Oxide Ceramic Materials and Nanocomposites with Mixed Conductivity by Using Mechanical Activation Route, Mater. Res. Soc. Symp. Proc. 2003, 755, DD.6.27.1–DD.6.27.6.

[104] Mancheva M, Iordanova R, Dimitriev Y, Mechanochemical synthesis of nanocrystalline ZnWO4 at room temperature. J. Alloys Compd. 2011, 509(1), 15–20.

[105] Presecnik M, Bernik S. Influence of a mechano-chemical treatment on the synthesis and characteristics of p-type thermoelectric $Ca_3Co_4O_9$ ceramic. J. Alloy. Compd. 2016, 686, 708–716.

[106] Carbajal-Ramos IA, Andrade-Gamboa JJ, Condo AM, Gennari FC. Formation of Cubic Li_2TiO_3 By Mechanical Activation and Its Transformation to Monoclinic Phase: Stability in Helium and Hydrogen Flows. Solid State Ion. 2017, 308, 46–53.

[107] Avvakumov EG, Dyakov BE, Strugova LI, et al. Mechanical activation of solid state reactions. 1V. Solid state reduction of cassiterite. Izvestiya SO AS USSR. Seriya khimicheskih nauk 1978, 14(6), 3–11(In Russian).

[108] Vossel' SA, Pomoshnikov EE, Poluboyarov VA, Anufrienko VF. EPR study of copper into TiO_2 implementation process during mechanical treatment. Kinetika I kataliz 1984, 25(6), 1501–1504.

[109] Prokof'ev VY, Kunin FV, Il'in AP, Yurchenko EN, Novgorodov VN. Application of mechanochemical methods for synthesis of cordierite support and catalysts. Zurnal prikladnoy khimii 1997, 70(10), 1655–1659(In Russian).

[110] Shirokov YG. Mechanochemistry in technology of mixed catalysts for CO conversion. Voprosy kinetiki I kataliza: Mezvuzovskii sbornik. Ivanovo, Ivanovskii khimiko-tehnologicheskii institut, 1984, 3–9 (In Russian).

[111] Naugol'ny EP. Mechanochemical synthesis of cupper-magnesium catalyst. Ph.D. Thesis, Ivanovo, Ivanovskii khimiko-tehnologicheskii institute, 1999.

[112] Naugol'ny EP, Smirnov NN, Shirokov YG. Mechanochemical initiation of solid state reactions in the mixtures of hydroxocarbonates of magnesium and cupper. Izvestiya vyshih uchebnyh zavedenii. Khimiya I khimicheskaya tehnologiya 1999, 6, 46–51 (in Russian).

[113] Parhi P, Singh SS, Ray AR, Ramanan A. Mechanochemically assisted room temperature solid state metathesis reaction for the synthesis of $MMoO_4$ (M = Ca, Sr and Ba). Bull. Mater. Sci. 2006, 29(2),115–118.

[114] Manova E, Paneva D, Kunev B, Estournès C, Rivière E, Tenchev K, Léaustic A, Mitov I. Mechanochemical synthesis and characterization of nanodimensional iron–cobalt spinel oxides. J. Alloys Compd. 2009, 485(1–2), 356–361.

[115] Harris VG, Sepelak V. Mechanochemically Processed Zinc Ferrite Nanoparticles: Evolution of Structure and Impact of Induced Cation Inversion. J. Magn. Magn. Mater. 2018, 465, 603–606.

[116] Makarenkov DA, Baranov DA, Nazarov VI. Technological Aspects of the Use of Mechanical Activation Effects in Energy-Saving Granulation Processes. Theor. Found. Chem. Eng. 2017, 51 (4), 537–542.

[117] Kalinkin AM, Balyakin KV, Kalinkina EV, Nevedomskii VN. Solid-State Synthesis of Nanocrystalline $BaZrO_3$ Using Mechanical Activation. Inorg. Mater. 2017, 53 (5), 496–501.

[118] Kalinkin AM, Usoltsev AV, Kalinkina EV, Nevedomskii VN, Zalkind OF. Solid-Phase Synthesis of Nanocrystalline Lanthanum Zirconate Using Mechanical Activation. Russ. J. Gen. Chem., 2017, 87 (10), 2258–2264.

[119] Kolbanev IV, Degtyarev EN, Streletskii AN, Kokorin AI. Paramagnetic Centers Created Under Mechanochemical Treatment of Mixed Molybdenum-Vanadium Oxides. Appl. Magn. Reson. 2016, 47 (6), 575–588.

[120] Hongbo Y, Meiling C, Xiuhui W, Hong G, Synthesis of Magnesium-Aluminum Layered Double Hydroxides By Mechanochemical Method and Its Solid State Reaction Kinetics. Arch. Metall. Mater., 2015, 60(2), 1455–1457.

[121] Šepelák V, Heitjans P, Becker KD. Nanoscale spinel ferrites prepared by mechanochemical route. J Therm Anal Calorim. 2007, 90(1), 93–97.

[122] Berchmans LJ, Myndyk M, Da Silva KL, Feldhoff A, Šubrt J, Heitjans P, K.D. Becker KD, V. Šepelák V. A rapid one-step mechanosynthesis and characterization of nanocrystalline $CaFe_2O_4$ with orthorhombic structure. J. Alloys Compd. 2010, 500 (1), 68–73.

[123] Da Silva KL, Šepelák V, Düvel A, Paesano A, Hahn H, Litterst FJ, Heitjans P, Becker KD. Mechanochemical-thermal preparation and structural studies of mullite-type $Bi_2(Ga_xAl_{1-x})_4O_9$ solid solutions. J. Solid State Chem. 2011, 184, 1346–1352.

[124] Cristóbal AA, Botta PM, Aglietti EF, Conconi MS, Bercoff PG, Porto López JM. Synthesis, structure and magnetic properties of distorted $Y_xLa_{1-x}FeO_3$: Effects of mechanochemical activation and composition. Mater. Chem. Physics 2011, 130(3), 1275–1279.

[125] Cherkezova-Zheleva Z, Blaskov V, Mitov I, Klissurski D, Radev D, Tsokov P. Mechanochemically Activated Synthesis of Nanostructured $NiFe_2O_4$. Inorg. Mater. 2011, 47 (5), 527–530.

[126] Šepelák V, Düvel A, Wilkening M, Becker KD, Heitjans P. Mechanochemical reactions and syntheses of oxides. Chem. Soc. Rev 2013, 42 (18), 7507–7520.
[127] Prokof'ev VY., Gordina NE. Comminution and Mechanochemical Activation in Oxide Ceramics Technology (Review). Glass Ceram 2012, 69(1–2), 65–70.
[128] Gancheva M, Naydenov A, Iordanova R, Nihtianova D, Stefanov P. Mechanochemically Assisted Solid State Synthesis, Characterization, and Catalytic Properties of $MgWO_4$. J. Mater. Sci. 2015, 50 (9), 3447–3456.
[129] Fabian M, Bottke P, Girman V, Duvel A, Da Silva KL, Wilkening M, Hahn H, Heitjansef P, Sepel ˇ ak V. A simple and straightforward mechanochemical synthesis of the far-from-equilibrium zinc alumin1ate, $ZnAl_2O_4$, and its response to thermal treatment. RSC Adv., 2015, 5, 54321–54328.
[130] Wilkening M, Düvel A, Preishuber-Pflügl F, da Silva K, Breuer S, Sepel ˇ ak V, Heitjansef P. Structure and ion dynamics of mechanosynthesized oxides and fluorides. Z. Kristallogr 2017; 232(1–3), 107–127.
[131] Avvakumov EG, Senna M, Kosova NV. Soft Mechanochemical Synthesis: A Basis for New Chemical Technologies. Springer Science & Business Media, 2007.
[132] Kharlamova T, Pavlova S, Sadykov V., Chaikina M, Krieger T Ishchenko A, Pavlyukhin Y, Petrov S, Argirusis C. Mechanochemical Synthesis of Fe-Doped Apatite-Type Lanthanum Silicates. Eur. J. Inorg. Chem. 2010, 4, 589–601.
[133] Isupova LA, Tsybulya SV, Kryukova GN, Boldyreva NN, Vlasov AA, Alikina GM, Ivanov VP, Sadykov VA, Yakovleva IS. Physicochemical and catalytic properties of La_{1-x}Cax $FeO_{3-0.5x}$ perovskites. Kinet. Catal. 2000, 41(2), 287–291.
[134] Isupova LA, Tsybulya SV, Kryukova GN, Alikina GM, Boldyreva NN, Vlasov AA, Snegurenko OI, Ivanov VP, Kolomiichuk VN, Sadykov VA. Mechanochemical Synthesis and Catalytic Properties of the Calcium Ferrite Ca2Fe2O5. Kinetics and Catalysis 2002, 43(1), 129–138
[135] Isupova LA, Yakovleva IS, Tsybulya SV, Kryukova GN, Boldyreva NN. Mechanochemical Synthesis of Perovskite Deep Oxidation Catalysts $La_{1-x}Ca_xFeO_{3-0.5x}$. Chemistry for Sustainable Development 2002, 10(1–2), 27–38.
[136] Isupova LA, Yakovleva IS, Gainutdinov II, Pavlyukhin YT, Sadykov VA. Mössbauer studies of the phase composition and microstructure of $La_{1-x}Ca_xFeO_{3-y}$ system as related to the reactivity of the surface and bulk oxygen. React. Kinet. and Catal. Lett. 2004, 81(2), 373–382
[137] Sadykov VA, Isupova LA, Yakovleva IS, Alikina GM, Lukashevich AI, Neophytides S. Reactivity of the surface and bulk oxygen of $La_{1-x}Ca_xFeO_{3-y}$ system with respect to methane oxidation. React. Kinet. and Catal. 2004, 81(2), 393–398
[138] Isupova LA, Sadykov VA, Solovyova LP, Andrianova MP, Ivanov VP, Kryukova GN, Kolomiichuk VN, Avvakumov EG, Pauli IA, Andryushkova OV, Poluboyarov VA, Rozovskii AY, Tretyakov VF. Monolith perovskite catalysts of honeycomb structure for fuel combustion. In: G. Poncelet et al ed. Preparation of catalysts VI. Scientific Bases for the Preparation of Heterogeneous Catalysts. Amsterdam, Elsevier Science, 1995, 637–645.
[139] Grenier J, Pouchard M, Hagenmuller P. Vacancy Ordering in Oxygen-Deficient Perovskite-Related Ferrites. Structure and Bonding 1981, 47, 1–25.
[140] Grenier JC, Darriet J, Pouchard M, Hagenmuller P. Mise en Evidence d'une nouvelle famille de phases de type perovskite lacunaire ordonnee de formule $A_3M_3O_8$ ($AMO_{2.67}$). Mat. Res. Bull 1976, 11, 1219–1226.
[141] Vallet-Regi M, Gonzalez-Calbet J, Alario-Franco M. Structural Intergrowth in the $Ca_xLa_{1-x}FeO_{3-x/2}$ System (0x1); An Electron Microscopy Study. J. of Solid State Chem 1984, 55, 251–261.
[142] Alario-Franco MA, M. Henche JR, Vallet M, Calbet JMG, .Grenier J, Wattiaux A, Hagenmuller P. Microdomain Texture and Oxygen Excess in the Calcium-Lanthanum Ferrite $Ca_2LaFe_3O_8$. J. of Solid State Chem 1983, 46, P23–40.

[143] Rao CNR, Gopalakrishnan J, Vidyasagar K. Superstructure, Ordered Defects and Nonstoichiometry in Metal Oxides of Perovskite and Related Structures. Indian Journal of Chemistry 1984, 23A, 265–284.

[144] Isupova LA, Tsybulya SV, Kryukova GN, Rogov VA, Yakovleva IS, Sadykov VA. Microheterogeneous solid solutions in perovskites: Formation, microstructure and catalytic activity. In: Mixed Ionic Electronic System for advanced energy system, Orlovskaya N., Browning M. Eds, Boston/Dordrecht/London, Kluwer Academic Publishers, 2003, 137–156.

[145] Sadykov VA, Bulgakov NN, Muzykantov VS, Kuznetsova TG, Alikina GM, Lukashevich AI, Potapova YV, Rogov VA, Burgina EB, Zaikovskii VI, Moroz EM, Litvak GS, Yakovleva IS, Isupova LA, Zyryanov VV, Kemnitz E, Neophytides S. Mobility and Reactivity of the Surface and Lattice Oxygen of Some Complex Oxides with Perovskite Structure. In: Mixed Ionic Electronic System for advanced energy system, Orlovskaya N., Browning M. Eds, Boston/Dordrecht/London, Kluwer Academic Publishers, 2003, 49–70.

[146] Isupova LA, Pavlyukhin YT, Rogov VA, Alikina GM, Tsybulya SV, Yakovleva IS, Sadykov VA. Charge and coordination states of iron cations in $La_{1-x}Me_xFeO_{3-y}$ (Me=Ca, Sr, Ba) prepared by mechanochemical route as related to oxygen reactivity. In: MRS Symposium Proceeding. Solid State Chemistry of Inorganic Materials V. (Symposium held 29 November–2 December, 2004, Boston, Massachusetts, USA). Jing Li, Nathaniel Brese, Mercouri Kanatzidis, Martin Jansen eds. Pennsylvania, MRS, Warrendale, 2005, 848, 511–516.

[147] Nadeev AN, Tsybulya SV, Gerasimov EY, Kulikovskaya NA, Isupova LA. Structural features of the formation of $La_{1-x}Ca_xFeO_{3-\delta}$ ($0 \leq x \leq 0.7$) heterovalent solid solutions. Journal of Structural Chemistry 2010, 51(5), 891–897.

[148] Yakovleva IS, Isupova LA, Tsybulya SV, Chernysh AV, Boldyreva NN, Alikina GM, Sadykov VA. Mechanochemical synthesis and reactivity of $La_{1-x}Sr_xFeO_{3-y}$ perovskites. J. Mater. Sci. 2004, 39, 5517–5521.

[149] Isupova LA, Yakovleva IS, Alikina GM, Rogov VA, Sadykov VA. Reactivity of $La_{1-x}Sr_xFeO_{3-y}$ ($x = 0–1$) perovskites in oxidation reactions. Kinetics and Catalysis 2005, 46(5),729–735.

[150] Isupova LA, Prosvirin IP. X-ray photoelectron spectroscopy investigation of perovskites $La_{1-x}Sr_xFeO_{3-y}$ (0 x < 1.0), prepared *via* a mechanochemical route1. Russian Chemical Bulletin, International Edition 2013, 62(7), 1564–1569.

[151] Nadeev AN, Tsybulya SV, Shmakov AN, Yakovleva IS, Isupova LA, Kryukova GN. High-temperature studies of $La_{1-x}Sr_xFeO_{3-\delta}$ solid solutions using synchrotron radiation. Journal of Structural Chemistry 2007, 48(6), 1105–1109.

[152] Nadeev AN, Tsybulya SV, Belyaev VD, Yakovleva IS, Isupova LA. Weakly bound oxygen and its role in stability of solid solutions $La_{1-x}Sr_xFeO_{3-\delta}$. Journal of Structural Chemistry 2008, 49 (6), 1077–1083.

[153] Nadeev AN, Tsybulya SV, Yakovleva IS, Isupova LA. Mobile oxygen form and structural stability of $La_{1-x}Sr_xFeO_{3-y}$ perovskites. Acta Cryst. 2008, Suppl. A.64, 520.

[154] Nadeev AN, Tsybulya SV, Gerasimov EY, Isupova LA. High-temperature phase transitions in the $La_{0.25}Sr_{0.75}FeO_{3-\delta}$ solid solution with a perovskite structure. Journal of Structural Chemistry 2009, 50(1), 108–113.

[155] Nadeev AN, Tsybulya SV, Kryukova GN, Yakovleva IS, Isupova LA. Vacancies ordering in $La^{3+}{}_{1-x}Ba^{2+}{}_xFeO_{3-\delta}$ perovskites. Z. Kristallogr. Suppl. 2007, 26, 381–386.

[156] Isupova LA, Nadeev AN, Yakovleva IS, Tsybulya SV. Mechanochemical synthesis and physicochemical properties of $La_{1-x}Ba_xFeO_{3-\delta}$ ($0 \leq x \leq 1$) perovskites. Kinetics and Catalysis 2008, 49(1), 133–137.

[157] Cherepanov VA, Gavrilova LY, Barkhatova LY, Voronin VI, Trifonova MV, Bukhner OA. Phase Equilibria in the La-Me-Co-O (Me=Ca, Sr, Ba) Systems. *Ionics*. 1998, 4, 309–315.

[158] Gavrilova LY, Cherepanov VA, Surova TV, Baimistruk VA, Voronin VI. Phase equilibria and oxygen nonstoichiometry in complex oxide phases of the La-Ca-Co-O system. Russian J. of Phys. Chem. 2002, 76(2), 150–156.

[159] Mastin J, Einarsrud MA, Grande T. Crystal Structure and Thermal Properties of $La_{1-x}Ca_xCoO_{3-\delta}$ ($0 \leq x \leq 0.4$). Chem. Mater. 2006, 18, 1680–1687.

[160] Kononyuk IF, Tolochko SP, Lutsko VA., Anishchik VM. Preparation and properties of $La_{1-x}Ca_xCoO_3$ ($0.2 \leq x \leq 0.6$). Solid State Chem 1983, 48, 209–2014.

[161] Isupova LA, Yakovleva IS, Sutormina EF, Gerasimov EY., Rogov VA. Catalytic activity in methane oxidation of $La_{1-x}Ca_xCoO_{3-\delta}$ ($x = 01$) perovskites prepared by mechanochemical route. Adv Mater Sci 2018 3(2):5p doi: 10.15761/AMS.1000141.

162] K. Tabata, I. Matsumoto, S. Kohiki. Surface characterization and catalytic properties of $La_{1-x}Sr_xCoO_3$. J. mater. Sci. 1987, 22, 1882–1886.

[163] Isupova LA, Sadykov VA, Ivanov VP, Rar AA, Tsybulya SV, Andrianova MA, Kolomiichuk VN, Petrov AN, Kononchuk OF. Catalytic and physico-chemical properties of $La_{1-x}Sr_xCoO_3$ perovskites. React. Kinet. Catal. Lett 1994, 53(1), 223–230.

[164] Tsybulya SV, Kryukova GN, Isupova LA, Shmakov AN, Cherepanova SV, Sadykov VA. Formation of a real structure of perovskites $(La_{1-x}Sr_x)CoO_{3-\delta}$ in the region of the morphotropic phase transition. Journal of Structural Chemistry 1998, 39(1),75–79.

[165] Isupova LA, Alikina GM, Tsybulya SV, Boldyreva NN, Kryukova GN, Yakovleva IS, Sadykov VA. Real structure and catalytic activity of $La_{1-x}Sr_xCoO_3$ perovskites. J. of Inorg. Mater 001, 3(6), 559–562.

[166] Isupova LA, Sadykov VA, Tsybulya SV, Kryukova GN, Ivanov VP, Petrov AN, Kononchuk OF. Effect of a structural disorder on the catalytic activity of mixed La-Sr-Co-Fe-O perovskites. React. Kinet. Catal. Lett 1997, 62(1), 129–136.

[167] Yakovleva I., Isupova LA, Rogov VA. Oxygen species and their reactivity in the mechanochemically prepared substituted perovskites $La_{1-X}\ Sr_{\ x}\ CoO_{3-Y}$ ($x = 0–1$). Kinetics and Catalysis 2009, 50(2), 275–283.

[168] Reller A, Thomas JM, FRS, Jeffersonn DA, Uppal MK. Superstructures formed by ordering of vacancies in a selective oxidation catalyst: grossly defective $CaMnO_3$. In: Proc R. Soc. Lond. A. 1984, 394, 223–241.

[169] Wollan EO, Koehler WC. Neutron Diffraction Study of the Magnetic Properties of the Series of Perovskite-Type Compounds $[(1-x)La,xCa]MnO_3$. Physical Review 1955, 100(2), 545–563.

[170] Topfer J, Goodenogh JB. $LaMnO_{3+}$ revisited. J. of Solid State Chem 1977, 130, 117–128.

[171] Rao CNR, Cheetham AK, Mahest R. Gigant Magnetoresistance and Related Properties of Rare-Earth Manganates and Other Oxide Systems. Chem. Mater 1996, 8, 2421–2432.

[172] Petrov AN, Tikhonova IL, Zuev AY. Phase Relations in the La(Sr)-Mn-Cu-O system and Oxygen Nonstoichiometry of Copper-Substituted Lanthanum Manganates. Solid Oxide Fuel Cells. Electrochemical Society. Stimming U., Singhal S., Tagawa H., and Lehnert W. Eds. The Electrochemical Society Proceedings Series. Pennington. NJ. 1997, 40, 927–936.

[173] Yue Wu, Tao Yu, Bo-sheng Dou, Chen-xian Wang, Xiao-rong Fan and Lian-chi Wang. A comparative study on perovskite-type mixed oxide catalysts A′xA1 − xBO3 − λ (A′ = Ca, Sr, A = La, B = Mn, Fe, Co) for NH3 oxidation. J. of Catalysis 1989, 120(1), 88–107.

[174] Filonova EA, Cherepanov VA, Voronin VI. A Study of the phase composition and crystal structure in the series of $LaCo_{1-x}Mn_xO_{3\pm\delta}$ solid solutions. Russian journal of physical chemistry A 1998, 72(10), 1706–1708.

[175] Faaland S, Knudsen KD, Einarsrud MA, ROrmark L, HOier R, Grande T. Structure, stoichiometry, and phase purity of calcium substituted lanthanum manganite powders. J. Solid State Chem 1998, 140, 320–330.

[176] Mitchell JF, Argyriou DN, Potter CD, Hinks DG, Jorgensen JD, Bader SD. Structural phase diagram of $La_{1-x}Sr_xMnO_{3+s}$. Relationship of magnetic and transport properties. Phys. Rev. B 1996, 54(9), 6172–6182.
[177] Mahendiran R, Tiwary SK, Kaychaudhuri AK, Ramakrishnan TV, Mahesh R, Rangavittal N, Rao C.N.R. Structure, electron-transport properties and giant magnetoresistance of hole-doped $LaMnO_3$ systems. Phys. Rev. B 1996, 53(6), 3348–3358.
[178] Isupova LA, Tsybulya SV, Kryukova GN, Alikina GM, Boldyreva NN, Yakovleva IS, Ivanov VP, Sadykov VA. Real structure and catalytic activity of $La_{1-x}Ca_xMnO_3$ perovskites. Solid State Ionics 2001, 141–142, 417–425.
[179] Yakovleva IS, Isupova LA, Rogov VA, Sadykov VA. Forms of oxygen in La1 – xCaxMnO 3 + δ (x = 0–1) perovskites and their reactivities in oxidation reactions. Kinetics and Catalysis 2008, 49(2), 261–270.
[180] Rebinder PA. Surface effects in disperse systems. Physicochemical mechanics. Moscow, Nauka, 1979 (In Russian).
[181] Nechiporenko SP. Physikochemical mechanics of disperse systems in technology of structural ceramics. Kiev, Naukova Dumka, 1968 (In Russian).
[182] Nechiporenko SP. Principal questions of processes of treatment and molding of ceramic pastes. Kiev, AN USSR, 1960 (In Russian).
[183] Kruglitskii NN. Basics of physicochemical mechanics. Kiev. Vysha shkola, 1976 (In Russian).
[184] Ur'ev NB. High-concentration disperse systems. Moscow, Khimia, 1980 (In Russian).
[185] Powell J, Assabumrungrat S, Blackburn S. Design of ceramic paste formulations for co-extrusion. Powder Technology 2013, 245, 21–27.
[186] Aranzabal A, Iturbe D, Romero-Sáez M, González-Marcos MP, González-Velasco JR, González-Marcos JA. Optimization of process parameters on the extrusion of honeycomb shaped monolith of H-ZSM-5 zeolite. Chemical Engineering Journal 2010, 162, 415–423.
[187] HYPERLINK https://www.sciencedirect.com/science/article/pii/S0272884200000705?via%3DihubAnanthakumar S, Menon ARR, Prabhakaran K, Warrier K.G. Rheology and packing characteristics of alumina extrusion using boehmite gel as a binder. Ceramics International 2001, 27(2), 231–237.
[188] Graczyk J, Gleissle W. Rheology and Extrudability of Alumina Paste for Catalysts. Ind Ceram. 1996, 16(3), 199–203.
[189] HYPERLINK https://www.sciencedirect.com/science/article/pii/S0955221902000456?via%3Dihub Das RN, Madhusoodana CD, Okada K. Rheological studies on cordierite honeycomb extrusion. Journal of the European Ceramic Society December 2002, 22(16), 2893–2900.
[190] Yurchenko EN, Prokof'ev VY, Il'in AP. Shirokov YG. Regulation of structural, mechanical and rheological properties of based on titanium dioxide molding pastes. Zurnal prikladnoii khimii 1995, 68(4), 607–612.
[191] Winstone G. Production of Catalyst Supports by Twin Screw Extrusion of Pastes. 2011. University of Birmingham.
[192] Blackburn S, Wilson DI. Shaping ceramics by plastic processing. Journal of the European Ceramic Society 2008, 28, 1341–1351.
[193] Benbow JJ, Bridgwater J. Measurement of paste yield by cone penetration. Chemical Engineering Science 1987, 42, 915–919.
[194] Eftekhari B, Yekta BE, Mahabad NA, Ebadzadeh T. Rheological study on cordierite paste during extrusion. Advances in Applied Ceramics 2007, 106(4), 161–164.
[195] Isupova LA, Alexandrov BY, Popovsky VV, Moroz EM, Litvak GS, Kyukova GN. Influence of preparation conditions on the properties of deep oxidation cupper oxide catalysts. Zurnal prikladnoi khimii 1988, 9, 1976–1980 (In Russian).

[196] Isupova LA, Alexandrov BY, Popovsky VV. Use of mechanochemical activation for preparation of gas purification oxide catalysts. In: Kataliticheskaya ochistka gasov. Thesis of 5 all-USSR conference, 1989, 3–5 (In Russian).
[197] Isupova LA, Sadykov VA, Tsybulya SV, Litvak GS, Kryukova GN, Burgina EB, Golovin AV. Development of Fe_2O_3-based catalysts of different geometries for environmental catalysis. Chemistry for Sustainable Development 2003, 11(1), 89–99.
[198] Isupova LA, Sadykov VA, Avvakumov EG, Kosova NV. Mechanical activation in the technology of high-temperature oxide catalysts. Khimiya v interesah ustoichivogo razvitiya 1998, 6(1–2), 207–210 (In Russian).
[199] Zolotovsky BP. Scientific basis of crystalline hydroxides mechanochemical and thermochemical activation in preparation of catalysts and supports. Dr.Sci. Thesis, Institut kataliza SO RAN, 1992 (In Russian).
[200] Shirokov YG, Il'in AP, Kirillov IP et al. Influence of mechanochemical treatment on quality of absorbent. Zurnal prikladnoii khimii 1979, 52(6), 1228–1233 (In Russian).
[201] Shirokov YG. Mechanochemistry in technology of mixed catalysts for CO conversion. Voprosy kinetiki I kataliza: Mezvuzovskii sbornik. Ivanovo, Ivanovskii khimiko-tehnologicheskii institut, 1984, 3–9 (In Russian).
[202] Shirokov YG. Role of mechanochemistry in the catalyst pastes preparation step. Izvestiya vyshih uchebnyh zavedenii. Khimiya I khimicheskaya tehnologiya 2001, 44(2), 3–14 (In Russian).
[203] Prokof'ev VY, Razgovorov PB, Il'in AP. Bases of physico-chemical mechanics of molded catalysts and sorbents. Moscow, Krasand, 2013 (In Russian).

7 Ceramometal supports and catalysts prepared through hydrothermal treatment of Al-containing powders

Ceramometals synthesized from aluminum-containing powders have been known for quite a long time [1–3]. Synthesis of such materials is based on the ability of aluminum to be partially oxidized by water under hydrothermal treatment (HTT) conditions with subsequent hydrolysis of the oxidation products, their precipitation and crystallization on the surface of metal particles [4]. Crystallization at the contact sites of metal particles leads to their cementation into a porous monolith.

There are several main approaches to the synthesis of porous ceramometals and their application as supports and catalysts.

a) Synthesis of Al_2O_3/Al ceramometals from powdered aluminum with subsequent deposition of the active component (AC) from a solution [5–11].
b) Incorporation of the AC into Al_2O_3/Al matrix via mixing the powdered AC as oxides or alloys with aluminum.

Details of ceramometal synthesis are illustrated in Fig. 7.1 [5, 6]. Powders are charged into special molds, which allow access of water and withdrawal of the produced hydrogen. It is possible to vary features of the aluminum powder, temperature, pressure and time of HTT. The mold with the loaded powder is then subjected to HTT. Details of the process are reported in [4, 7–10]. A firm porous monolith is produced upon HTT; it is taken out of the mold and calcined. Size and form of the resulting monoliths can be widely varied depending on the design.

7.1 Ceramometal supports and catalysts based upon Al_2O_3/Al

7.1.1 Influence of the nature of aluminum powder on the microstructure, textural and mechanical properties of Al_2O_3/Al ceramometals

In Al_2O_3/Al ceramometals, aluminum is evenly distributed in the oxide matrix. A typical relief of the fracture face of Al_2O_3/Al cermet prepared by oxidation of aluminum powder by water at 100 °C followed by calcination in air at 550 °C is shown in Fig. 7.2 [11]. Aluminum particles are covered by a porous oxide film.

Analysis of the ceramometal macrostructure (Fig. 7.3a–c) shows that it substantially depends on the type of the aluminum powder. The loosest packing of the monoliths is typical for ceramometals produced from PAP-2 powder.

Averaged characteristics of the obtained materials are also very different (Table 7.1) [11]. The average macropore diameter determined by the Darcy method

https://doi.org/10.1515/9783110587777-007

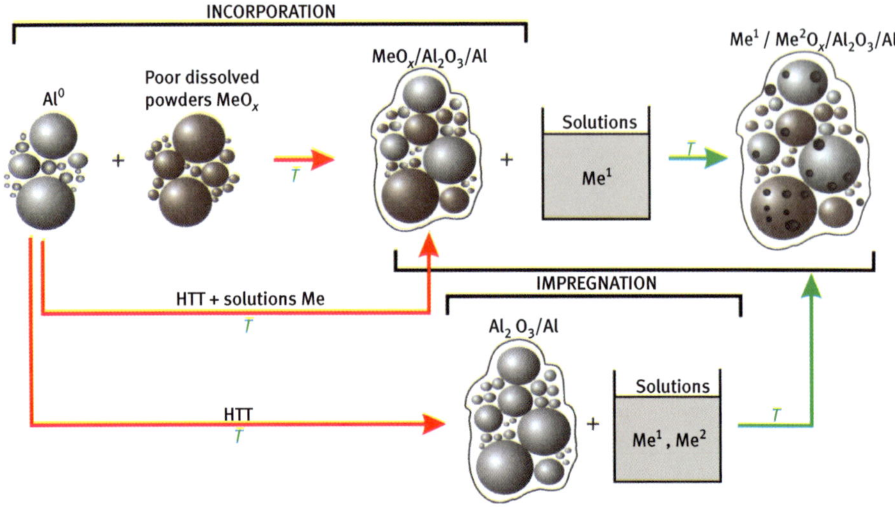

Fig. 7.1: Main stages of the preparation of ceramometal catalysts with aluminum powder [5, 6].

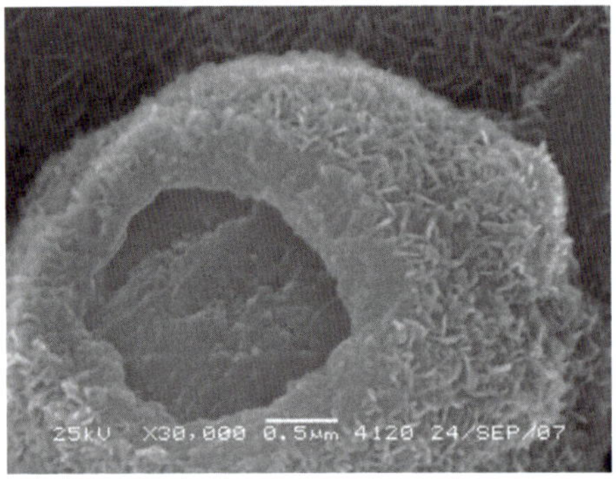

Fig. 7.2: SEM micrograph of Al_2O_3/Al ceramometal prepared from ASD-4 powder: (a) aluminum core and (b) porous alumina matrix [7].

increases from 1 to 22 μm in the series of ceramometals ASD-4 < ASD-1 < PAP-2. The maximum pore size varies in a similar order. The permeability coefficients of obtained materials also differ by more than a factor of 50 correlating with the average pore size (Table 7.1).

Comparison of the macropore structure parameters (Table 7.1) with a quantitative analysis of the macrotexture (Fig. 7.3) shows that a decrease in the average size of ceramometal Al_2O_3/Al particles decreases the average size of macropores and permeability. The average size and shape of particles in cermets are determined by the particles of the aluminum source powder used for the synthesis. According to the data obtained by the Koulter method, the average particle size changes as follows: ~33 μm

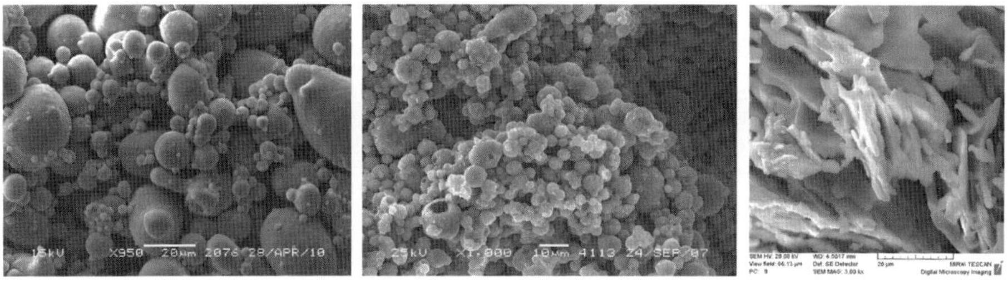

(a) (b) (c)

Fig. 7.3: SEM micrographs of porous Al_2O_3/Al composites prepared from different Russian commercial-grade aluminum powders: (a) ASD-1, (b) ASD-4 and (c) PAP-2 [11].

Table 7.1: Macropore structure, permeability and mechanical properties of Al_2O_3/Al composites prepared from different aluminum powders [11].

Source of aluminum powder	Average particle size of Al^0 (μm)	Permeability coefficient, $K \times 10^{-13}$ (m^2)	Maximum macropore size (μm)	Average macropore size (μm)	Crushing strength (σ) (MPa)	Al_2O_3 (wt.%)	Cermet porosity (ε) (%)
ASD-1	33*(25)**	~3.5	12.5	6.5	11	~24	~36
ASD-4	12(5)	0.4	4.5	~ 1	12	~44	~57
PAP-2	~5.4, 35	20	~63.5	22	23	~94	~42

*Coulter method
**According to SEM

(ASD-1), ~12 μm (ASD-4), ~5 and ~35 μm (PAP-2) (Table 7.1). This trend qualitatively matches the results obtained by scanning electron microscopy (SEM) (Table 7.1, Fig. 7.3).

The differences in the shape and average sizes of porous cermet particles are quantitatively expressed as the loading density of aluminum powder in a die before HTT. The loading density is largely determined by the size and shape of aluminum particles. The flat shape of partially aggregated PAP-2 particles results in a substantially lower filling density compared to round particles, about 0.3–0.4 g/cm^3. Round particles typical of ASD-1 and ASD-4 aluminum source provide a denser packing, 1.6–1.8 g/cm^3 for ASD-4 and about 1.3–1.4 g/cm^3 for ASD-1. Apparently, different particle packing is also preserved in the obtained ceramometal monoliths (Fig. 7.4). As a result, the average size of macropores in PAP-2 monolith is substantially larger. The composite obtained from ASD-4 has the smallest pore size (Table 7.1).

The degree of aluminum conversion to hydroxide, determining the fraction of oxide in Al_2O_3/Al cermets, has a comparatively minor effect on parameters of the macropore structure (Table 7.1). This is the result of relatively mild conditions of HTT (~100 °C, 1 bar, 4 h). At higher temperature and pressure of HTT in an

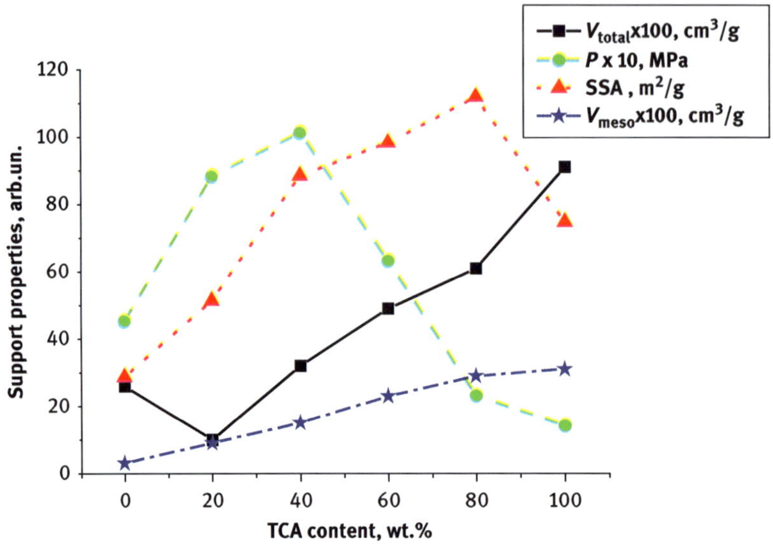

Fig. 7.4: Some textural and mechanical properties of supports with varied Al hydroxide content [5].

autoclave, the influence of the conversion is much higher [5–7]. So, permeability of the composite materials prepared using hydrothermal aluminum oxidation can be substantially varied. This is very important for internal diffusion limitations of the catalysts.

One of the most remarkable properties of composites prepared by cementing aluminum-containing powders under HTT conditions is the presence of developed nano (micro-, meso-)porous structure formed by primary aluminum oxide nanoparticles and their aggregates. It is important because nanoporosity (nanopore volume) provides improved water capacity and increase of the AC content loaded through impregnation as well as the AC dispersion. These particles are formed during thermal decomposition of aluminum hydroxides obtained from aluminum metal particles at the HTT stage. The values of the aluminum oxide specific surface areas (SSAs) reported in parenthesis (Table 7.2) show that the SSAs of the oxide differ substantially for ceramometals prepared from ASD-1 and ASD-4 despite the fact that their total SSAs are similar. This result is largely related to different ratios of the aluminum hydrothermal oxidation (HTO) rate to the aging rate of the HTO products under HTT conditions [7]. The particle dimensions should increase due to recrystallization of nanoparticles when the relative aging rate increases, whereas the SSA should decrease. For ASD-1 and ASD-4 powders the aging rates determined by the external HTT conditions were approximately equal. Meanwhile, the rates of Al consumption at the second diffusion-controlled stage of hydrothermal oxidation were substantially

Table 7.2: Parameters of the nanoporous structure of Al_2O_3/Al composites [11].

Aluminum powder	Specific surface area of Al_2O_3/Al (Al_2O_3) (m^2/g)	Nanopore volume, V (cc/g)	Nanopore diameter (nm)
ASD-1	52 (221)	0.05	3.8
ASD-4	54 (318)	0.05	4.7
PAP-2	121 (~121)	0.16	4.7

different. For ASD-4 this rate was higher by two orders of magnitude [11]. Therefore, the smallest hydroxide (oxide) particles were obtained for this ceramometal. For PAP-2 the lowest oxidation rate was observed in the diffusion region. This led to a significant growth of the primary particle dimensions and decrease of the alumina SSA in comparison with the other samples.

The nanopore structure of ceramometals was increased via incorporation of aluminum hydroxide, which is the alumina precursor. Figure 7.4 illustrates variation of main properties of granulated Al_2O_3/Al supports with the AlOH content. As is seen, the total pore volume goes through the minimum while crushing strength varied almost contrariwise. As compared to supports prepared from pure components, blending increases crushing strength (up to ~10 MPa). Taking into account that the total pore volume is 2–3 times higher than that of micro- and mesopores, one can conclude that macroporosity presumably influenced the mechanical properties of supports.

These pores are formed by the space between the aluminum (d ~ 20–25 μm) and AlOH aggregate (d ~ 9 μm) particles comprising the powdered blends. At the low AlOH content, its particles fill interstitials between larger aluminum ones. Hence, more dense packing occurs leading to the growth of the crushing strength of the granules. At higher AlOH content its particles pull out aluminum particles, which lead to less dense packing of blends and decrease of the crushing strength.

7.1.2 Catalytic properties of $Cr_2O_3/Al_2O_3/Al$ composites in dehydrogenation of light alkanes

Figure 7.5 illustrates conversion (X) and selectivity (S) of isobutane to isobutene dehydrogenation for $CrO_x/Al_2O_3/Al$ catalysts versus AlOH content in ceramometal. The maximum conversion was observed for the sample containing 50% AlOH. The catalysts were prepared through impregnation of Al_2O_3/Al supports. This sample is also characterized by the highest selectivity to isobutene. The developed macropore structure of the support granules with 3–5 mm size excludes problems of internal diffusion limitations typical for granulated catalysts. Overall, this provides for a high isobutene conversion that is close to the equilibrium one at this temperature and unusually high (>94%) selectivity to the dehydrogenation product.

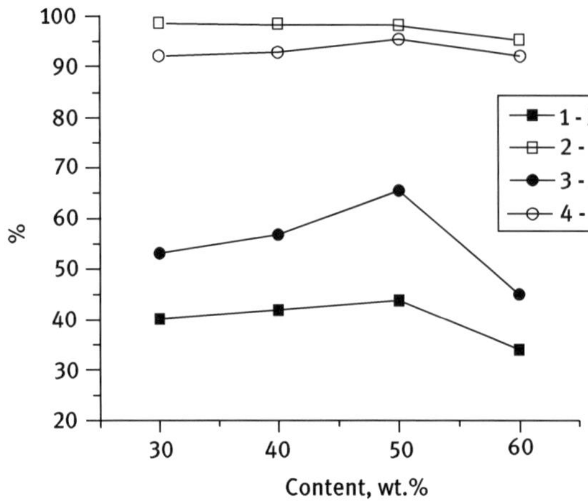

Fig. 7.5: Dependence of isobutane conversion X (1,3) and selectivity to isobutene S (2,4) at 540 °C (1,2) and 590 °C (3,4) on AlOH content (wt.%) in the initial blends used for preparation of granulated composites $CrO_x/Al_2O_3/Al$ [7].

7.1.3 Catalytic properties of $FeZrH/Al_2O_3/Al$ catalysts in Fischer–Tropsh synthesis

The method used for preparation of porous ceramometals from aluminum mixtures with other powdered components makes it possible to incorporate powdered ACs with a high catalytic activity while preserving their main properties. For instance, hydrogenated iron–zirconium alloys have relatively high productivity in Fischer–Tropsch synthesis of liquid C_{5+} hydrocarbons, which are used as synthetic fuels [12]. However, their application in catalysis is complicated by the fact that these alloys are rather fragile, whereas the use of small particles in a fixed bed reactor leads to significant pressure drop. The attempts to compact small grains of hydrogenated iron–zirconium alloys by mechanical pressing in a mixture with powdered aluminum did not yield promising results [13]. Granulation of the AC with the average particle size ~0.375 mm in a porous Al_2O_3/Al cermet led to a fourfold increase of the catalytic activity compared to AC with the particle size ~2.5 mm (Fig. 7.6.) [14]. This effect is, most likely, caused by developed macroporous structure of the composite ceramometals. However, further decrease of the AC particle size in the ceramometal by an order of magnitude did not lead to a substantial increase of the AC productivity normalized to unit mass, especially in production of liquid hydrocarbons. The lack of effect in this case is caused by the influence of the aluminum HTT products that decorate the AC surface blocking the active sites and, thus, leveling the effect of AC dispersion (Fig. 7.7.). In spite of alumina deposition on the surface, a high activity of AC is provided by a high permeability of this porous layer toward reagents and FTS products due to developed ultra-macroporosity. Qualitatively, the

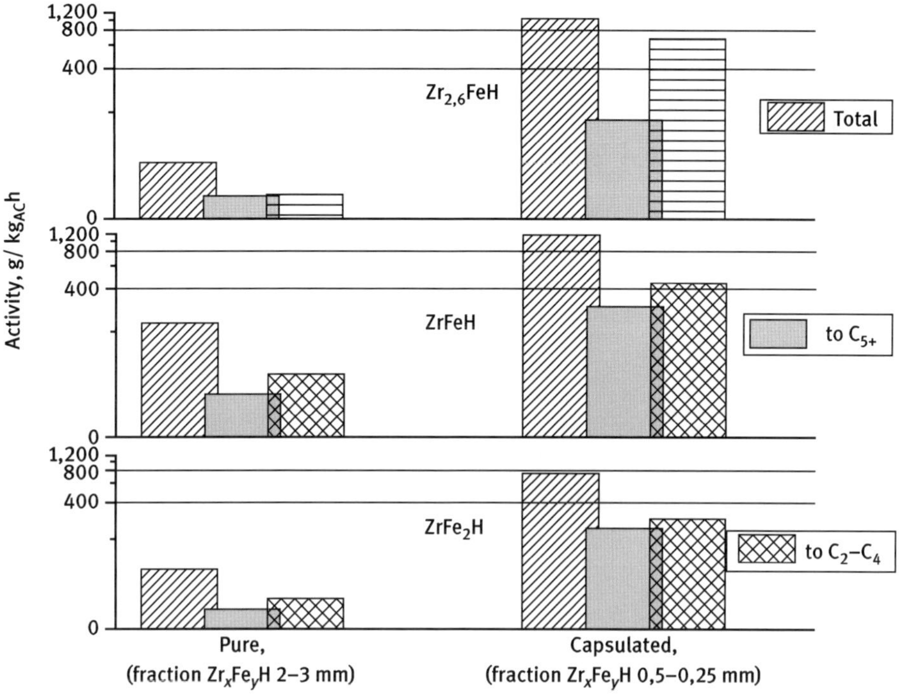

Fig. 7.6: Effect of incorporation of hydrogenated ZrFe intermetallides with varied Fe:Zr ratio on their activity to different FTS products [15].

pores with diameters up to tens of micrometers are visible in SEM images of composite "$ZrFeH_{0.5}$"/Al_2O_3/Al catalyst (Fig. 7.7a).

Comparison of the textural characteristics of catalysts with their catalytic performance in Fischer–Tropsch synthesis revealed for alumina-containing composites prepared from different aluminum powders and alumina precursors is presented in Fig. 7.8. There is a clear tendency for decreasing selectivity to C_{5+} products and increasing that to methane with the increase of the micropore and mesopore volume and SSA (Fig. 7.8). Some deviation can be explained by the effect of the average pore diameter which could differ for composites with the same integral pore volume. Observed relation between selectivity and SSA could be explained by variation of the concentration of active sites (surface concentration of iron) as well as by the impact of diffusion characteristics of porous alumina, both factors are known to be of great importance for the FTS selectivity. Detailed analysis of these factors has been earlier presented for cobalt-containing FTS catalysts [16]. Probably, the surface of alumina strongly affects the FTS characteristics through readsorption of reaction intermediates leading to chain termination, while readsorption on the active sites situated on the surface of AC leads to the chain growth. Another factor of importance

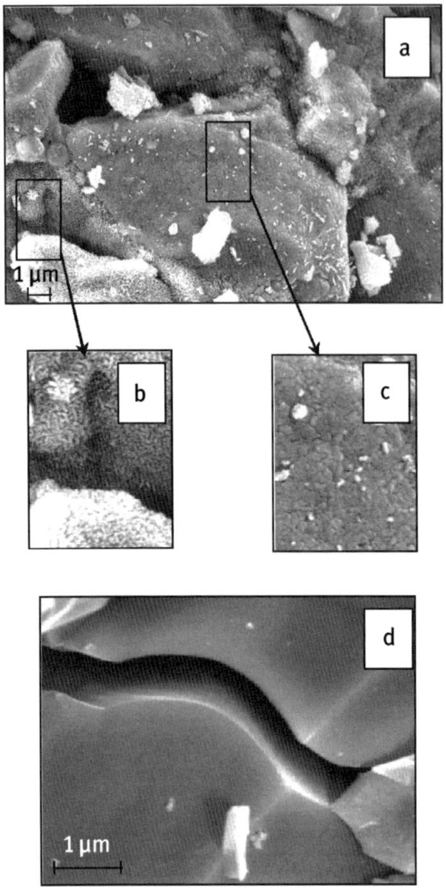

Fig. 7.7: SEM micrographs of $ZrFeH_x/Al_2O_3/Al$ catalyst prepared through HTO and calcination at 350 °C (a), $ZrFeH_{0.5}$ after the same treatment (d), inserts (b) and (c) – surface of incorporated AC (b) – with porous alumina, (c) – without porous alumina.

could be filling of micropores by water and liquid hydrocarbons formed in FTS or a nonuniform distribution of active sites inside the composite catalysts. In all cases, the microporosity of alumina has a negative effect on the chain growth reaction. So, the alumina in ceramometal matrix is less useful for FTS contrary to impregnated AC in dehydrogenation.

7.2 Ceramometal catalyst based upon MeAlO/MeAl

An essential restriction of Al_2O_3/Al ceramometals is stability of metallic aluminum, the melting point of which does not exceed 660 °C. So, a method based on the use of aluminum-containing powdered alloys was further developed for the synthesis of MeAlO/MeAl ceramometals. Such alloys obtained by mechanochemical treatment of aluminum with powdered metals, similar to neat aluminum, retained the ability

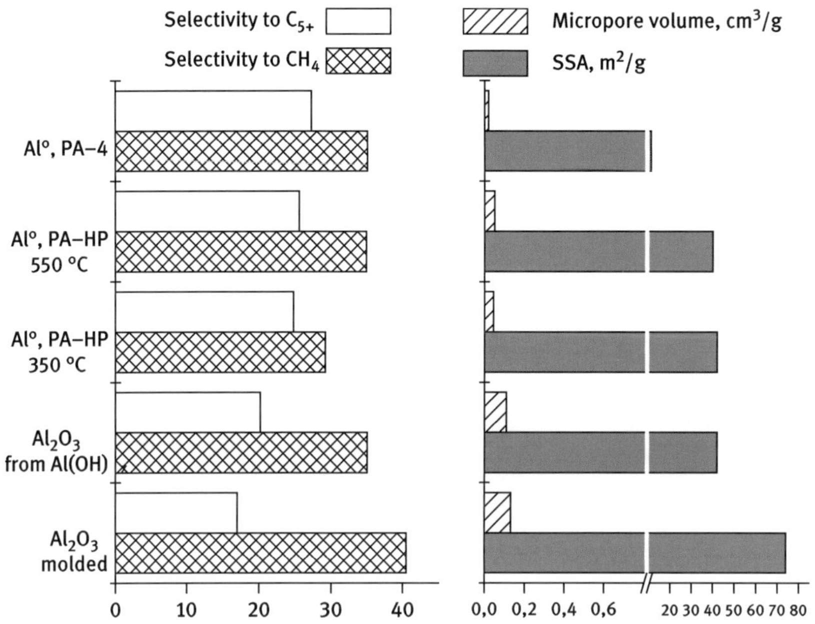

Fig. 7.8: Influence of microporous structure of alumina on the selectivity of composite catalysts based upon hydrogenated ZrFe intermetallides.

to cementation into porous monoliths under HTT conditions [9, 10]. A weakly developed mesoporous structure is typical for MeAlO/MeAl ceramometals; so it was developed by incorporation of materials having a more developed mesopore structure and subsequent impregnation similar to stages (a) and (b). In the general form, possible variants of ceramometal synthesis from alloys and variants of the produced monoliths are displayed in Fig. 7.9.

The most important stage in the preparation of MeAlO/MeAl ceramometals is their mechanochemical alloying (MA). This procedure substantially affects the properties of resulting ceramometals but is virtually unknown for specialists in classical methods of catalyst preparation. As the MA time is extended, it facilitates changes in the phase composition and dispersion of metal powders. Microstructure of the particles also changes from layered to more homogeneous [17–20]. This microstructure of metal particles is retained in the cores of produced ceramometals. Similar to Al_2O_3/Al ceramometals, the obtained composites have the egg-shell microstructure with the oxide shell and metallic cores. Similar also is the bi-disperse pore structure with nanopores concentrated in the oxide shell and ultramacropores formed by cavities between metal particles of the initial precursor powder with the size of some tens of micrometers. The absence of metallic aluminum significantly increases thermal stability and mechanical strength of such ceramometals.

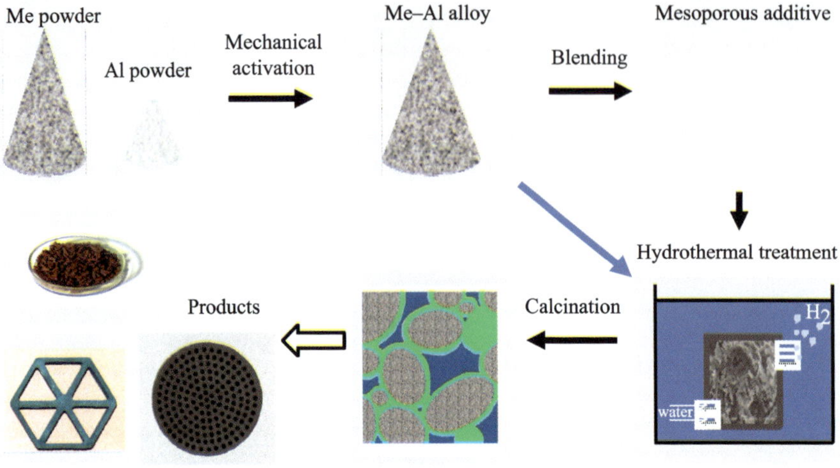

Fig. 7.9: Main stages of the preparation of ceramometal catalysts with Me–Al powder.

7.2.1 Ceramometals CuAlO/CuAl and CuFeAlO/CuFeAl as low-temperature WGS catalysts

For CuAlO/CuAl ceramometals, with extension of the MA time of powdered Cu–Al precursors, the concentration of Al_4Cu_9 intermetallic goes through the maximum. The concentration of the solid solution $Cu_{1-x}Al_x$ is nearly constant, except for the sample activated for 6 min. The concentration of aluminum in the solid solution decreases monotonically with the time of activation (excluding 6 min point), while CuO concentration goes through the minimum (Table 7.3) [21]. Remarkable is the absence of crystallized forms of mixed copper–aluminum oxides according to X-ray diffraction (XRD), although they are observed by energy-dispersive X-ray analysis, nuclear magnetic resonance and X-ray photoelectron spectroscopy [21].

For CuFeAlO/CuFeAl ceramometals, which were synthesized from MA products during oxidative treatments, metallic phases Al_4Cu_9 and Al_4Cu (found only for CuFeAl-6), metallic copper and trace amounts of iron, as well as the oxide phases – tenorite CuO and spinel close to $AlFe_2O_4$ – were found [22]. The variation in the content of all phases in ceramometals is presented in Fig. 7.10. In general, changes in the composition of metallic components (Fig. 7.10) are similar to those typical for the metallic precursor CuFeAl [20]. The increase in spinel content and decrease in CuO amount with the time of MA is related to more pronounced interaction of the powdered metal components during ball milling. This favors mixed oxide formation in the course of oxidative ceramometal synthesis.

Overall, the key feature of this synthesis method is a large amount of X-ray amorphous phases for precursors in the powdered metal alloys obtained by MA [19, 20].

Table 7.3: XRD phase analysis of CuAlO/CuAl ceramometals (wt.%) [21].

Phase	Time of MA (min)			
	3	6	9	12
$Cu_{1-x}Al_x$ solid solution	49% ($x = 0.14$)	62% ($x = 0.10$)	48% ($x = 0.12$)	53.0% ($x = 0.06$)
Al_4Cu_9 intermetallic	12%	20%	27%	0%
CuO	39%	18%	25%	47%

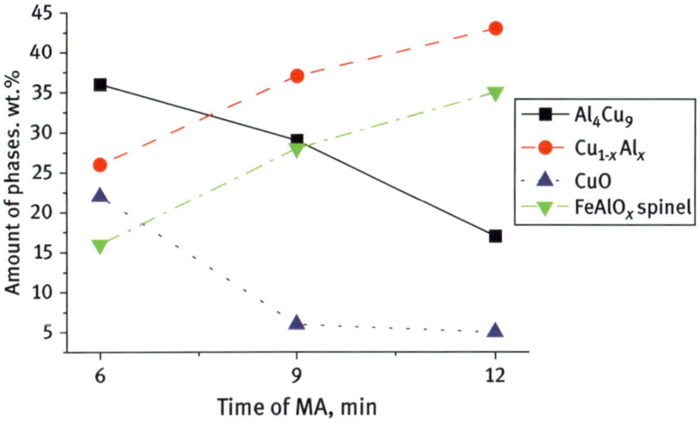

Fig. 7.10: Content of different phases in CuFeAlO/CuFeAl ceramometals estimated by X-ray diffraction analysis (TOPAS software) versus the time of preliminary mechanochemical alloying of the CuFeAl precursor.

A large amount of X-ray amorphous phases in the oxide film is typical of ceramometals [21, 22].

As the MA time of precursor is extended, the textural-mechanical properties of CuAlO/CuAl ceramometals change also nonmonotonically (Fig. 7.11). This is caused by various factors: changes in the phase composition of precursors, dispersion of the particles and their microstructure.

Specific catalytic activity of ceramometals CuAlO/CuAl and CuFeAlO/CuFeAl is higher than that of the conventional CuZnAl (IK-4–25) catalyst (Fig. 7.12) because of a higher surface concentration of copper and stabilization of more active centers (copper clusters) on the surface [21, 22].

The specific catalytic activity changes similarly to the concentration of Al_4Cu_9 intermetallide and contrariwise to the surface concentration of copper, which may be related to the formation of surface clusters that are more active than crystalline copper (Fig. 7.12.) [21].

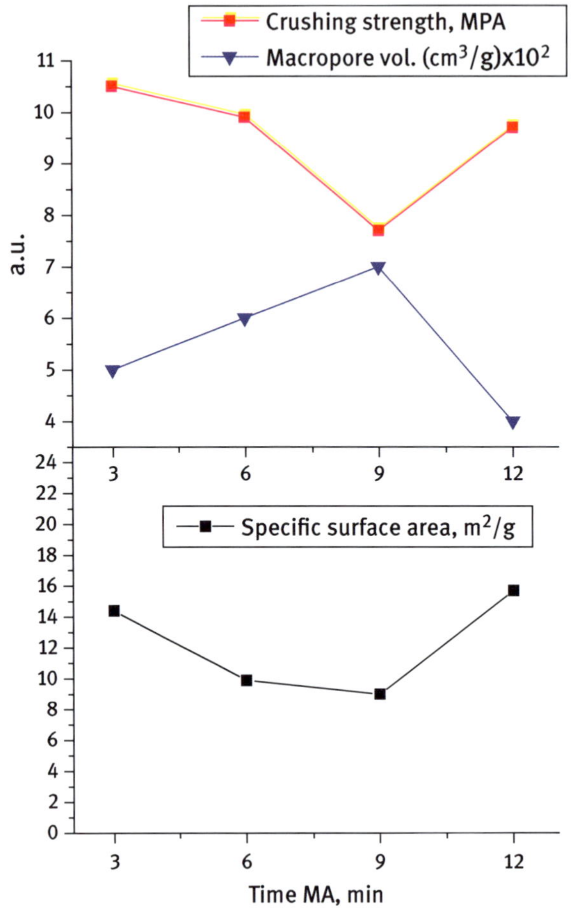

Fig. 7.11: Variation of the crushing strength, macropore volume and specific surface area of CuAlO/CuAl ceramometals with the time of preliminary MA of Cu–Al precursor.

The presence of metal particles with the size up to some tens of microns determines two additional features of ceramometal composites: a developed network of ultramacropores with the size up to 10 μm and a high bulk density as compared to conventional oxide catalysts employed in WGS (Fig. 7.13). The latter factor is of particular importance for estimating the effective activity of unit volume of the catalytic bed.

Indeed, the activity per the volume unit of a fine fraction of cermet was lower as compared to that of the oxide catalyst, whereas an opposite trend was observed for the coarse fraction (Fig. 7.13). Activity of the 3 × 3 × 5 mm cermet granules ($k = 5.1\ s^{-1}$) exceeded by more than 20% the activity of granulated oxide catalyst ($k = 4.4\ s^{-1}$). Thus, when going from a fine to a coarse fraction, activity of the cermet decreases by 35%, and activity of the oxide catalyst by 52%. Note also that the density of ceramometals

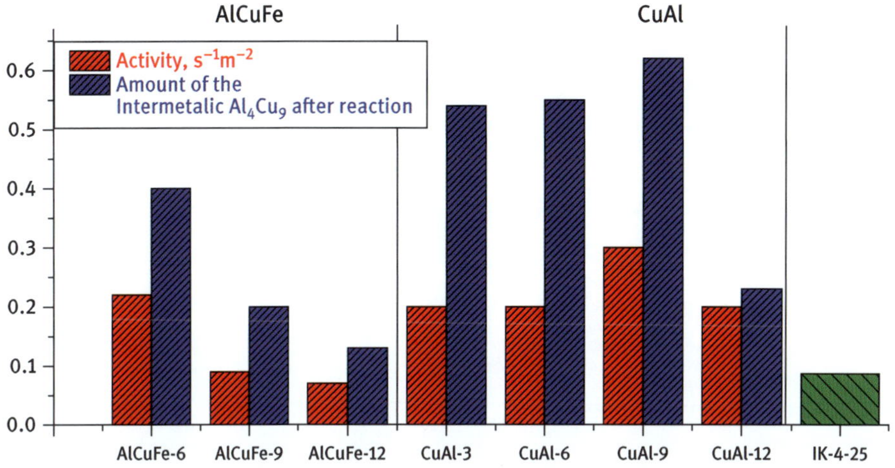

Fig. 7.12: Specific catalytic activity in WGS and amount of intermetallic Al_4Cu_9 in ceramometals. Comparison with CuZnAl oxide catalyst (IK-4–25) [21, 22]. Reaction test conditions: (240 °C. CO : H_2O : H_2 = 8 : 42 : 50. P = 1 bar, GHSV = 3,000–4,000 h^{-1}).

(~4 g/cm^3) is substantially higher than that of CZA (~2 g/cm^3) (Fig. 7.13), which compensates their lower activity per the mass unit. These factors are important arguments, which demonstrate that ceramometal catalysts are promising for WGS. Leaching that promotes formation of additional nanostructures based on copper oxide can be an additional factor increasing the concentration of surface active sites (Fig. 7.13).

7.2.2 Ceramometal Al_2O_3/CoAlO/CoAl as a support for dehydrogenation catalyst under MW irradiation

The CoAlO/CoAl ceramometal synthesized from the mechanochemically treated blend containing 20% of aluminum according to the scheme displayed in Fig. 7.9 had a high strength but a low porosity (Table 7.4). The metallic part consisting of cobalt and aluminum alloys was resistant to oxidation in air even at 900 °C due to a dense alumina interface separating the metallic part from the oxide shell (Fig. 7.14) [23]. The mesopore structure was developed via incorporation of alumina, which served as a mesoporous component [23].

This made it possible to increase the porosity by more than an order of magnitude, and SSA – by two orders of magnitude (Table 7.4). As shown in Fig. 7.14, alumina globules in Al_2O_3/CoAlO/CoAl composite are surrounded by ceramometal CoAlO/CoAl skeleton that maintains a strength of ~5 MPa, which is satisfactory for catalysis. The dehydrogenation catalyst was synthesized by impregnation with an iron-containing solution. For comparison, a catalyst was prepared also by

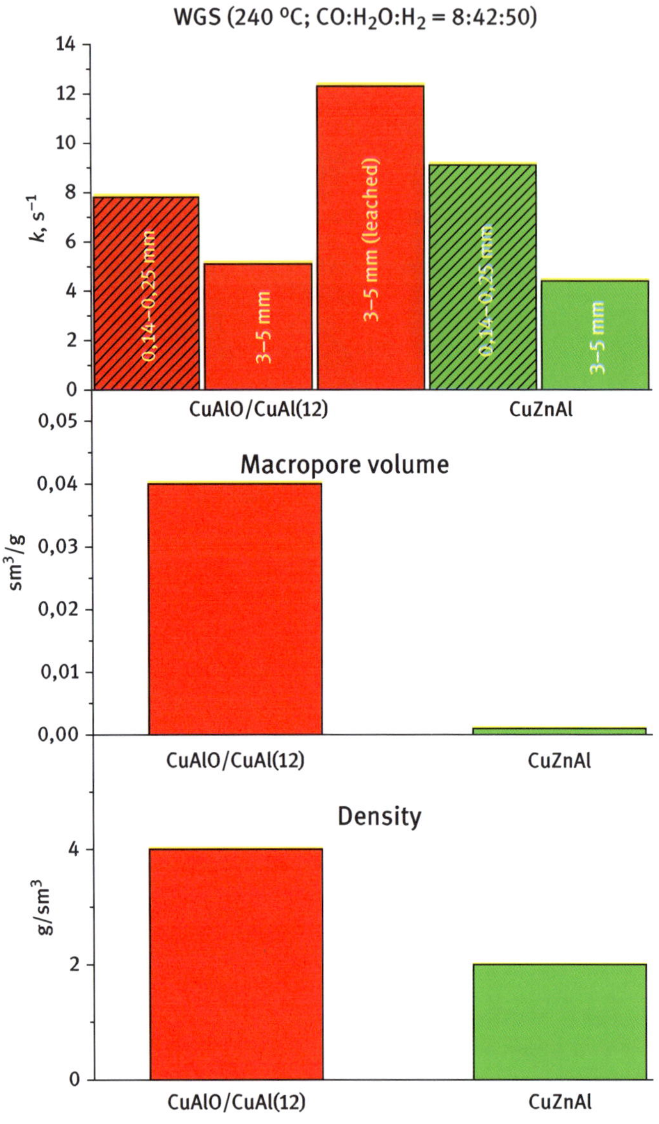

Fig. 7.13: Volume activity of WGS catalysts (the model of a plug-flow reactor), macropore volume and apparent density of granules of ceramometal and oxide catalysts.

Table 7.4: Textural and mechanical properties of CoAlO/CoAl and Al_2O_3/CoAlO/CoAl ceramometals.

Ceramometal	Crushing strength (MPa)	Porosity (%)	Specific surface area (m^2/g)
CoAlO/CoAl	70	5	1
Al_2O_3/CoAlO/CoAl	5	60	122

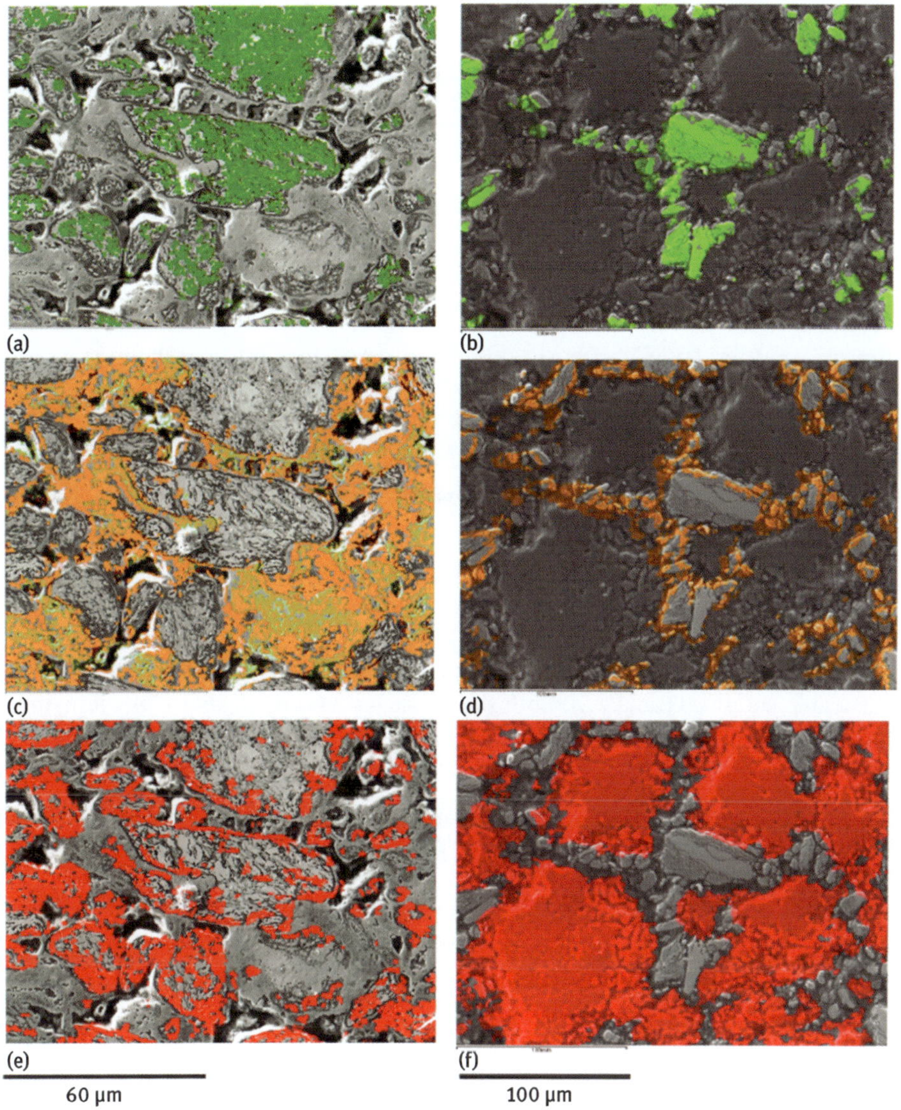

Fig. 7.14: Spatial distribution of phases with different stoichiometry (mapping) according to SEM and INCA data for ceramometals. *a, c, e* – CoAlO/CoAl; *b, d, f* – Al_2O_3/CoAlO/CoAl. *a, b* – metal cores; *c, d* – oxide matrix; *e* – interface; *f* – alumina globules [23].

impregnation of conventional γ-Al_2O_3 with the same solution. Both catalysts were examined with respect to microwave irradiation.

It is found that the main dielectric parameters of the ceramometal catalyst differ from those of the conventional catalyst: $\varepsilon = 2.03$, tg δ – 0.016 for FeO_x/Al_2O_3 catalyst versus $\varepsilon = 3.06$, tg $\delta = 0.065$ for the FeO_x/Al_2O_3/CoAlO/CoAl catalyst (Table 7.5).

Table 7.5: MW absorption and catalytic dehydrogenation data [23].

Sample	MW absorption			Activity (C_6H_6 and C_6H_{12} formation)	
	ε	tg δ	η (%)	K (mg/s)	Selectivity (%)
Fe_2O_3/Al_2O_3	2.03	0.013	~1.5	~0.00007	5–10
Fe_2O_3/ (Al_2O_3) CoAlO/CoAl	3.06	0.065	~26	~0.00018	30–40

It means that the MW absorption by the ceramometal is substantially higher than by pure porous alumina. So, the ceramometal can be used as an efficient absorber of microwave power in MW cavities when they are placed into the maximum of the electric component of the electromagnetic field. The difference in MW absorption between two catalysts is greater when samples are located in the MW cavity in the maximum of the magnetic component. The measured efficiency, η, of MW power-to-heat conversion for γ-Al_2O_3-based catalyst is about 1.5%, whereas for Al_2O_3/CoAlO/CoAl-based catalyst for the same sample weight (500 mg) η = 26%. The ceramometal catalyst is found to be more efficient for endothermal hexane dehydrogenation when heated using the MW irradiation compared to the conventional porous oxide catalyst (Table 7.5). Both the average rate of C_6H_6 and C_6H_{12} formation and selectivity for these products show an increase. The influence of the CoAlO/CoAl skeleton on the activity is negligible due to its low SSA.

7.2.3 Ceramometal CuO/Al_2O_3/FeAlO/FeAl as the combustion catalyst

As in the case of cobalt-containing ceramometals, iron-containing composites also have a poorly developed mesoporous structure [24]. To develop the mesopore structure of ceramometal, the incorporation of their precursor like powdered TSEFLAR™ (CTA) product of centrifugal thermal treatment of gibbsite – commercial grade IC-02–76 [25] was made through blending of CTA with the products of MA. Catalytic activity in CH_4 deep oxidation was studied for CuO/Al_2O_3/FeAlO/FeAl plates in the temperature range of 400–800 °C using feed 3% CH_4 + 6.5% O_2 in N_2 at the feed rate 10 L/h [26].

In Fig. 7.15., XRD diffraction patterns of separate components – FeAlO/FeAl (prepared from the product of Fe + Al mechanical alloying) and CTA product along with that of ceramometals with incorporated alumina are presented. CTA product contains both the δ- and γ-alumina phases due to a rather short (~1 h) time of calcination under air at 900 °C. Four basic ceramometal phases of Fe (Fe–Al alloy) along with corundum phase are present. For alumina-modified ceramometal the phase composition is described by superposition of those for constituent materials.

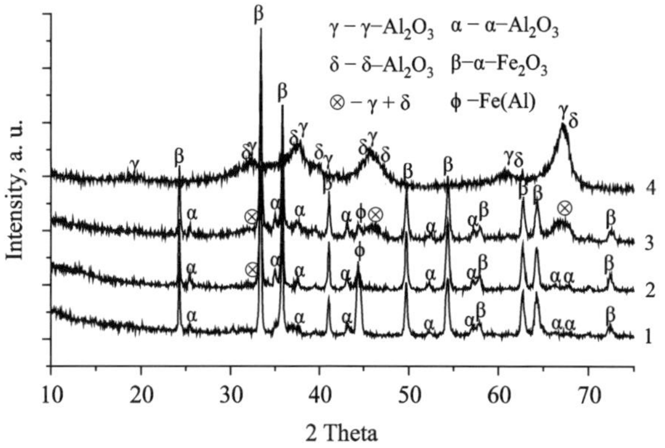

Fig. 7.15: XRD diffraction pattern of Al_2O_3/FeAlO/FeAl cermet supports prepared from powdered blends containing different amounts of CTA: 1 – 0 wt.%, 2 – 20 wt.%, 3 – 50 wt.%, 4 – pure CTA, TCA calcined at 900 °C [15].

The fraction of mesopores in cermets apparently increases with the amount of added CTA product (Fig. 7.16). Macroporous structure changes as well due to increasing the volume of ultramacropores (Fig. 7.16). SSA increases with alumina content while the crushing strength decreases (Fig. 7.17), though being nearly constant (~10 MPa) in the range of alumina content 20–50 wt.%.

The catalytic oxidation of methane over ceramometal monoliths CuO/Al_2O_3/FeAlO/FeAl with different alumina (and, hence, CuO) content revealed that activity increases with the increase of alumina content from 10% to 30% (Fig. 7.17). Apparently, the main

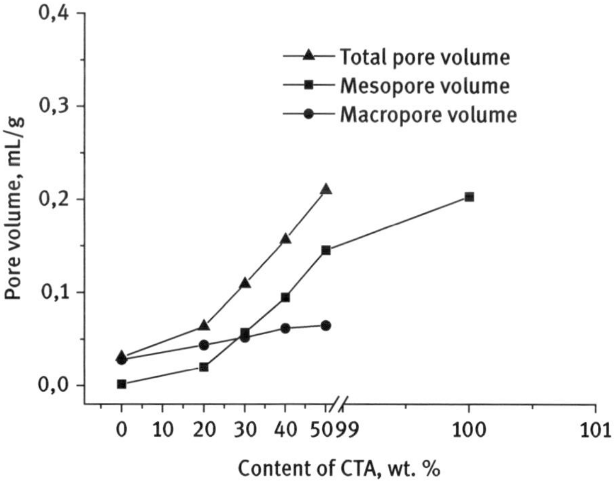

Fig. 7.16: Total, nano and macropore volume of CuO/Al_2O_3/FeAlO/FeAl cermet support prepared from powdered blends with different CTA content [15].

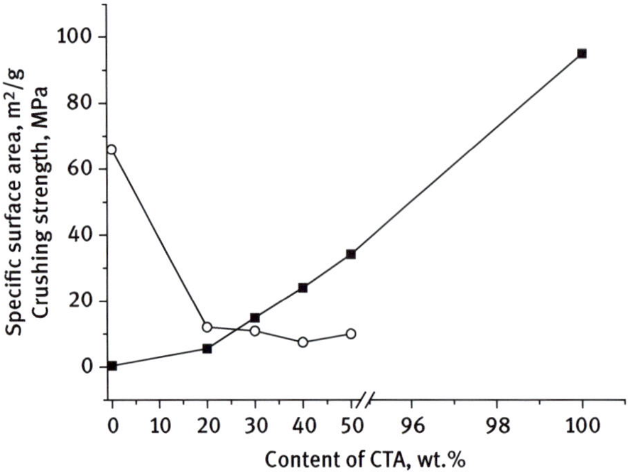

Fig. 7.17: Specific surface area and crushing strength of cermet supports prepared from powdered blends with different CTA content [15].

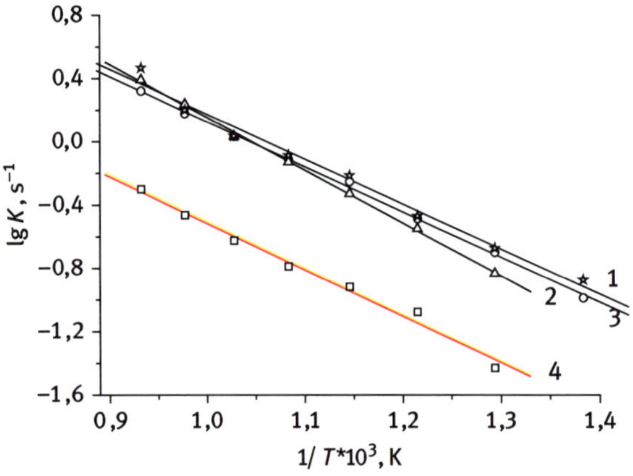

Fig. 7.18: Temperature dependence of the efficient first-order rate constant of CH_4 total oxidation for catalysts $CuO/Al_2O_3/FeAlO/FeAl$ with different CTA content (wt.%): 1–50%, 2–40%, 3–30%, 4–10%. Feed composition: 3 vol.% CH_4 + 6.5 vol.% O_2 in Ar, contact time 0.26 s.

factor here is the increase of CuO content, which is the active phase here. The further increase of alumina (and, as the result, copper oxide content) has not improved activity. Hence, in terms of the crushing strength (Fig. 7.18), activity and price, the optimum could correspond to the catalyst with the CTA content of ca. 30 wt.%.

It should be emphasized that the catalytic experiments were carried out on a plate. In spite of a 1–2 mm gap between the wall and the catalyst, the macropore structure of the plate ensured quite a good activity toward the oxidation of methane.

Another example of the application of ceramometal catalysts is fuel combustion in a fluidized bed. Attrition resistance is of special importance for the catalytic oxidation of carbon in a fluidized bed. Ceramometal monoliths $CuO/Al_2O_3/FeAlO/FeAl$ shaped as honeycombs are promising for this purpose. Details of the tests are presented in Chapter 8.

7.3 Conclusions

Synthesis of porous ceramometal composites from metallic aluminum-containing powders allows obtaining various catalytically active materials with essentially different composition and shape for a wide range of catalytic processes. At present, studies are performed mostly on laboratory scale; a sole exception is the combustion of coal, for which pilot tests were carried out.

The synthesized ceramometals have a developed macropore structure with the pore size reaching some tens of micrometers, which provides quite a high permeability of the granulated composites. The nanoporous component is concentrated in the oxide shell. It can be substantially increased via incorporation of mesoporous powders. The presence of metallic particles enhances the absorption of microwave power, increases the loading density per unit volume and provides a higher thermal conductivity in comparison with porous oxide materials. Overall, such ceramometals may be quite efficient for the processes requiring a high heat and mass exchange in compact catalytic devices.

References

[1] Kuznetzova, LL., Ananin, VN., Pashis, AV., & Belyaev, VV. Studies of composite catalysts of nickel on metal-ceramic substrates. React. Kinet. Catal. Lett. 1991, 43, 545–552.

[2] Kuznetsova, LI., Zaikovskii, VI., Ziborov, AV., & Plyasova, LM. Morphology and structure of composite catalysts of nickel on metal-ceramic substrate. React. Kinet. Catal. Lett. 1991, 43, 553–558.

[3] Tikhov, SF., Sadykov, VA., Potapova, YuV., Salanov, AN., Kustova, GN., Litvak, GS., Zaikovskii, VI., Tsybulya, SV., Pavlova, SN., Ivanova, AS., Rozovskii, AY., Lin, GI., Lunin, VV., Ananyin, VN., & Belyaev, VV. The study of the formation of supports and catalysts based upon Al_2O_3/Al cermets. Stud. Surf. Sci. Catal. 1998, 118, 797–806.

[4] Tikhov, SF., Sadykov, VA., Potapova, YuV., Yudaev, IV., Lapina, OB., Tsybulya, SV., Salanov, AN., & Titkov, AI. Hydrothermal synthesis of mechanically strong porous monoliths $MeO_x/Al_2O_3/Al$. Mater. Res. Soc. Symp. Proc. 2005, 878E, Y4.10.1-Y.4.10.6.

[5] Tikhov, SF., Romanenkov, VE., Sadykov, VA., Parmon, VN., & Ratko, AI. Physico-chemical principles of the synthesis of porous composite materials through the hydrothermal oxidation of aluminum powder. Kinet. Catal. 2005, 46, 641–685.

[6] Tikhov, SF., Romanenkov, VE., Sadykov, VA., Parmon, VN., & Ratko, AI. Porous composites on the base of oxide-aluminum cermets (synthesis and properties). Publ. House of SB RAS, "Geo" Branch, Novosibirsk, Russia, 2004. [In Russian]

[7] Tikhov, SF., Potapova, YuV., Fenelonov, VB., Sadykov, VA., Salanov, AN., Tsybulya, SV., & Melgunova, LF. Porous metalloceramics Al_2O_3/Al prepared by oxidation of powdered aluminum in hydrothermal conditions. III Reactivity of aluminum, specificity of the reaction mechanism by water vapor, and microtexture of cermets. Kinet. Catal. 2003, 43, 322–334.

[8] Tikhov, SF., Potapova, YuV., Sadykov, VA., Fenelonov, VB., Yudaev, IV., Lapina, OB., Salanov, AN., Zaikovskii, VI., & Litvak, GS. Synthesis of alumina through hydrothermal oxidation of aluminum powder conjugated with surfactant-directed oriented growth. Mater. Res. Innovations 2005, 9, 431–446.

[9] Tikhov, S., Sadykov, V., Ratko, A., Kuznetsova, T., Romanenkov, V., & Eremenko, S. Kinetics of aluminum powder oxidation by water at 100 °C. React. Kinet. Catal. Lett. 2007, 92, 83–88.

[10] Kim, J., & Lee, D. Synthesis and properties of core-shell metal-ceramic microstructures and their application as heterogeneous catalysts. Chem. Cat. Chem 2014, 6, 2642–2647.

[11] Tikhov, SF., Pakhomov, NA., Nemykina, EI., Salanov, AN., Sadykov, VA., Romanenkov, VE., & Pietiushyk, TYe. Porous ceramic matrix Al_2O_3/Al composites as supports and precursors for catalysts and permeable materials. In: Cuppoletti J., ed. Metal, ceramic and polymeric composites for various uses. INTECH, Croatia, 2011, 195–210.

[12] Rozovskiiv, AYa. Main paths of the methane and synthesis gas processing. State of art and outlook. Kinet. Catal. 1999, 41, 358–371.

[13] Vitnova, LA., Mordovin, VP., Kliger, GA., Bogolepova, EI., Kurkin, VI., Shyukin, AN., Marchevskaya, EV., & Slivinskii, EV. Tendencies in promotion of the iron containing intermetallide Fischer-Tropsch catalysts. Russ. J. Petrochem. 2002, 42, 111–118.

[14] Tikhov, SF., Kuz'min, AE., Bespalko, YuN., Kurkin, VI., Sadykov, VA., Bogolepova, EI., Tsybulya, SV., Kalinkin, AV., Mordovin, VP., Salanov, AN., Zaikovskii, VI., & Shavorsky, AA. ZrFe intermetallides for Fischer-Tropsch synthesis: pure and encapsulated into alumina-containing matrices. Stud. Surf. Sci. Catal. 2007, 163, 153–175.

[15] Tikhov, SF., Sadykov, VA., Valeev, KR,., Salanov, AN., Cherepanova, SV., Bespalko, YuN., Ramanenkau, VE., Piatsiushyk, YaYa., & Dimov, SV. Preparation of porous ceramometal composites through the stages of mechanical activation and hydrothermal partial oxidation of Me-Al powders. Catal. Today 2015, 246, 232–238.

[16] Iglesia, EI., Reyes, SC., Madon, RJ., & Soled, SL. Selectivity control and catalyst design in the Fischer-Tropsch synthesis: sites, pellets and reactors. Adv. Catal. 1993, 39, 221–302.

[17] Tikhov, SF., Usoltsev, VV., Sadykov, VA., Pavlova, SN., Snegurenko, OI., Gogin, LL., Vostrikov, ZY., Salanov, AN., Tsybulya, SV., Litvak, GS., Golubkova, GV., & Lomovskii, OI. CrAl alloy-based cermet monolith with polymodal pore structure for partial oxidation of methane to synthesis gas. Stud. Surf. Sci. Catal. 2006, 162, 641–648.

[18] Kim, JS., Kwon, YS., Golubkova, GV., Lomovskii, OI., Dudina, DV., Dovlitova, LS., Malakhov, VV., Tikhov, SF., Usoltsev, VV., & Sadykov, VA. Formation of intermetallic phases during mechanical alloying and annealing of Cr + 20 wt % Al mixtures. Inorg. Mater. 2008, 44, 677–681.

[19] Dudina, DV., Lomovsky, OI., Valeev, KR., Tikhov, S.F., Boldyreva, NN., Salanov, AN., Cherepanova, SV., Zaikovskii, VI., Andreev, AS., Lapina, OB., & Sadykov, VA. Phase evolution during early stages of mechanical alloying of Cu-13wt.% Al powder mixtures in a high-energy ball mill. J. Alloys Compd. 2015, 629, 343–350.

[20] Tikhov, SF., Valeev, KR., Salanov, AN., Cherepanova, SV., Boldyreva, NN., Zaikovskii VI, V.A., Sadykov, VI., & Dudina, DV. Lomovsky OI, Romanenkov ., Pyatsyushik EE, Phase formation

during high-energy ball milling of the33Al-45Cu-22Fe (at.%) powder mixture. J. Alloys Compd. 2018, 736, 289–296.

[21] Tikhov, S., Minyukova, T., Valeev, K., Cherepanova, S., Salanov, A., Kaichev, V., Saraev, A., Andreev, A., Lapina, O., & Sadykov, V. Design of micro-shell Cu–Al porous ceramometals as catalysts for the water–gas shift reaction. RSC Adv. 2017, 7, 42443–42454.

[22] Tikhov, SF., Minyukova, TP., Valeev, KR., Cherepanova, SV., Salanov, AN., Shtertser, NV., & Sadykov, VA. Design of ceramometal CuFeAlOx/CuFeAl composites and their catalytic potential for water gas shift reaction. Mater. Chem. Phys. 2019, 221, 349–355.

[23] Tikhov, SF., Andreev, AS., Salanov, AN., Cherepanova, SV., Lapina, OB., Sadykov, VA., Tanashev, YuYu., & Bolotov, V.A. Ceramic matrix composites prepared from CoAl powders. J. Mater. Sci. 2016, 51, 10487–10498.

[24] Usoltsev, V., Tikhov, S., Salanov, A., Sadykov, V., Golubkova, G., & Lomovskii, O. Properties of porous $FeAlO_y/FeAl_x$ ceramic matrix composite influenced by mechanical activation of FeAl powder. Bull. Mater. Sci. 2013, 36, 1195–1200.

[25] Pinakov, VI., Stoyanovskiy, OI., Tanashev, YuYu., Pikarevsky, AA., Grinberg, BE., Dryab, VN., Kulik, KV., Danilevich, VV., Kuznetsov, DV., & Parmon, VN. TSEFLAR™ – the centrifugal flash reactor for rapid thermal treatment of powdered materials. Chem. Eng. J. 2005, 107, 157–61.

[26] Tikhov, SF., Simonov, AD., Yazykov, NA., Dubinin, YV., Usoltsev, VV., Yakovlev, VA., Sadykov, VA., Salanov, AN., Suprun, EA., & Parmon, VN. Catalytic combustion of brown coal particulates over ceramometal honeycomb catalyst. Catal. Sustain. Energy 2012, 1, 82–89.

8 Catalytic combustion of fuels on oxide catalysts in the fluidized state

Combustion of fuels is the most important process for generation of energy (heat and electricity). Conventional combustion, which is based on high-temperature flame combustion, is carried out at high temperatures, 1,200–1,600 °C. This leads to considerable contamination of the atmosphere with harmful wastes (nitrogen and sulfur oxides, CO, soot, aromatics, etc.). Very stringent requirements are imposed also on heat resistance of the equipment. When the process is performed in a fluidized bed with an inert material, the process temperature can be lowered to 800–1,000 °C [1, 2]. This allows decreasing the concentration of harmful wastes released into the atmosphere, particularly the "atmospheric" nitrogen oxides. In addition, the indicated temperatures prevent ash melting and reduce slagging and corrosion of the equipment. The fluidized mode increases the heat transfer coefficient, which makes it possible to use low-grade fuels with a high content of ash and moisture [1, 2]. However, combustion of such fuels is accompanied by the release of quite a large amount of harmful substances, and stringent requirements to heat resistance of materials are retained. Fuel burns not only in the fluidized bed but also over the bed, which makes it necessary to increase dimensions of the reactors [3, 4].

Unconventional method for catalytic combustion of various fuels and wastes in a fluidized catalyst bed has been developed at the Boreskov Institute of Catalysis SB RAS. The catalytic combustion of fuels mainly eliminates drawbacks associated with combustion of fuels in the fluidized bed of an inert material. This technology is based on four principles:

- the use of a deep oxidation catalyst;
- combustion in the fluidized catalyst bed;
- combination of heat release and heat removal in a joint fluidized bed;
- combustion without a significant air excess [5–10].

The distinctive feature of catalytic combustion in comparison with other types of burning is that fuel is oxidized predominantly on the surface of solid catalysts without formation of naked flame or with its minimum contribution [4]. Fuel components are oxidized by the catalyst surface oxygen, and reduced surface is regenerated by the gas-phase oxygen. As compared to conventional combustion methods, the presence of a catalyst mitigates requirements to thermochemical properties of structural materials of the apparatus, reduces heat losses through the walls, simplifies control of the process, and prevents secondary endothermic processes leading to the formation of toxic substances. In addition, the use of a catalyst reduces the explosion hazard and increases the energy intensity of the processes. As a result, the dimensions and metal consumption of structures can be considerably reduced. The use of a catalyst allows also decreasing the concentration of nitrogen oxides, both the

https://doi.org/10.1515/9783110587777-008

thermal (from air nitrogen) and fuel (from nitrogen-containing components of fuel) ones [5–10].

This section presents published results of laboratory, pilot and industrial testing of the catalytic combustion processes in a fluidized catalyst bed. Such processes are especially promising for local power generation in the northern territories with a weakly developed infrastructure and thinly scattered population, where provision of gas supply is economically inefficient, and transportation and storage of liquid fuel are expensive.

8.1 Main types of the fuels being used, their characteristics and products of their incomplete combustion (oxidation)

Residual oil and diesel fuel are widely used as fuels for small heat supply units. However, for distant communities, owing to transportation problems, a promising source of fuel may be crude oil, which has quite a high calorific value, up to 42,000 kJ/kg [11]. In the Russian Federation, oils with a high sulfur content are in wide use. Examples of two types of such sulfur-containing oils, sulfur light oil (SO) and heavy sulfur oil (HSO), are listed in Table 8.1. High sulfur content substantially hinders their combustion in flame burners. Starting of oil burners is complicated by explosions and fires. Large amounts of soot, CO, nitrogen oxides and certainly sulfur oxides are emitted with flue gases [12, 13]. Two groups of nitrogen-containing compounds in sulfur-containing oils are distinguished: nitrogenous bases and weakly basic nitrogenous compounds [11]. Nitrogenous bases are present mostly in the low and medium fractions and are represented by alkyl and cycloalkyl derivatives of pyridine and quinoline. Heavy fractions contain benzoquinoline and benzoacridine. Neutral nitrogen compounds include indole and carbazole derivatives, cyclic amides of aromatic amino acids and porphyrins. Noteworthy is the high content of vanadium in heavy oils, up to 6 kg/t, as well as nickel, cobalt, chromium and rubidium. In light oils, vanadium and nickel content is low, so nitrogen-containing compounds in such oils are not bound to metals, in distinction to heavy oils [11, 13].

Table 8.1: Main characteristics of Russian petroleum [12, 13].

Petroleum type	Density at 20 °C (g/cm^3)	Weight content (%)				
		C	H	O	N	S
Sulfur light	0.840	84.60	12.58	0.65	1.5	0.66
Heavy high sulfur	0.872	84.72	12.53	0.16	0.29	2.30

The combustion of sulfurous oil in the fluidized bed of an inert material (quartz sand) with a small excess of oxygen does not lead to a complete burnout of oil; its conversion does not exceed 80%. As the bed height is increased, CO concentration reaches 8,800 ppm (Fig. 8.1). Methane is detected in flue gases; its concentration at the reactor outlet is up to 0.15%. Therefore, SO_2 is absent both at the reactor outlet and over the entire height of the fluidized bed. This is caused by the presence of alkali and alkali-earth metals in quartz sand, which are capable of interacting intensively with sulfur dioxide to form sulfates and sulfites of the corresponding metals. The absence of nitrogen oxides is associated with their reduction to molecular nitrogen, which occurs because the reaction mixture contains high concentrations of CO and organic products of the incomplete burning of fuel along the full height of the fluidized bed of an inert material (see Fig. 8.2). The complete combustion of oil in the fluidized bed of an inert material can be achieved at temperatures above 850 °C. In this case, there is a sharp growth in the concentrations of nitrogen oxides in flue gases, which are formed due to the oxidation of nitrogen-containing compounds in oil and the oxidation of nitrogen by air oxygen. The concentration of nitrogen oxides in the flue gases can reach 1,000 ppm or more [12].

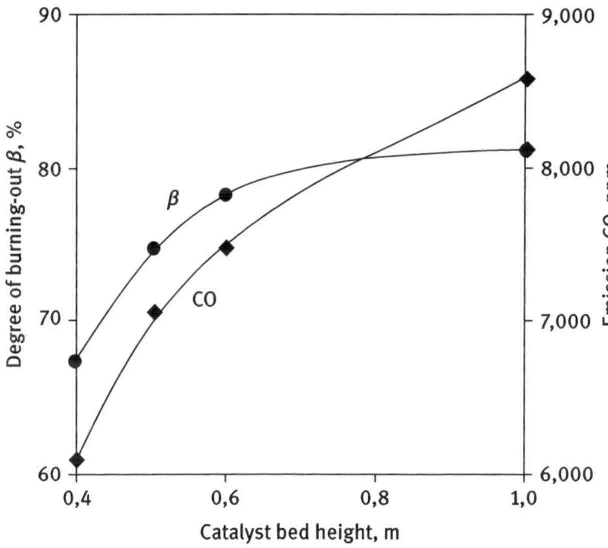

Fig. 8.1: Changes in burnout degree of sulfurous oil over the height of an inert material layer in the reactor at 700 °C [12].

As the price of oil and its products is growing, coal becomes a promising fuel for heat supply units. Brown coal is the most used type of combustible coals. A typical Kansk–Achinsk coal organic matter composition is (weight %): C – 71.5; H – 5.0; N – 1.0; O – 22.1; S – 0.4; ash content in the dried brown coal is 10.1%. At the same

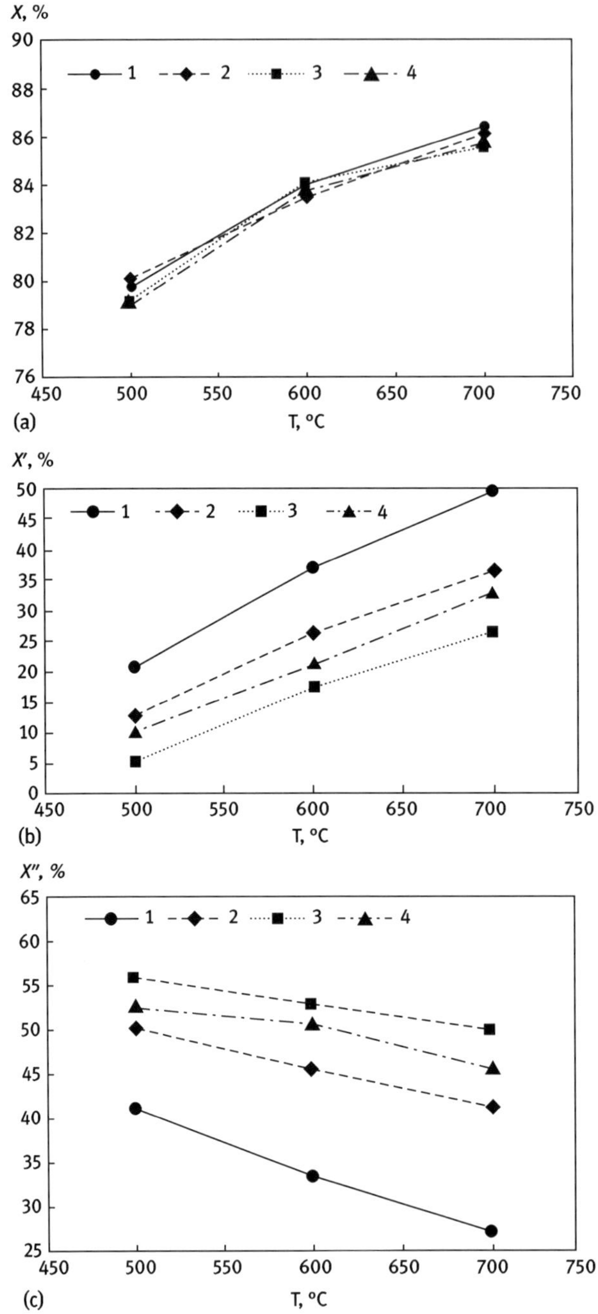

Fig. 8.2: Degree of conversion of wood to volatiles versus the pyrolysis temperature (a), degree of conversion of wood to noncondensable gases (b), degree of conversion of wood to condensable gases (tar and water) (c): (*1*) IK-12-73 catalyst; (*2*) IK-12-72 catalyst; (*3*) Al_2O_3; (*4*) sand [20].

time, the catalytic combustion in a fluidized bed can be performed using other types of solid fuel as well as wastes from pulp and paper industry, which are difficult to use in conventional boiler plants (Table 8.2.) The listed data indicate that these types of fuels contain large amounts of volatile components and incombustible substances that produce ash.

Table 8.2: Ash and volatiles content and degree of burnout of some carbon-containing materials in the demo catalytic heat generating unit [14].

Material	Volatiles content (wt%) *	Ash content (wt%)	Degree of burnout (wt%)
Anthracite	10.1	17.5	80.4
Coal G	41.7	25.0	94.7
Coal–water slurry	41.7	25	97.4
Coal B	47.0	28.3	98.6
Shale**	65.7	49.0	97.2
Peat	70.4	24.3	99.2
Lignin sludge	78.7	19.0	99.4
Saw dust	85.2	0.95	99.9

*The volatiles content data refer to the combustible mass of the carbon-containing materials.
**Shale was burned in a benchmark unit.

A detailed analysis of the products of oil shale processing (heavy coal-tar products, HCTP), which can also be used as a fuel [15] is made in Tables 8.3 and 8.4.

Table 8.3: Working mass composition of heavy coal-tar products (HCTP) F1 and F2 [15].

HCTP samples	Working mass composition (%)					Ash (wt%)	Wt (%)***	Density (kg/m^3)
	C	H	O	S	N			
F1	40.0	4.1	7.0	1.3	0.1	39.8	7.7	1,050
F2	40.5	3.9	8.3	1.2	0.1	34.8	11.2	1,250

Wt is the total moisture in heavy HCTP, %;
**The oxygen content was determined from the balance.
***The amount of total moisture is taken by the mass losses during drying in a desiccator over KOH.

As mentioned in Tables 8.3 and 8.4, these products have high contents of sulfur and ash. The nitrogen content is also quite high The lowest heat of combustion of

Table 8.4: Composition of dry ash free (daf) mass of heavy coal-tar products (HCTP) F1 and F2 [15].

HCTP samples	Total output of volatiles (daf) (%)	Composition of dry ash mass (%)				
		C	H	O	S	N
F1	84.2	76.2	7.8	13.3	2.5	0.2
F2	86.5	75.0	7.2	5.4	2.2	0.2

such fuels varies within 17,000–16,974 kJ/mol [15]. More detailed estimates of combustion heat of various fuels can be found in [16].

The composition of ash residue of the oil shale processing products is shown in Table 8.5. The content of such detrimental components as phosphorus, chlorine and sulfur in ash is not high. However, the presence of chlorine-containing compounds may lead to the formation of chlorinated benzodioxins and benzofuranes, which are especially dangerous [17, 18].

Table 8.5: Ash contents of heavy coal-tar products (HCTP) (%) [15].

Components	F1	F2
Na_2O	0.91	0.53
MgO	5.31	5.48
Al_2O_3	17.44	16.89
SiO_2	45.72	45.86
P_2O_5	0.37	0.31
SO_3	4.17	3.41
Cl	0.07	0.18
K_2O	4.71	3.41
CaO	14.78	17.51
TiO_2	0.56	0.54
Fe_2O_3	5.49	4.27
ZnO	0.16	0.05
SrO	0.07 .	–
BaO	0.06.	–
Other compounds	0.18	1.56

It should be noted that the oxidation of solid fuels commonly proceeds in two steps, owing to the presence of volatile substances emitted from a solid at high temperatures. The first step includes release of volatiles and their combustion in the gas phase. In the second step, the carbon residue interacts with the gas-phase components.

This may be a slow oxidation of coke by gas-phase oxygen. Endothermic processes can also proceed, for example, the interaction of carbon dioxide with coke and the formation of carbon monoxide [4, 19].

A detailed study of pyrolysis in the presence of various catalytic and inert additives is reported in [20].

It was shown that the total degree of wood pyrolysis to char is mainly determined by the bed temperature. With increasing temperature, the mass content of condensable substances (water, tar, acids) increases (Fig. 8.2a), and the amount of noncondensable gases (CO_2, CO, H_2 and CH_4) (Fig. 8.2b) increases independently from additives. The amount of ash and water, on the contrary, decreases as the temperature is raised (Fig. 8.2c). For the wood used in the work [20], the total yield of volatile substances determined by a standard procedure was 84.7% of the combustible mass. Raising the temperature and using the complete oxidation catalysts that contain copper and chromium oxides (IK-12-72 and IK-12-73) facilitate development of the indicated trend in comparison with inert materials (sand and alumina) (Fig. 8.2). These data validate possibility of endothermic processes of steam and carbon dioxide reforming of hydrocarbons. The complete oxidation catalysts promote a rapid oxidation of hydrocarbons and their "switching out" from the endothermic process. The complete oxidation proceeds even faster in the presence of air oxygen.

The catalytic combustion in a fluidized bed can be applied also for remediation of the sludge from municipal sewage. Thus, the annual amount of sewage sludge formed in the Russian Federation is ca 2.5 million tons calculated on a dry basis. Many special facilities are needed to bury it for 100–150 years. However, the sludge contains a large amount of combustible organic components (Table 8.6.)

Table 8.6: Elemental composition of the organic part of a dried sewage sludge from different sources [21].

Town	Content (wt%)						
	C	H	N	S	Cl	P	O
Novosibirsk	36.0	4.8	2.3	0.6	0.1	0.2	16.1
Omsk	36.4	4.7	2.4	0.5	0.2	0.2	16.1
Moscow	35.8	4.7	2.2	0.6	0.2	0.3	16.0

The mineral part of sewage sludge includes aluminum and silicon oxides (60–70%), calcium and magnesium oxides (10–15%), iron oxides (6–7%) and much smaller amounts of other oxides [21]. Noncatalytic combustion of such wastes results in the formation of not only carbon, nitrogen and sulfur oxides, but also benzopyrenes and dioxins [22].

Thus, a wide scope of liquid and solid fuels can be used for combustion in local heat generating units. However, their application in conventional units is hindered by a large amount of impurities. Some impurities can significantly deteriorate the environment. Other impurities have nonoptimal calorific value and decrease the efficiency of the main fuel oxidation reaction. Catalytic combustion makes it possible to significantly neutralize the detrimental effects of this process.

8.2 Examples and characteristics of catalytic combustion processes in a fluidized bed

A typical design of the industrial fluidized bed reactor is shown in Fig. 8.3 [10]. The reactor (1) has a detachable air distributor (2). The air is introduced sideways (3) and distributed through a circular gap (4) into horizontal perforated tubes (5),

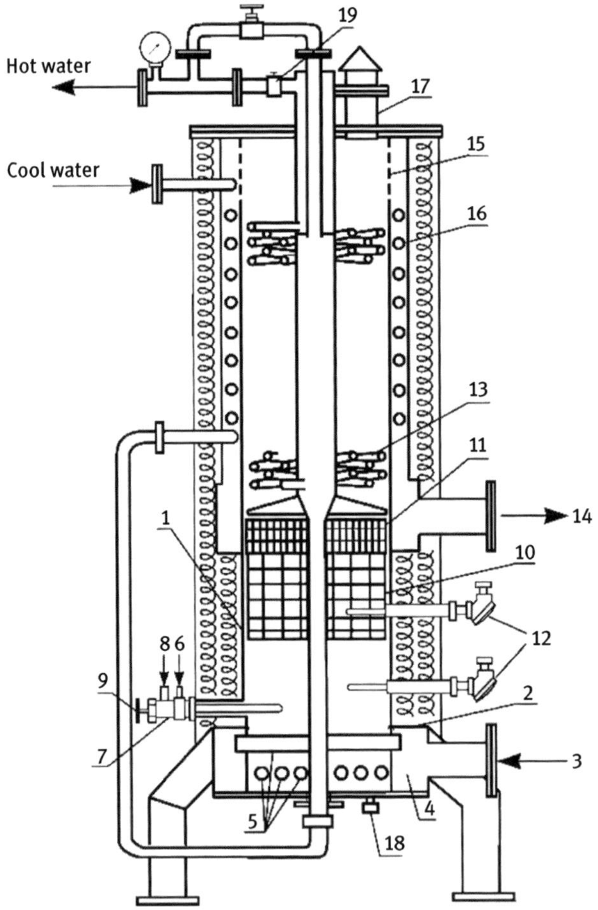

Fig. 8.3: Layout of CHSU reactor [10].

which work as a gas-distributing grid. Fuel (6) is sprayed into catalyst bed through injectors (7) by air (8). The injectors are fitted with a needle (9) for cleaning them during the operation. Starting from the height of 250 mm from the gas-distributing grid, the fluidized bed is staged by wire grids (10) with a mesh size of 30–30 mm^2 and a block (11) of six grids of mesh size 10–10 mm^2. Free fluidized bed in the injectors zone serves for fuel distribution and mixing with air. The fuel burns out in this zone approximately by 80%. Complete oxidation of the fuel occurs in the staged bed (10). Thermocouples (12) are introduced both in the staged and free fluidized bed sections.

An important feature of this reactor is the non-isothermal fluidized bed [5]. That is, the bottom part of the bed is kept at the optimum temperature for the fuel oxidation, while the top part of the bed is kept at any preset temperature (Fig. 8.4). A heat exchanger (13) with staggered-corridor tubing [6] provides the maximum coefficients of bed-to-tubes heat transfer. Such configuration makes more random a possible heterogeneity (large bubbles) and prevents the formation of catalyst "caps" on the upper tube surface, in contrast to staggered tubing. Nearly 10% of the heat exchanger submerges in the bed, and the remaining overbed part acts as economizer. Flue gases (14) pass through a baffle gauze (15) to prevent particles carry over from the reactor, and then through a heat exchanger (16) to cool the body and decrease heat losses. The reactor is fitted with a relief valve (17), a socket (18) to discharge a catalyst, and a safety valve (19) at the hot water pipe [10]. It should be noted that smaller reactors, which were used in laboratory and pilot testing, commonly had a less complicated design [12, 13, 20].

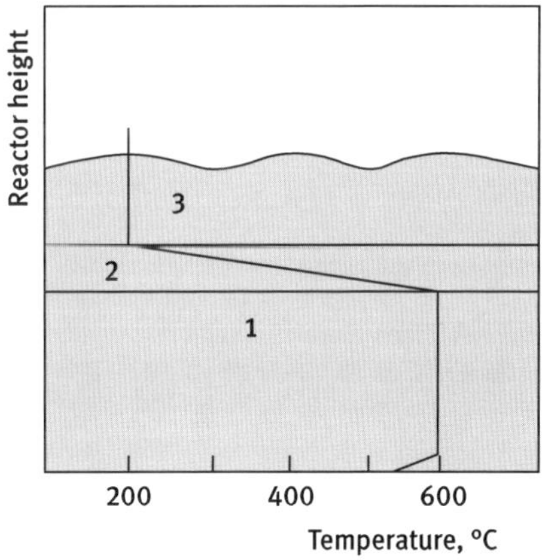

Fig. 8.4: Temperature profile along the height of the fluidized catalyst bed: (1) zone of fuel combustion in the zone of free fluidized bed (isothermal zone); (2) zone of fluidized bed with packing (nonisothermal zone); (3) zone of free fluidized bed (isothermal) [10].

The combustion of sulfurous oils was investigated in a laboratory setup with a vibratory fluidized bed in a reactor with the diameter 40 mm and SWCHKZ-1 catalyst (550–700 cm^3); the height of a vibratory fluidized bed was 950–1,000 mm; air flow rate, 2.6–2.9 m^3/h; and oil flow rate, 150–350 g/h (air excess 3.1–1.4) [12, 13]. As shown in Fig. 8.5, when the temperature over the height of a fluidized bed (Fig. 8.3) is raised from 500 to 700 °C, the degree of fuel burnout (β) increases from ~84% to ~99%.

As the temperature is raised, the concentration of CO decreases and approaches zero at the reactor outlet. The concentration of nitrogen and sulfur oxides, on the contrary, increases to 45 and 25 ppm, respectively (Fig. 8.5). The emission of nitrogen oxides is much lower than the theoretical value (from the nitrogen content in oil). It means that nitrogen oxides are reduced by hydrocarbons and CO over the catalyst. In addition, the presence of the catalyst makes it possible to lower the maximum temperature to 750 °C. This does not lead to the formation of thermal nitrogen oxides from air even at an increased air excess coefficient (Fig. 8.6). The content of sulfur oxides below the theoretical value (150 ppm) is caused by their adsorption on the catalyst.

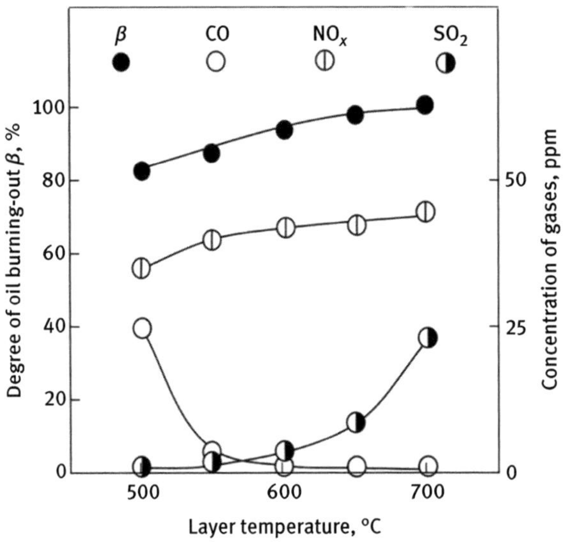

Fig. 8.5: Temperature dependences of sulfur crude oil burnout (β) in a fluidized bed of aluminum–copper–chromium catalyst and the concentrations of CO, SO_2 and NO_x [12].

An increase in the oxygen concentration due to the growing air excess coefficient also facilitates a decrease in the outlet concentration of SO_2 (Fig. 8.7). The produced SO_3 has a higher adsorption ability on the catalyst, thus decreasing the concentration of sulfur oxides in the laboratory experiments [13]. However, as the time of laboratory experiment is extended, the outlet concentration of SO_2 increases to theoretical values for the indicated fuel. This is caused by saturation of the catalyst with sulfur oxides [11, 13].

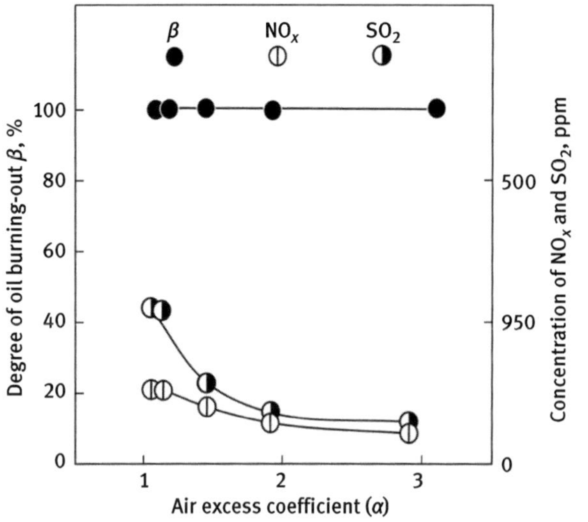

Fig. 8.6: The degree of sulfur crude oil burnout β and the concentrations of SO_2 and NO_x at the reactor outlet in dependence on the amount of excess air α at 700 °C [12].

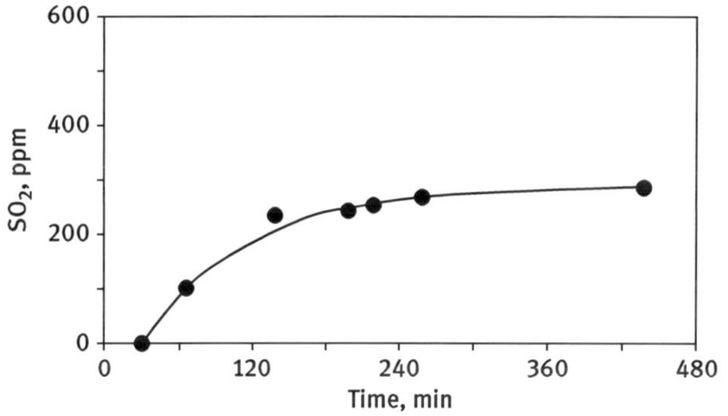

Fig. 8.7: The SO_2 concentration at the reactor outlet during the SCHKZ-1 catalyst sulfation during HSO combustion without absorbent at a temperature of 700 °C, oil consumption of 260 g/h, and air excess of 1.9× [13].

The concentration of sulfur oxides can be decreased by additional loading of calcium compounds (oxide or carbonate) into the fluidized bed (Fig. 8.8) [12, 13]. This is accompanied by a substantial decrease in the outlet concentration of CO. Sulfur-containing compounds also exert a detrimental effect on the reduction of fuel nitrogen oxides [12, 13].

In distinction to oil, the burnout degree of solid fuels can vary from ~80% for anthracite to ~99% for sawdust (Table 8.2). A considerable heat transfer from the

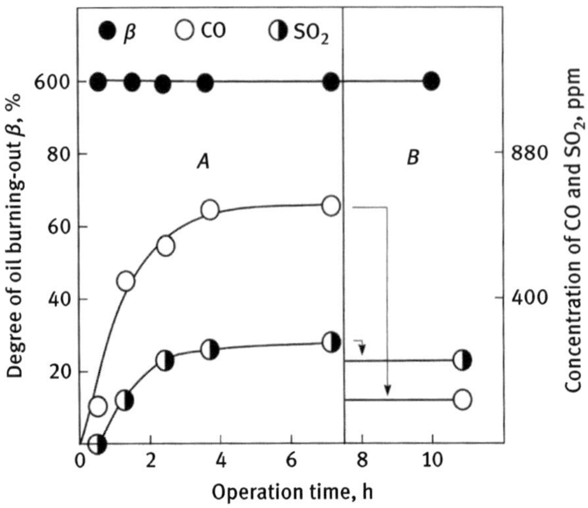

Fig. 8.8: The degree of high-sulfur crude oil burnout (*β*) and the concentrations of CO and SO_2 as a function of the catalyst operation time at 700 °C: (*A*) petroleum combustion without calcite; (*B*) oil combustion in the presence of calcite (calcite consumption, 11 g/h) [12].

Table 8.7: Technical characteristics of the catalytic heat generating unit KTU-2 [23].

Characteristics	Values
Reactor diameter (without heat insulator)	0.42 m
Reactor height	2.43 m
Heat exchanger surface area	3.3 m^2
Fuel utilization efficiency	93%
Temperature in combustion zone	600–700 °C
Temperature of flue gases after cyclone	100 °C
Temperature of water at the outlet	95 °C
Water consumption upon its heating by 50 °C	8.0 m^3/h
Fuel consumption (at a calorific value of 5,000 kcal/kg)	86 kg/h
Air flow rate	500 m^3/h
Installed power of electrical equipment:	
Electrical heater	20 kW
Water pump	4 kW
Fuel pump	0.27 kW
Air blower	4 kW
Catalyst loading	150 m^3

catalyst grains to the gas-coal flow provides a rapid heating of carbon particles. In the pores of fuel particles, there occur rapid gas heating and expansion, evaporation of moisture, and destruction of coal matter with a release of considerable amounts of volatile organic substances (up to 50% of organic mass of coal). This is followed by gasification of the coke residue by oxygen, carbon dioxide and water with the formation of complete and incomplete oxidation products. Volatile organic substances and incomplete oxidation products (CO and H_2) are oxidized to CO_2 and H_2O directly on the catalyst. It is commonly accepted that the burnout degree of coal weakly depends on the catalyst activity and is determined mostly by the particle size ratio of the catalyst and coal, temperature, and air excess coefficient [33]. However, upon combustion of a solid fuel in the fluidized bed of an inert material, the burnout degree of coal at 700–750 °C is much lower as compared to the catalytic combustion. The cause of such a difference is that upon combustion of a solid fuel in the fluidized bed of an inert material, coke burnout is the limiting process. This is related to the delivery of oxygen to the surface of coal particles. The use of a deep oxidation catalyst removes diffusion limitations; as a result, the burnout degree of coal is substantially increased.

Testing of the modified heat generating unit KTU-2 with a power up to 500 kW (Table 8.7) in the combustion of Kansk–Achinsk brown coal using the catalyst IK-12-73 showed that in the temperature range of 640–680 °C the burnout degree was 93.8%. The content of wastes in flue gases was as follows: CO – 0.05 vol.%; NO_x – 74 mg/m^3; SO_2 – 51.3 mg/m^3 and dust – 130 mg/m^3 [23].

In some works [41–44], deep oxidation of methane and propane in a fluidized bed of CuO/γ-Al_2O_3 catalyst was studied upon variation of the fluidized bed height, feed rate, concentration of reagents, and temperature. It was shown that the concentration of CH_4 at the reactor outlet is below 10 ppm at the feed rate 0.4 m/s and temperature 700 °C (the inlet concentration of CH_4 was 3%). Within the two-phase fluidization model, a kinetic model describing the experimental data has been developed.

The important characteristics of fuel combustion processes for heat generation are their technical and economic indices. A comparison of some indices with non-catalytic setups is shown below. According to Tables 8.8 and 8.9, at a comparable power of heat generating units with grate furnace (NEC) and layer furnace (KV), the catalytic heat supply unit (CHSU) decreases not only the emission of hazardous substances but also the dimensions of the unit by a factor of ca. 20, the process temperature by a factor of 1.5 and fuel consumption due to its more efficient utilization.

Extrapolation of the obtained experimental data to the heating units with a higher power shows that in comparison with the unit based on an inert material in the fluidized bed, the use of the catalyst also considerably decreases the dimensions and the content of harmful wastes. An increase in the degree of fuel combustion enhances the efficiency (Table 8.10).

The efficiency of a heat generating unit with constant dimensions can be improved by varying the catalyst loading. Thus, upon combustion of a liquid fuel, the useful thermal capacity increased more than threefold when the catalyst loading was raised from 70 to 130 L (Fig. 8.9). The increase in capacity was accompanied by an increase in the fuel utilization efficiency from 75% to 93% (Fig. 8.10). This occurs mostly due to a decrease in heat losses with flue gases (Fig. 8.10). The capacity can be controlled for solid fuels, too [21].

Table 8.8: Comparison data for heat generators using flare combustion (NEC boiler, St. Petersburg, RF) and catalytic combustion (CHSU) of solid fuels [14].

Parameter	NEC boiler	CHSU
Thermal power, kW (Gcal/h)	230 (0.2)	230 (0.2)
Fuel utilization efficiency (%)	75	93
Heat generator dimensions (mm):	3,000	400*
Length	1,400	3,000
Width	1,800	0.45
Height	7.56	88
Boiler volume (m^3)	94	700
Fuel consumption rate for brown coal with a heating	1,100	50–250
value of 2,440 kcal/kg (kg/h)	to 2,000	50–300
Firebox temperature (°C)	to 1,000	1–50
Toxic substance content of the flue gas (mg/m^3):	to 500	
NO_x		
CO		
SO_x		

*Apparatus diameter

Table 8.9: Operating data of hot-water boiler KV-50 with layer furnace produced at Cherepanovo Boiler Plant (Novosibirsk region, RF) and catalytic heat supply units (CHSU) [10].

Parameter	KV-50	CHSU
Heat productivity (kW)	50	50
Temperature of output water (°C)	95	95
Fuel consumption (Q = 5,000 kcal/kg) (kg/h)	11.9	9.9
Power efficiency (%)	72–78	87–93
Boiler volume (m^3)	2.85	0.09
Power consumption blower (kW)	0.3	0.65
Catalyst load (kg)	–	10
Content of toxic compounds in flue gases (mg/m^3):		
NO_x	Up to 2,000	50–150
CO	Up to 1,000	50–100
SO_x	Up to 500	1–50

The following results were obtained for industrial operation with the spherical catalyst SCHKZ-1 and quartz sand taken in a 1:4 ratio at a total loading of 650 kg. The degree of attrition did not exceed 0.2 wt % per day for the catalyst, and 0.50 wt % per day for sand. The degree of burnout reached 94–97% for the low-grade G coal (Kuznetsk Basin). The concentration of nitrogen oxides in flue gases did not

Table 8.10: Comparison of working characteristics of "Tsukisima Kikai" furnace (Japan) for combustion of solid wastes in a fluidized bed of sand (base) operating at Ust-Ilimsk timber and new furnace based on catalytic combustion with an approximate capacity of 5 Gcal/h [10].

Parameter	Base[a]	Catalytic[b]
Furnace productivity (t/h):		
Dry sludge	2.3	2.3
Evaporated moisture	11.2	11.2
Wet sludge	13.5	13.5
Furnace dimensions:		
Hearth area (m^2)	32.2	8.2
Furnace volume (m^3)	510.0	41.0
Specific load of furnace by solid wastes		
Per hearth area (kg/h m^2)	71.5	280.0
Per furnace volume (kg/h m^3)	4.5	56.1
Specific load by evaporated moisture		
Per hearth area (kg/h m^2)	348.0	1,365.0
Per furnace volume (kg/h m^3)	22.0	273.0
Power consumption by blower (kW h/t ads[c])	271.1	190.0
Consumption of equivalent fuel (t/t ads[c])	0.405	0.219
Specific heat of combustion of dry sludge (kcal/kg)	3,420	3,420
Sand load (t)	100	–
Catalyst load (t)	–	12.2
Unit weight (t)	90	5.9
Content of toxic compounds in flue gases (mg/m^3):		
NO_x	Up to 800	50
CO	Up to 1,500	60
SO_x	Up to 1,000	5

[a]Operating data of the existing plant.
[b]Characteristics were obtained by extrapolation of operating data for the plant with dry sludge capacity 1 t/h.
[c]Absolutely dry substance.

exceed 100–200 mg/m^3, and carbon oxide – 100–300 mg/m^3. In comparison with the existing boiler plant, the monthly consumption of coal decreased by a factor of 4.5 [21]. Overall, technical and economical advantages of the catalytic combustion of fuels in a fluidized bed are evident. However, the key technological problem of such a process is the catalyst.

8.3 Catalysts for fluidized bed processes and their deactivation

During the operation in a fluidized bed, catalysts (Fig. 8.11) are subjected to chemical, thermal and mechanical action [28]. Chemical interaction includes the influence of reaction medium, which can lead to phase transformations of the active component

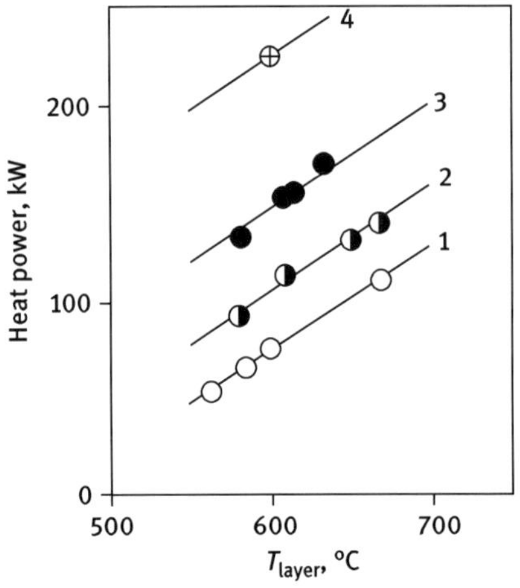

Fig. 8.9: Useful heat power of CHU on liquid fuel in dependence on the temperature and amount of catalyst loaded in the reactor: 1 – catalyst loading 70 t, 2 – catalyst loading 80 t, 3 – catalyst loading 100 t, 4 – catalyst loading 130 t [21].

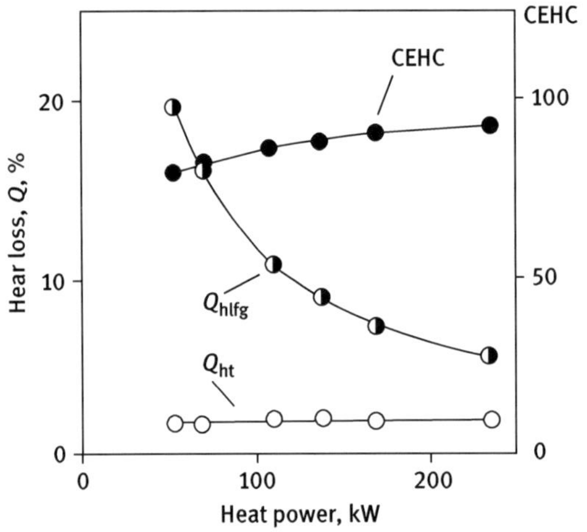

Fig. 8.10: Heat loss (Q_n) and coefficient of efficiency of fuel heat capacity (CEHC) in dependence on CHU heat power: Q_{ht} – heat loss of incomplete fuel combustion; Q_{hlfg} – heat loss with flue gas [21].

and substrate, deposition of polymerized by-products or coke (in case of a stoichiometric mixture), and poisonous effect of sulfur as well as alkali and other metals that are present in poorly refined fuels. Thermal action is related to a long-term operation of a catalyst in the temperature range of 700–1,000 K and possible short-term overheats of individual catalyst particles (or monoliths) up to 1,300–1,400 K. The short-term overheats are caused by nonuniform distribution of fuel over a gas-distributing

Fig. 8.11: A typical view of a spherical catalyst used in fluidized bed catalytic combustion [30].

grate with a considerable heat release on individual grains or monoliths. Mechanical action in the fluidized bed mode includes attrition of the catalyst grains and monoliths as well as splitting and destruction of monoliths due to collision of the catalyst particles with each other, with the sand particles and with the surface of heat exchange tubes. If a fixed bed of monolithic catalysts is used for coal combustion, coal particles can also possess abrasive properties. The listed factors are closely interrelated and lead to catalyst deactivation [28, 29].

The following requirements are imposed on the deep oxidation catalysts for a fluidized bed:

- quite a high catalytic activity;
- spherical shape and narrow grain size distribution;
- high mechanical strength and attrition resistance; and
- stability of the above listed parameters during the operation.

The activity of catalysts in complete oxidation reactions should provide a high degree of fuel burnout, a low ignition temperature, and low concentrations of harmful emissions (CO, NO_x, CH_x). Taking into account the cost, most promising are the catalysts based on oxides of period 4 transition metals. The initial stage in the development of catalysts for catalytic heat generator (CHG) included testing of the model oxidation catalysts containing transition metal oxides, which were synthesized by impregnation of alumina beads, in a pilot facility for combustion of residual oil at nearly a stoichiometric oxygen content in air [28]. After testing, the catalysts were examined in the model reaction of butane oxidation under gradientless conditions [31]. It was found that activity of all the catalysts in the test reaction after testing for 10 ÷ 25 h decreased by a factor of 5 ÷ 10 and then remained virtually constant (Fig. 8.12).

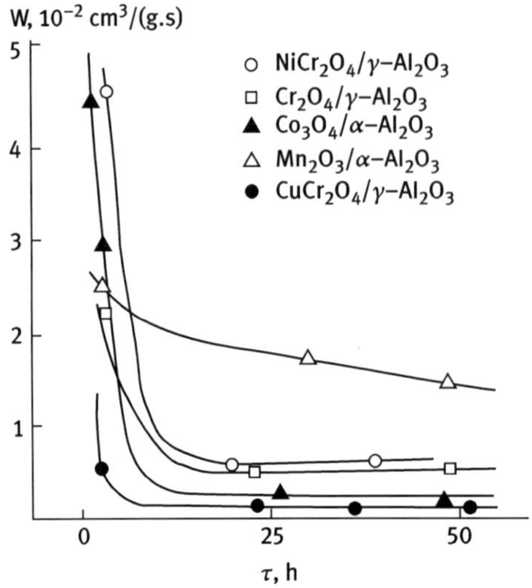

Fig. 8.12: Activity of catalysts in the reaction of complete oxidation of butane versus the CHG operation time [27].

An important characteristic of environmental safety is the residual concentration of CO. In noncatalytic combustion, the residual concentration of CO is usually high [13, 32]. In catalytic combustion of residual oil over an oxide catalyst, the initially low residual concentration of CO slowly increases, reaching 1,500–2,000 ppm after 200–300 h of testing (Fig. 8.13). Thus, catalytic combustion of fuels on oxide catalysts is accompanied by slow processes leading to their deactivation.

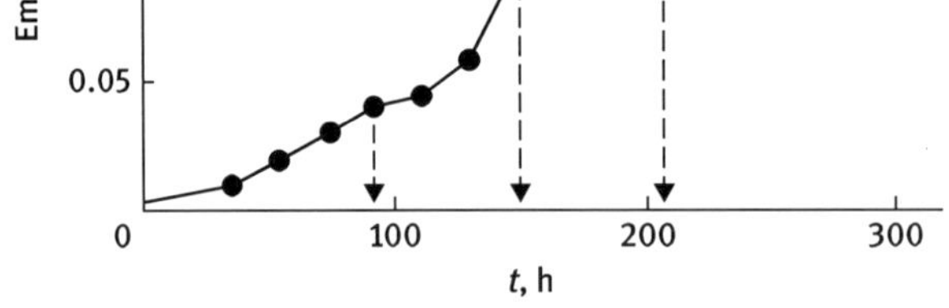

Fig. 8.13: CO content at the exit of the CHG pilot facility versus its operation time: arrows show the instants of addition of fresh catalyst portions [27].

The primary factors leading to a loss in activity include sintering, which decreases the specific surface area of catalysts by a factor of 2–3 just in the first 10–25 h of operation in CHG. This process is irreversible. Combustion of liquid fuels in stoichiometric mixtures was accompanied by carbon build-up, the amount of which reached 1 wt.%. Annealing in oxygen resulted in a partial restoration of activity. Combustion of sulfur-containing fuels leads to the formation of sulfur oxides, which interact with the catalyst to yield various sulfur-containing compounds. Thus, upon combustion of sulfur residual oil M100, the concentration of sulfates (mostly aluminum ones) increases to 1–4 wt.% in the first hours of catalyst operation and then virtually does not change [28]. Accumulation of sulfur compounds was observed also upon combustion of brown coal [33]. A loss in activity of various oxide catalysts in the joint reaction of complete oxidation of phenol and hydrogen sulfide in the presence of water vapor (60–70%) was studied in detail in [32] (Table 8.11). The process led to the formation of both the sulfates and more weakly bound adsorbed sulfur oxide compounds. Overall, the indicated factors are responsible for the changes in activity observed in the first tens of hours. Sulfation of the catalyst upon combustion of sulfur-containing oils increases CO emission during 4 h. After that, the outlet CO concentration remains approximately constant (Fig. 8.8) [12].

Table 8.11: Activity of catalysts in the oxidation of phenol in the presence of hydrogen sulfide (composition of the initial mixture: C_6P_5OH 4.1 × 10^{-5} mol/L; H_2S 2.5 × 10^{-5} mol/L; H_2O 63%; air 37%; GHSV 28,000 h^{-1}) [34].

Catalyst	Specific surface area (m^2/g)		Temperature of 90% phenol conversion (°C)	
	Before reaction	After reaction	Without H_2S	With H_2S
CuO/Al_2O_3 (IK-12-1)	125	50	315	380
$CuCr_2O_4/Al_2O_3$ IK-12-2	87	79	330	350
CuO	0.4	2.4	335	410
Cr_2O_3	29	21	340	405
Co_3O_4	27	9	285	395
V_2O_5	5	6	330	375
Fe_2O_3	26	11	355	415
$CuCr_2O_4$	11	10	325	360

According to Table 8.11, the mixture that is most resistant to sulfur is the active component containing copper chromite spinel. Thus, much efforts were made to develop and test the oxide catalysts based on chromium oxides [14, 35–37].

Some properties of the chromium oxide catalysts are listed in Table 8.12.

Thermal stability of the IK-12-72 catalyst is limited by 850 °C. The use of magnesium chromite as the active component allows obtaining catalyst with a high mechanical strength and thermal stability up to 1,000 °C. However, in some cases, activity of the $MgCr_2O_4/\gamma\text{-}Al_2O_3$ catalyst (IK-12-72) was insufficient to completely exclude the emission of CO. The presence of magnesium and copper chromites in the active component makes it possible to obtain the catalyst that possesses high catalytic activity, mechanical strength and thermal stability (up to 1,000 °C). The IK-12-73 catalyst with the composition $Cu_xMg_{1-x}Cr_2O_4/\gamma\text{-}Al_2O_3$ complies with the main requirements to deep oxidation catalysts for a fluidized bed. Thermal stability of this catalyst is attributed to the formation of a solid solution of copper chromite in magnesium chromite. Attrition is observed for all the catalysts in a fluidized bed. Taking into account that any chromium III oxides contain an admixture of chromium VI oxides, dust particles of the catalyst are quite toxic. In addition, the production of such catalysts is also toxic [36].

In this connection, of interest are the catalysts based on iron oxides [30, 35, 36, 39]. In comparison with the catalysts containing chromium oxides, the iron oxide catalysts are able to considerably decrease nitrogen oxides emission upon combustion of fuels [35]. However, the residual emission of CO and the ignition temperature of such catalysts are noticeably higher [35, 40] (Table 8.13).

The values listed in Table 8.13 were obtained by testing for several hours. More prolonged changes in the catalyst properties, which slowly increase the emission of CO in flue gases (Fig. 8.13) for hundreds of hours, are caused most likely by the long-term changes in the composition of active component and the nature of active sites. For copper-chromium oxide catalysts, slow changes in activity were also observed in the test reaction of butane oxidation after testing the catalysts in combustion of a liquid fuel (solar oil) [40, 41] (Table 8.14). A detailed study of model Cu-Cr oxide catalysts supported by impregnation on aluminum oxides of different phase composition revealed that such systems have a very complicated structure. They can include many amorphous and disperse structures containing individual oxides like CrO_3 or CuO and their compounds with alumina [43–45]. Some of the indicated structures can be detected by XRD, while others are amorphous. Highly active and readily reducible structures were found on the surface of copper-chromium oxides; they decompose in the reaction medium during a long-term combustion of fuels [45]. After fuel combustion for 70–80 h, the activity of such oxides referred to their unit surface area decreases by a factor of 14–15 [42]. The interaction with the reaction medium upon combustion of organic substances under industrial conditions produces changes in the real/defect structure of the surface, which can be studied using some spectral methods that are sensitive to the surface state: XPS, EXAFS, IR spectroscopy of adsorbed probe molecules, and others. Thus, it was shown for the $CuO/\gamma\text{-}Al_2O_3$ system that not only the copper oxide particles are present on the alumina surface, but also the highly reactive metastable Cu–O clusters, the structure

Table 8.12: Chemical and textural-mechanical properties of chromium oxide catalysts [36–38].

Catalyst (Specs. "TU")	Composition		Textural properties		Abrasiveness, (60 min) (%)	Density	Crushing strength (MPa)
	Chemical (wt.%)	Phase	SSA (m^2/g)	Pore vol. (cm^3/g)			
IK-12-72 (TU 6-09-5505-88)	–	$MgCr_2O_4$, Al_2O_3	180	~0.4	0.43 (day)	1.03	20–25
IK-12-73 (TU 6-68-102-89)	Cr_2O_3 (1.7) CuO (1.7) MgO (3.6) Al_2O_3 (93)	Spinel, $Mg_xCu_yCr_2O4$, γ-Al_2O_3	106	0.27	5%	1.02	20–25
SCHKZ-1 (TU 2171-005-5144844-2002)	Cr_2O_3 (6.5) CuO (3.5) Al_2O_3 (90)	Spinel, $CuCr_2O_4$, γ-Al_2O_3	115	0.30	3%	–	–

Table 8.13: Comparative data on activity and stability of chromium oxide and iron oxide catalysts in fluidized bed combustion of brown coal [39].

Catalyst	Temperature of 50% conversion of CO (°C)	Emission of CO (ppm) (700 °C)	Degree of attrition (%) (ASTM D-40558)
SCHKZ-1	245	320	2.2
IK-12-73	275	700	0.4
IK-12-74A	375	2,800	0.9
FeO_x/Al_2O_3	308	1,000	0.8

Table 8.14: Deactivation of $CuCr_2O_4/Al_2O_3$ catalyst after testing in diesel fuel combustion [40, 41].

Catalyst	Time-on-stream (h)	Activity*10^{-2}, cm^3 butane/g · s	Ref.
$CuCr_2O_4$ (5%)/Al_2O_3	0	4.67	[21]
	2	0.25	
	43	0.14	
$CuCr_2O_4$ (30%)/Al_2O_3	0	3.62	[22]
	8	0.25	
	80	0.06	

of which differs from the CuO copper oxide (tenorite) phase. Slow destruction of such clusters and formation of the aluminate phase in the reaction medium deteriorate the catalyst activity [46].

Stability of catalysts implies several factors:

- stability of their catalytic activity during the operation, which may decrease under the combustion conditions (high temperature and redox medium), and interaction of the active component with catalytic poisons (sulfur oxides, alkali metals, etc.);
- stability of the catalyst mechanical strength and attrition resistance; the degree of attrition should be not lower than 0.5% per day;

For the heating season in Siberia and Far East regions, the catalyst service life should be not shorter than 6 months.

For conventional chromium-containing catalysts, attrition is the determinative and limiting factor of their irreversible loss in a fluidized bed. In addition, essential are not only the high mechanical strength of grains, but also their spherical shape and narrow grain size distribution. These factors can decrease the catalyst loss during the operation in the fluidized bed mode.

Table 8.15: Service life of $CuCr_2O_4/Al_2O_3$ catalyst at different temperatures of fuel combustion [51].

Operating temperature (K)	Service life (time)	Wearability (%/day)
970	~3 years	0.1
1,000	~142 days	0.7
1,020	36 days	2.8
1,050	105 h	23

At the initial stage of implementation of the CHG process, the catalyst wear was as high as 2–5% per day. So, the catalyst deactivation due to its destruction and carry-over from reactor strongly exceeded other deactivation factors. The resistance to wear was estimated by various methods. The well-known method for evaluating the strength of granulated materials and describing the attrition kinetics is the use of ASTM International standards (the US international organization that develops and publishes voluntary standards for materials, products, systems and services): crushing strength of a grain (ASTM D-4179, ASTM D-6175); crushing strength of a layer of granulated particles (ASTM D-7084-4); and attrition resistance (ASTM D-4058-96, ASTM D-5757-95) [46–50]. For the initial (up to 100 h) step of catalyst operation in CHG, quite a good correlation was found between strength and crumb content upon attrition. As a rapid method for measuring the mechanical strength of spherical grains, it was recommended to measure the crushing strength under static conditions, which makes it possible to reliably test the mechanical properties of supports and catalysts. Catalysts with a strength below 10 MPa usually have an anomalously high rate of wear, up to 5% a day. The strength within 10–35 MPa provides an approximately constant stationary rate of attrition (0.22%/day). The catalyst wearability significantly depends also on the process temperature. Thus, in [52] it was found that the rate of wear increases with raising the temperature. Service life of the supported copper–chromium catalyst, which is very sensitive to the temperature parameter, was estimated at 970–1,050 K (Table 8.15.)

For the strongest catalysts additionally containing magnesium oxides, changes in the strength from 30.5 to 24.1 MPa were observed after 120 h of operation in CHG, and up to 21.6 MPa after 216 h [28]. It means that the strength loss rate dropped from 0.053 to 0.027 MPa/h. The data obtained made it possible to use one catalyst lot only for a single heating season (6 months). So, under standard operating conditions of industrial CHG, the catalyst was periodically added to compensate its wear. A similar procedure was used in coal combustion under fluidized conditions with the addition of sand as a heat carrier [11]. This decreased not only the consumption of a spherical catalyst, but also the degree of its attrition (Fig. 8.14).

The subsequent development of procedures and methods of catalyst synthesis [21] made it possible to use the catalyst IK-23-73 (Katalizator Ltd., Novosibirsk). The

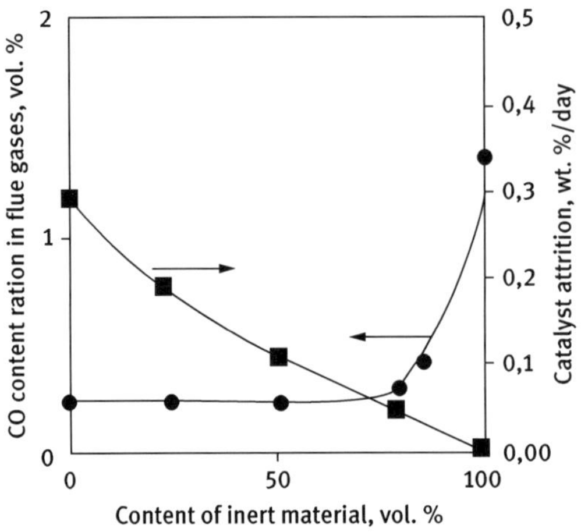

Fig. 8.14: The effect of sand addition on the catalytic and strength properties of aluminum-copper-chromium catalyst in the oxidation of diesel fuel [10].

catalyst was synthesized as alumina grains of 1.5–2 mm in diameter with the deposited active component (up to 10 wt.%) – copper, magnesium and chromium mixed oxides. The bead catalyst (Shchelkovo Catalyst Plant Ltd., Moscow) containing alumina-supported copper and chromium oxides is currently used in industry (Table 8.13).

8.3.1 Ceramometal honeycomb catalysts for fuel combustion in a fluidized bed

Along with abrasiveness, the fluidized catalyst bed has also a disadvantage related to the presence of gas bubbles, where a breakthrough of unburned fuel is possible and flare combustion of fuels can proceed by the conventional mechanism, that is, with the formation of toxic products (CO, NO_x, SO_2). To eliminate this phenomenon, the fluidized bed should be arranged using small-volume packing that destroys gas bubbles, but increases the catalyst attrition degree [21].

The indicated drawbacks can be attenuated with the use of a stationary catalytic packing mounted instead of grids (Fig. 8.3); as a result, fuel combustion can be performed at temperatures below 800 °C and advantages of fuel combustion in a fluidized catalyst bed are retained. The catalytic packing is made as honeycomb monoliths, grids or Raschig rings. Elements of the catalytic packing are deep oxidation catalysts operating in the kinetic or internal diffusion region. Laboratory and bench testing of various packings in combustion of liquid and solid fuels showed that the presence of a catalytic packing in a fluidized bed of an inert material allows reaching the same degree of fuel burnout and the same content of toxic substances in flue gases as in the case of fuel combustion in the organized fluidized catalyst bed [53]. Along with the high catalytic activity toward fuel combustion, the packing elements should have a sufficient specific surface area. The important additional requirements are high mechanical strength and attrition resistance of a packing as well as a high thermal stability and thermal conductivity.

The recently developed samples of honeycomb cermet catalysts generally comply with these requirements [29, 33]. Such catalysts are the highly porous $CuO/Al_2O_3/FeAlO/FeAl$ composites with the specific surface area of 5–19 m^2/g and the developed system of ultramacropores containing pores with the size up to several tens of micrometers. Features of the pore structure of cermets in comparison with other porous catalysts are displayed in Fig. 8.15. For conventional oxide supports and catalysts, the volume of macropores and their fraction in the total volume are insignificant [33, 40]. The micro- and mesoporous components are completely absent in the monolithic supports, including honeycomb structures, foam materials and others. To increase these components, various "washcoating" procedures are employed. Such coatings are readily abraded with sand and powdered solid fuels in the fluidization mode. The presence of open ultramacropores in cermets substantially decreases diffusion resistance in the porous composite.

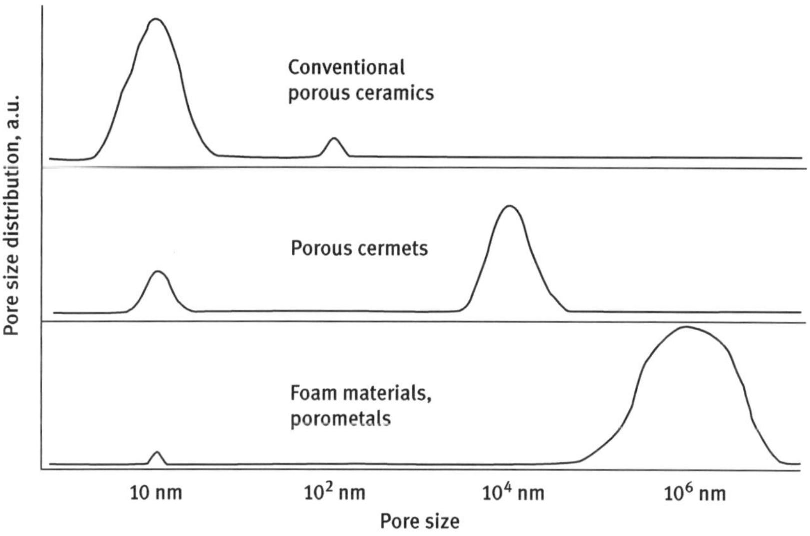

Fig. 8.15: Features of the pore structure of different supports and catalysts [33].

Microstructure of such catalysts includes metallic particles of Fe and FeAl alloys as well as the alumina globules with deposited copper oxide, which are uniformly distributed in the FeAlO matrix (Fig. 8.16). The presence of metallic particles in the cermet skeleton (Fig. 8.16c) considerably enhances the mechanical strength of cermets, which reaches 25–26 MPa.

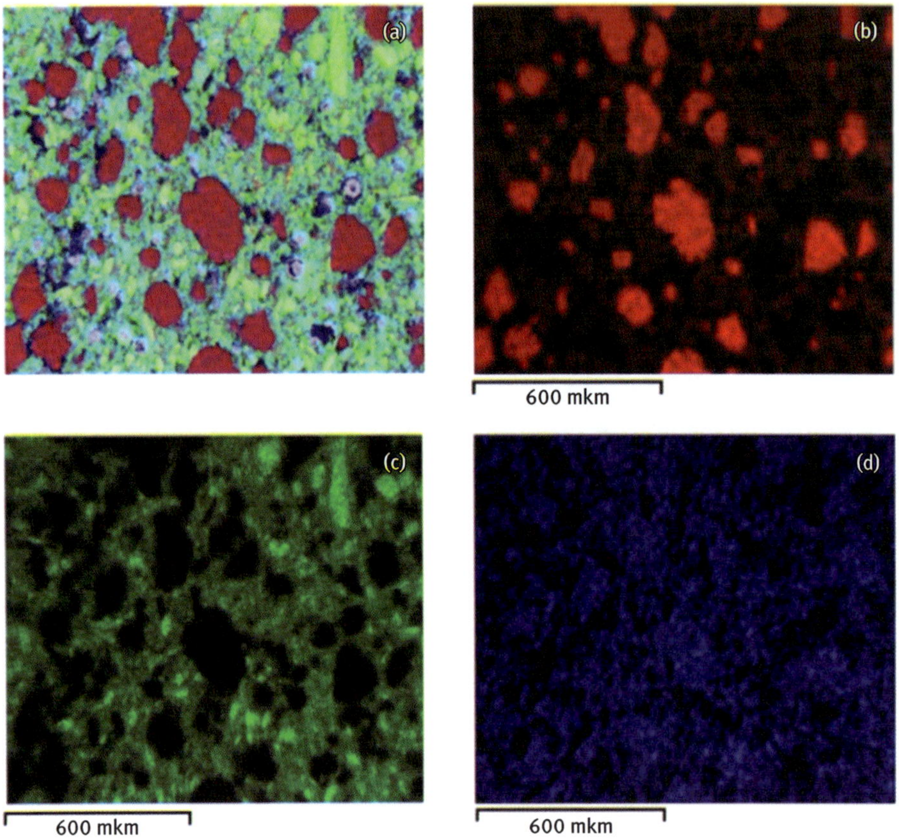

Fig. 8.16: Distribution of different components (phases) according to mapping of the polished surface of cermet support Al_2O_3/FeAlO/FeAl: a – general view, b – pure alumina (red), c – FeAlO/FeAl matrix (green and salad), d – oxygen (blue). Blue and violet spots on (a) are ultramacropores [33].

The cermet catalysts synthesized as honeycomb monoliths (Fig. 8.17) were mounted in a tubular reactor with the internal diameter 78 mm according to Fig. 8.18.. At the first stage, testing was carried out in a pilot facility for 100 h. Kansk–Achinsk coal (see above) was used as a solid fuel. Sand loading was 3.5 L. Coal consumption was 0.68 kg/h; air flow was 10 m^3 (STP)/h and the air excess coefficient $\alpha = 2.65$. The volume portion of the catalytic package was 24%, total mass – 2.1 kg. The testing showed that the conversion of coal increased from 91–95% to 98–98.5%, and the concentration of CO and sulfur oxides decreased substantially (Fig. 8.19) [33].

Long-term (~4,000 h) testing of the monoliths in a heat generating unit of a boiler plant with a power of 3 Gcal/h produced by TermoSoft-Sibir Ltd. was carried out in Kulunda, Altai Region. The strength of cermet catalysts was shown to decrease by ca. 30% (Fig. 8.20a). The honeycomb structures completely retained

Fig. 8.17: Cermet honeycomb structures before (left) and after (right) the service life testing [29].

Fig. 8.18: General view of cermet honeycombs divided by stainless steel rings before loading into reactor [33].

their shape (Fig. 8.17) and weight of the monoliths remained virtually unchanged. Taking into account data on the strength–wear relation obtained for ceramic catalysts, it can be predicted that such catalysts could operate for one or two seasons without replacement.

Changes in the activity of cermet catalysts were estimated from the rate constant of octane oxidation in the excess of oxygen. The dynamics of the activity of cermet catalyst versus the time on stream in combustion of brown coal was found to be similar to that in combustion of residual oils (Fig. 8.20b): deactivation in the initial period (34–42% depending on temperature) strongly exceeds deactivation in other periods (~18%) relative to the initial activity. The loss in the activity is not so

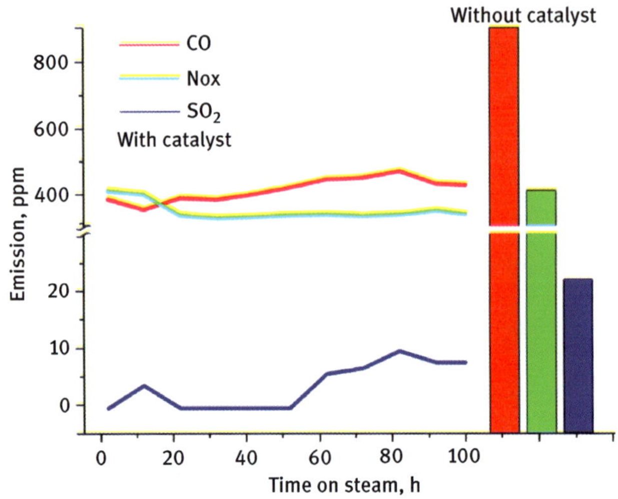

Fig. 8.19: Concentrations of CO, nitrogen and sulfur oxides at the reactor outlet versus the time of pilot testing. Concentrations of the same substances during noncatalytic combustion of brown coal are shown for comparison purposes.

pronounced as in the case of ceramic catalysts. Deactivation of cermet catalysts is caused nearly by the same factors as in the case of combustion of residual oils. A decrease in the specific surface area from 19 to 9 and then to 5 m^2/g is observed. In addition, a high (up to 7 wt.%) sulfur content, caused by the formation of surface sulfates or sulfites, was found in samples after testing. It seems interesting that sulfur concentrates mostly in the subsurface layer with a thickness of ca. 1 μm. It seems that exactly the sulfur-containing compounds concentrated in the surface layer together with the developed macropore structure provide a good preservation of activity of cermet catalysts as compared to ceramic catalysts, which commonly do not possess the developed macropore structure.

8.4 Conclusions

Catalytic combustion of fuels in a fluidized bed has significant advantages over noncatalytic combustion, especially in a fixed bed. The presence of catalysts substantially enhances the technological, economical and environmental characteristics of the process in comparison with noncatalytic ones. It becomes possible to use different wastes for heat generation. The major problem hindering the development of the technology of catalytic combustion of fuels in a fluidized bed is insufficient stability of catalysts under reaction conditions. Therefore, the rate of mechanical wear is often higher than the rate of chemical deactivation of catalysts. This

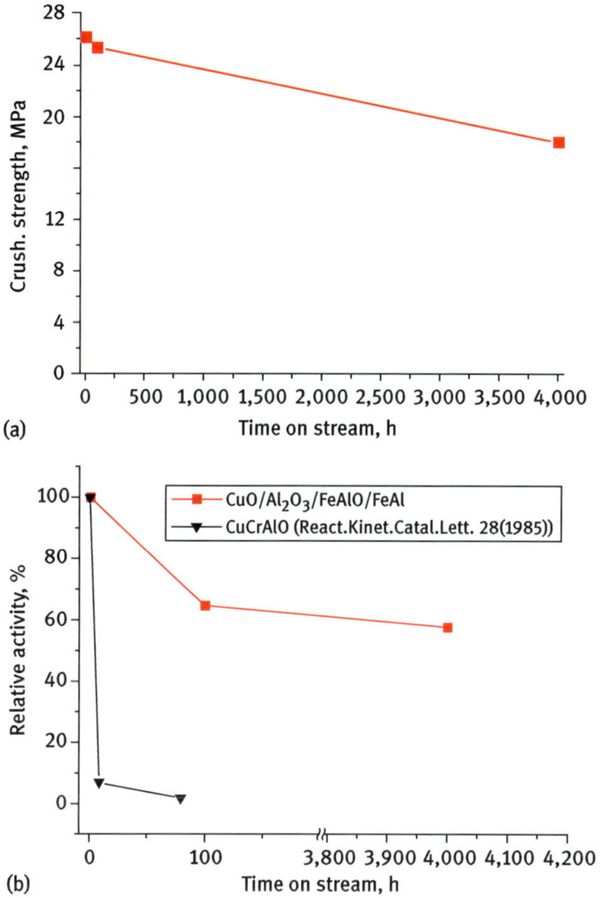

Fig. 8.20: Changes in the strength (a) and relative activity of cermet catalysts (b) after testing in combustion of brown coal [29].

problem can be solved with the use of radically new methods and materials, such as cermets produced as honeycomb structures with a large free cross section.

References

[1] Basu P. Combustion and gasification in fluidized beds, CRC Tailor &Francis Group, Boca Raton, Florida, 2006.

[2] Simeon NO. Fluidized bed combustion, Marcel Dekker Inc., NY, 2004, 581.

[3] Adanez J, Abad A, Garcia-Labiano F, Gayan P, de Diego LF.: Progress in Chemical-Looping Combustion and Reforming technologies. Progress in energy and combustion science 2012, 38, 215–82.

[4] Simonov AD, Fedorov IA, Dubinin YuV, Yazykov NA, Yakovlev VA, Parmon VN. Catalytic heat generators for industrial application. Catalysis in Industry 2012, 3, 50–7. [In Russian].

[5] Boreskov GK. Heterogeneous Catalysis, Nauka, Moscow, 1986) [in Russian].

[6] Boreskov GK, Levitskii EA, Ismagilov ZR. Burning of fuels and catalytic heat generators. Zhurnal Fiziko-Khimicheskogo Obshchestva (J. Russian Phys. Chem. Soc.). 1984, 29, 389–398 [In Russian].

[7] Parmon VN, Ismagilov ZR, Kerzhentsev MA. Catalysis for energy production. In *Perspectives* In: Catalysis (Chemistry for 21st Century), Blackwell Scient. Publ., Oxford–Boston, 1992, 337–57.

[8] Ismagilov ZR, Kerzhentsev MA. Fluidized bed catalytic combustion. Catal. Today 1999, 47, 339–46.

[9] Chemistry for Energy Future. In: Parmon VN, Tributsch H, Bridgwater AV, Hall D.O. eds. Oxford, Blackwell Sci., 1999.

[10] Simonov AD, Yazykov NA, Vedyakin PI, Lavrov GA, Parmon VN. Industrial experience of heat supply by catalytic installation. Catal. Today 2000, 60, 139–45.

[11] Ryabov VD. Oil and gas chemistry, "Forum", INFRA-M, Moscow, 2014, 6–193 [in Russian].

[12] Dubinin YuV, Simonov AD, Yazykov NA, Yakovlev VA. Burning sulfur crude oil in a fluidized catalyst bed. Catalysis in Industry 2015, 7, 314–20 [In Russian].

[13] Yazykov NA, Dubinin YuV, Simonov AD, Reshetnikov SI, Yakovlev VA. Features of sulfur oils combustion in fluidized bed. Chem. Eng. J. 2016, 283, 649–55.

[14] Simonov AD, Fedorov NA, Dubinin YuV, Yazykov NA, Yakovlev VA, Parmon VN. Catalytic heat-generating units for industrial heating. Catalysis in Industry 2013, 5, 42–9.

[15] Yazykov NA, Simonov AD, Aflyatunov AS, Dubinin YuV, Selishcheva SA, Yakovlev VA, Stepanenko AI. Combustion of shale heavy coal-tar products in a boiling layer of a catalyst. Chemistry for Sustainable Development 2017, 25 313–21.

[16] Yazykov NA, Simonov AD, Yakovlev VA. Assessment of the low heat values of plant biomass, peat and fossil coal based on technical analysis, Chemistry for Sustainable Development 2017, 25, 465–471.

[17] Fedorov LA, Myasoedov BF. Dioxins: analytical chemical aspects of the problem, Russ. Chem. Rev. 1990, 59, 1063–92.

[18] Prokof'ev AK. Determination of polychlorinated dibenzo-*p*-dioxins, dibenzofurans, biphenyls and chlorine-containing pesticides in the environment. Russ. Chem. Rev. 1990, 59, 1051–62.

[19] Simonov AD, Yazykov NA. Catalytic methods for remediation of solid organic wastes and biomass, Khimicheskaya promishlennost (Chemical industry) 1996, 3, 47–53 [In Russian].

[20] Yazykov NA, Trachuk AV, Dubinin YuV, Simonov AD, Yakovlev VA. Pyrolysis of wood in vibro-fluidized beds of catalysts and inert materials. Combust. Explos. Shock Waves 2013, 49, 608–13.

[21] Simonov AD, Yazykov NA, Trachuk AV, Yakovlev VA. Combustion of communal wastewater sludge in the fluidized bed of a catalyst, International Scientific Journal for Alternative Energy and Ecology (ISJAEE) 2010, 86, 61–6.

[22] Bernadiner M.N., Shurygin A.P. Fire processing and disposal of industrial wastes. Moscow, Khimiya, 1990 [In Russian].

[23] Parmon VN, Ismagilov ZR, Kirillov VA, Simonov AD. Catalytic heat generating units for solving environmental and energy problems. Catalysis in Industry 2002, 3, 20–28 [In Russian].

[24] Iamarino M, Chirone R, Lisi L, Pirone R, Salatino P, Russo G. $Cu/\gamma\text{-}Al_2O_3$ catalyst for the combustion of methane in a fluidized bed reactor. Catal. Today 2002, 75, 317–24.

[25] Iamarino M, Chirone R, Pirone R, Russo G, Salatino P. Catalytic combustion of methane in a fluidized bed reactor under fuel-lean conditions. Combust. Sci. Technol. 2002, 174, 361–75.

[26] Iamarino M, Salatino P, Chirone R, Pirone R, Russo G. Catalytic combustion of methane and propane in a fluidized-bed reactor. Proceedings of the Combustion Institute 2002, 29, 827–34.
[27] Simonov AD, Yazykov NA, Aflyatunov AS, Fedorov IA, Yakovlev VA, Parmon VN. Operating experience of hot-water boilers with catalytic combustion of liquid and solid fuels in the fluidized bed. Int. Sci. J. Altern. Energy Fuel 2014, 19(159), 71–85.
[28] Ismagilov ZR. Investigation of Deactivation and Changes in the Texture and Mechanical Properties of CHG Catalysts during Burning of Various Fuels. In: Problems of Catalyst Deactivation. Part 1. Nature of Changes in the Specific Catalytic Activity. Novosibirsk, Boreskov Institute of Catalysis, Siberian Branch, Acad. of Sci. of the USSR, 1985, pp. 74–96 [In Russian].
[29] Parmon VN, Simonov AD, Sadykov VA., Tikhov SF. Catalytic Combustion: Achievements and Problems. Combust Explos Shock Waves 2015, 51, 143–150.
[30] Fedorov AV, Yazykov NA, Ermakov DYu, Bulavchenko OA, Yakovlev VA. A study of spherical grains of the $CuO–Fe_2O_3–Al_2O_3$ catalyst obtained by droplet molding for combustion of brown coal in a fluidized bed. Catalysis in Industry 2017, 17, 324–30 [In Russian].
[31] Popovskii VV, Sazonov VA, Chermoshentseva GK, Panarina TL, Iliseeva LF. Comparative tests of catalysts in high- depth oxidation reaction. In Catalytic Cleaning of Gases. Part 2, Boreskov Institute of Catalysis, Siberian Branch, Acad. Sci. USSR, Novosibirsk, 1981, 80–92.
[32] Ciambelli P; Palma V; Tikhov SF, Sadykov VA, Isupova LA, Lisi L. Catalytic activity of powder and monolith perovskites in methane combustion, Catal. Today 1999, 47, 199–207.
[33] Tikhov SF, Simonov AD, Yazykov NA, Dubinin Yu, Usoltsev V, Yakovlev V, Sadykov V, Salanov A, Suprun E, Parmon V. Catalytic Combustion of Brown Coal Particulates over Ceramometal Honeycomb Catalyst. Catal. Sustain. Energy 2012, 1, 82–89.
[34] Simonov AD. Catalytic remediation of flue gases, wastewater and solid wastes from the production of cellulose by sulfate method. Chem. Sustain. Development 1998, 6, 277–92.
[35] Ismagilov ZR, Shkrabina RA, Barannik GB, Dobrynkin NM, Sazonov VA, Kerjenzev MA, Kirichenko OA, Aleksandrov VYu. Novel catalysts and catalytic processes for environmental protection. Russ. Chem. J. 1993, 3, 48–55.
[36] Fedorov AV, Yazykov NA, Ermakov DYu, Kaichev VV, Simonov AD, Yakovlev VA. Investigation of iron-containing deep oxidation catalysts for a fluidized bed. Catalysis in Industry 2015, 2, 61–67 [In Russian].
[37] Sazonov VA, Kimkhai ON. Industrial catalyst for gas purification II, Novosibirsk, Boreskov Institute of Catalysis SB AN USSSR, 1989 [In Russian].
[38] Dubinin Yu, Yazykov N, Simonov A, Yakovlev V, Saraev, A, Kaichev V, Bulavchenko O, Oschenko, Mokrinskii, Yermakov. Investigation of catalysts for deep oxidation of organic substances in a fluidized bed. Catalysis in Industry 2013, 4, 68–76 [In Russian].
[39] Fedorov AV, Ermakov DYu, Kaichev VV, Bulavchenko OA, Yakovlev VA. Investigation of physicochemical and catalytic properties of mixed oxides $CuO–Fe_2O_3–Al_2O_3$ in deep oxidation reactions. Catalysis in Industry 2017, 17, 315–323 [In Russian].
[40] Arendarskii DA, Paukshtis EA, Ismagilov ZR, Yurchenko EN. Optical spectroscopic studies of $CuCr_2O_4$/-Al_2O_3. Catalyst deactivation under operation in catalytic heat generators. React. Kinet. Catal. Lett. 1985, 28, 195–201.
[41] Arendarskii DA, Pashis AV, Shepelin AP, Ismagilov ZR. XPS studies of Cu-Cr catalyst deactivation in catalytic heat generators. React. Kinet. Catal. Lett. 1985, 28, 211–17.
[42] Ismagilov ZR, Arendarskii DA, Kirichenko OA, Barannik GB, Moroz EM, Ushakov VA, Malakov VV, Boldyreva NN. Study of the reactions and catalysts of combustion of fuels. 4. Genesis of the phase composition of applied aluminum-copper-chromium oxide catalysts. Kinet. Catal. 1989, 30, 806–14.

[43] Arendarskii DA, Ismagilov ZR, Barannik GB, Ovsyannikova IA, Goldenberg GI, Rogov VA, Pashis AV. Study of the reactions and catalysts of combustion of fuels. 5. Distribution of the active component of a supported aluminum-copper-chromium oxide catalysts. Kinet. Catal. 1990, 31, 1040–45.
[44] Arendarskii DA, Ismagilov ZR, Barannik GB, Rogov VA, Paukshtis EA. Study of the reactions and catalysts of combustion of fuels. 6. Role of the elements of the active component of an aluminum-copper-chromium oxide catalysts in the reaction of oxidation of CO. Kinet. Catal. 1990, 31, 1046–50.
[45] Tikhov SF, Sadykov VA, Kryukova GN, Paukshtis EA, Popovskii VV, Starostina TG, Kharlamov GV, Anufrienko VF, Poluboyarov VF, Razdobarov VA, Bulgakov NN, Kalinkin AV. Microstructural and spectroscopic investigations of the supported copper-alumina oxide system: nature of aging in oxidizing reaction media. J. Catal. 1992, 134, 506–24.
[46] Standard Test Method for Single Pellet Crush Strength of Formed Catalysts and Catalyst Carriers, ASTM International, West Conshohocken, PA (2011, A. D4179-11, at http:/www.astm.org)
[47] Standard Test Method for Radial Crush Strength of Extruded Catalyst and Catalyst Carrier Particles, ASTM International, West Conshohocken, PA, (A. D6175-03(2013), at http: 2013, www.astm.org/)
[48] Standard Test Method for Determination of Bulk Crush Strength of Catalysts and Catalyst Carriers, ASTM International, West Conshohocken, PA, (A. D7084-04(2009) 2009, at http: www.astm.org /)
[49] Standard Test Method for Attrition and Abrasion of Catalysts and Catalyst Carriers, ASTM International, West Conshohocken, PA, (A. D4058-96(2015), 2015, at http:www.astm.org/)
[50] Standard Test Method for Determination of Attrition of FCC Catalysts by Air Jets, ASTM International, West Conshohocken, PA, (A. D5757-11(2017) 2017, at http:www.astm.org/)
[51] Kirichenko OA. Development of approaches to the prediction of wear resistance of CHG catalysts. In: Stability of granulated supports and catalysts, Novosibirsk, Institute of Catalysis SB AS USSR, 1989, 76–83 [In Russian].
[52] Yazykov NA, Simonov AD, Mishenko TI, Aflyatunov AS, Smolin SV, Parmon VN. Fuel combustion in the fluidized bed of an inert material with an unmovable catalytic small-volume package. Chemistry for Sustainable Development 2003, 11, 1–6.

9 Ammonia oxidation to NO_x in nitric acid production

Nonconcentrated nitric acid is produced via the catalytic oxidation of ammonia with air oxygen followed by adsorption of the formed nitrogen oxide by water [1–8]. Therefore, the yield of the target product is determined to a great extent by activity and selectivity of the employed catalyst and conditions of its operation. At present, the prevalent catalyst for ammonia oxidation in the nitric acid production consists of gauzes that are made of platinum alloys with rhodium or with rhodium and palladium. During operation, the surface of platinum gauzes corrodes; as a result, the catalyst platinoid filament increases in diameter, the filaments become spongy and opaque, the surface area increases up to 20-fold, and platinum is lost as volatile oxides (chemical losses) due to carryover of the catalyst particles (mechanical losses). In a run over the gauzes, platinoid losses can reach two-thirds of the initial loading. In [9–12], corrosion of a platinum catalyst is attributed to the increased surface diffusion in the catalytic process and formation of volatile platinum oxides, which are condensed on more cold regions of the wire or reactor. According to Salanov et al. [13, 14], the observed changes are related to an increase in the local temperature due to the catalytic reaction proceeding on oxygen coordination defects, which increases surface diffusion and platinum transfer from the regions of defects (dislocations and intergrain boundaries). As a result, defect-free crystallites and pores are formed on the initially smooth but defect wire surface.

Since platinoids are quite expensive, for example, in February 2018, ~ \$1,000–1,020 for 1 oz (troy ounce – a unit of mass measurement equal to 31.1034768 g), many studies were aimed at finding less expensive oxide catalysts for ammonia oxidation. Results of the studies are summarized in [15–32].

Figure 9.1 shows the temperature dependence of conversion for some oxide catalysts (contact time $\tau = 2.5 \cdot 10^{-2}$ s, linear velocity 0.2 m/s). The testing was performed in laboratory setups at atmospheric and elevated pressures (up to 1.0 MPa).

Extensive studies of various oxide catalysts in the ammonia oxidation reaction demonstrated that although there are systems with high activity and stability (primarily the oxides containing 3d cations, which are often modified to enhance their activity and thermal stability), none of them can start operation at temperatures 200–250 °C, which is important for launching the reactor. So, the two-step oxidation of ammonia was proposed, where a part of platinoid catalytic gauzes (inefficient downstream gauzes in the gas flow (Figs. 9.2 and 9.3) [7, 8, 34]) is replaced by the oxide catalyst that is made of inexpensive and available materials and does not contain precious metals.

Iron oxide catalysts with the pellet/grain size of 4–6 mm have been developed for two-step systems. It was shown that on the iron–chromium catalyst NK-2U at ammonia concentration 10.5–11.3% and temperature 800 °C the yield of NO is 93.

https://doi.org/10.1515/9783110587777-009

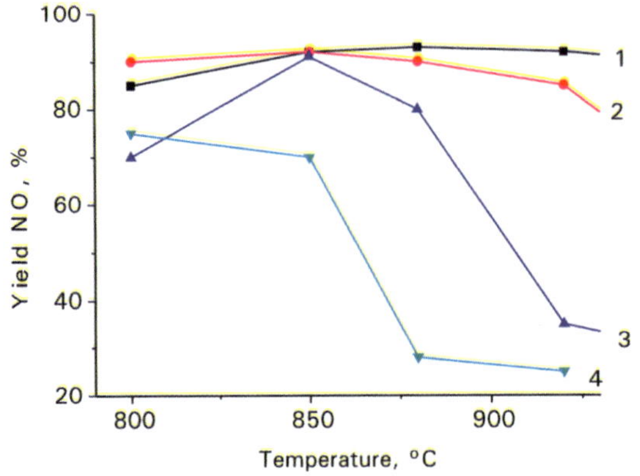

Fig. 9.1: Temperature dependence of NO yield on various nonplatinum catalysts [33]. 1 – 80% Fe_2O_3, 20% Al_2O_3; 2 – 86% Fe_2O_3, 7% Cr_2O_3, 7% Al_2O_3; 3 – 65% Fe_2O_3, 15% Cr_2O_3, 20% Al_2O_3; 4 – 85 % Co_3O_4, 15% Al_2O_3.

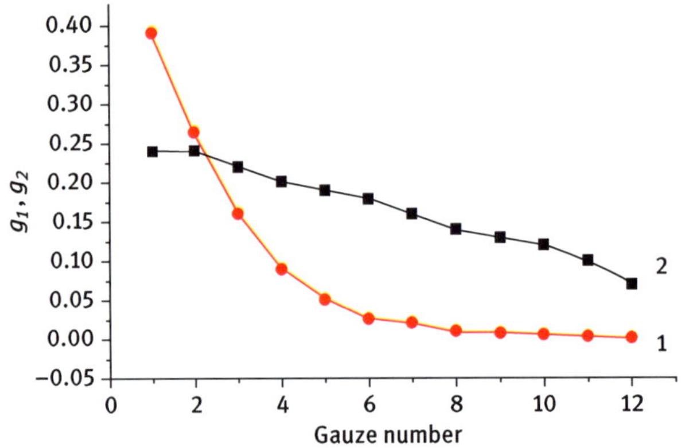

Fig. 9.2: The fraction (q) of converted NH_3 (1) and losses of platinoids (2) versus the location of a gauze in the stack [35].

3–98.4%, whereas on a more stable iron–aluminum catalyst at 890–900 °C the yield is 94.4% on the average, which virtually coincides with the nitrogen oxide yield on the platinoid catalyst [20, 36–40].

Starting from the 1960s, the two-step oxidation of ammonia (a platinum–rhodium gauze + a layer of the NK-2U oxide catalyst) has been employed virtually in all nitric acid processes in the former USSR where ammonia was converted at atmospheric pressure [7, 34], namely, in 225 ammonia oxidation plants or 450 reactors with the working diameter 2.9 m. Therefore, the initial mass of the loaded platinoids decreases by 50%, and irrecoverable losses – by 20–35% without lowering the nitrogen oxide yield relative to the design value.

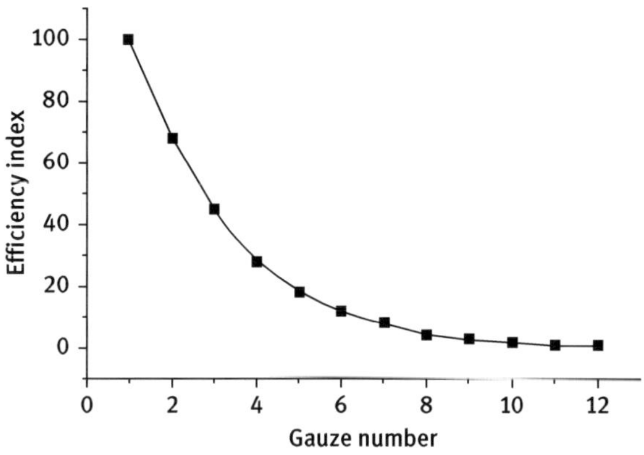

Fig. 9.3: The performance of a Pt gauze in dependence on its location in the stack [35].

In the 1970s, the two-step system comprising 9 (instead of 12) platinoid gauzes and a layer of pelletized or granulated iron–aluminum catalyst was proposed for high-pressure aggregates (up to 1.0 MPa [19]). The application of this system decreased the initial mass of the platinum-rhodium catalyst by 20–25% and irrecoverable losses of platinum by 15% without lowering the nitrogen oxide yield. However, industrial application of the granulated oxide catalyst in UKL-7 plants was associated with some difficulties because the catalyst produced dust and its loading into the catalytic reactor required a special basket. Centralized production of such baskets was not arranged and handcrafted baskets were imperfect, which ultimately led to the spillage of catalyst grains into the pot and to the gas flow bypasses due to burnout and rupture of the nichrome gauze of the basket bottom [41].

In the 1990s, a two-step catalytic system where the pelletized catalyst bed was replaced by a monolithic honeycomb catalyst with unidirectional channels has been developed at the Boreskov Institute of Catalysis SB RAS and industrially implemented in UKL-7 plants. The developed system allowed eliminating the drawbacks of irregular granulated bed: nonuniform thickness of the bed and the related gas flow bypasses, as well as a nonuniform velocity field both in the bed itself and in the platinoid gauzes of the first step. The system has been successfully operated for more than 15 years in nitric acid plants with conversion at a pressure of 0.716 MPa; it provides a decrease in the mass of platinum-rhodium catalyst by 25–33% relative to the initial stack, at least a 15% decrease in losses of platinoids, and a guaranteed average conversion of ammonia to nitrogen oxides at the level corresponding to a stack of the neat platinum-rhodium catalyst [8, 41–46]. It should be noted that a 100–120 mm layer of the pelletized catalyst (at a mass of 250–350 kg) has a resistance of 0.1–0.12 atm (1,000–1,200 mm H_2O), whereas the honeycomb catalyst has nearly a 100-fold lower resistance depending on the channel size and bed height [41].

The ABC flow in the channel is in the laminar mode and has a low mass transfer coefficient [47]. Geometrical characteristics of the honeycomb catalyst should prevent the ammonia breakthrough at a reduced number of gauzes.

Calculations made in [48] for monolithic catalysts with different selectivity, channel density and height (Fig. 9.4) showed that, for example, three gauzes removed from a standard stack (12 gauzes), which are used in UKL-7 plant to ensure the average NO yield of 93%, can be replaced by a monolithic nonplatinum catalyst with the bed height 50 mm and channel density 120 cpsi (at S = 90%), 170 cpsi (at S = 80%) or 270 cpsi (at S = 70%). If a monolithic catalyst with the bed height 100 mm is used (or two 50 mm beds of a monolithic catalyst), then three gauzes can be replaced by monoliths with the channel density 50 cpsi (at S = 90%), 70 cpsi (at S = 80%) or 120 cpsi (at S = 70%). It means that a greater density or height of channels can ensure the required NO yield even at a lower selectivity of the monolithic catalyst; therefore, a pressure drop on the monolithic catalyst bed will be insignificant.

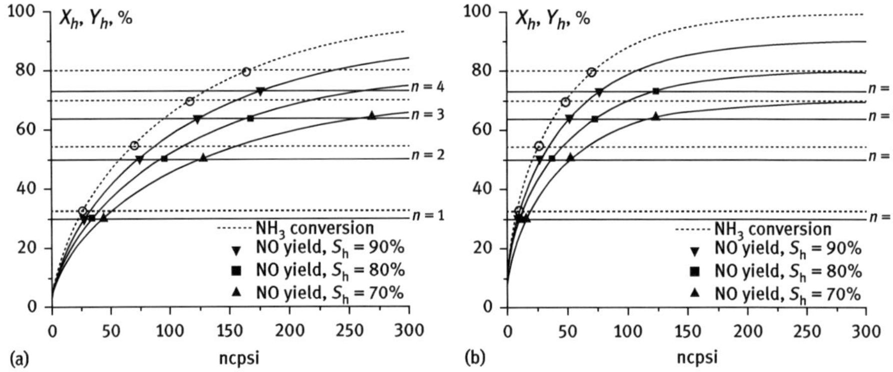

Fig. 9.4: NH_3 conversion (X_h) and NO yield (Y_h) versus monolith channel density (N_{cpsi}) for monoliths with different selectivity (S_h). (a) one monolith layer and (b) two monolith layers. Horizontal lines demonstrate the NH_3 conversion (dotted line) and the NO yield (solid line) on n removed gauzes with S_g = 0.92 [48].

Two types of monolithic catalysts are distinguished – the catalysts with monolithic supports and the bulk ones. In [49], characteristics of the oxide catalysts supported on cordierite monoliths are considered; it is noted that content of the active component decreases in testing, which can explain the observed deterioration of activity. It is accepted that the bulk catalysts commonly containing more than 50% of the active component are more stable. The main methods used to prepare bulk catalysts are pelleting and mixing. Mixing is more suitable for the preparation of monolithic catalysts: the active component is mixed with a binder and other additives (which affect the rheological properties of the resulting pastes and the structural-mechanical properties of the catalysts) with subsequent extrusion molding of the obtained plastic masses [42, 50–51].

It is well known that the oxides of such transition metals as cobalt, nickel, manganese and iron as well as the rare-earth oxides are quite active and selective in the oxidation of ammonia to nitrogen oxides [7, 33]. Nevertheless, analysis of the literature data revealed that individual oxides cannot be used as a basis for industrial catalyst due to a loss of activity at the working temperatures owing to recrystallization and sintering. When choosing the promising systems, toxicity and carcinogenicity of nickel and cobalt oxides should also be taken into account. Thus, these studies were carried out using the systems based on manganese and iron oxides modified with oxides of alkaline earth and rare-earth elements to obtain the perovskite-like structures. Aluminum oxides and aluminosilicates served as the structural modifiers to enhance thermal stability of the catalysts. For each gross composition of the catalyst, the effect of specific surface area, porosity and real structure on the catalytic performance of samples was studied by varying the synthesis conditions of initial oxides, features of the raw material and thermal treatment modes [42, 50–51].

Table 9.1 shows that the catalysts based on lanthanum manganite are more selective as compared to ferrite catalysts; therefore, the replacement of the reactive raw material by industrially produced concentrates of rare-earth elements and manganese ore does not lead to a significant decrease in selectivity. Iron oxide catalysts are also highly selective.

Table 9.2 lists the results of testing for the iron oxide and lanthanum manganite catalysts as fragments of monoliths having different channel density, height 4–20 mm and diameter 8 mm. The monolith shape varied by extrusion of the catalyst mass through spinnerets of different configuration (a hexagonal prism with triangular channels and a square prism with square channels). The wall thickness varied from 0.2 to 2 mm, and the channel side length, from 1.8 to 4.5 mm.

Testing performed on the catalysts having a similar chemical composition but different shape (samples 3, 5, 6, 7) confirmed the dependence of NO yield on the size of catalyst channels and testing temperature because the reaction proceeds in the external diffusion mode. It seems that catalysts with the wall thickness less than 0.5 mm are not required for testing at 800–850 °C; however, high-temperature testing demonstrated that the use of thinner walls is promising. The data obtained (samples 3, 5, 6, 7) showed that it is reasonable to decrease the wall thickness from 2 to 0.8 mm.

The following conclusions can be made from the experiments with variation of the active component features, its content in the catalyst, type of electrolyte and thermostabilizing additive, and raw material:

1. Other conditions being equal, the introduction of a thermostabilizing additive in the iron oxide catalyst (samples 5 and 11–14) increases selectivity at 950 °C, especially in the case of Ni-modified additive.

Table 9.1: Testing* of catalyst samples in the catalytic system with one platinum gauze.

No.	Catalyst	Testing conditions		NO yield**		Ammonia breakthrough ***
		T (°C)	*V* (m/s)	1	2	
1	(DyY)MnO_3	925	1.25	95.3	86.8	Not detected
2	(DyY)FeO_3	920	1.25	92.5	83.5	Not detected
3	(DyY)$Mn_{0.3}Fe_{0.7}O_3$	920	1.25	91.7	83.1	0.0057
		900	1.72	90.8	74.9	0.015
		905	0.85	94.0	90.7	0.014
		900	2.26	91.6	68.6	
4	a-Fe_2O_3	900	1.25	95.8		
6	(LaCe)FeO_3	920	1.25	93.4	86.9	
7	$Ca_2Fe_2O_5$	940	1.25	93.7	84.7	Not detected
8	$CaMnO_3$	900	1.72	93.6	88.2	0.037
9	(LaCe)MnO_3	890	1.25	93.4	88.3	
		920	0.85	94.0	93.8	
		900	1.72	88.4	85.4	
10	Catalyst no. 9 was tested as a monolith	890	1. 25	94.7		
		900	0.85	-93.5		
		890	1.72	92.5		
				89.2		
11	Catalyst no. 9 after commercial testing	940	1.25	90.7	86.3	

Catalysts contain the active component and 12% of alumina; samples 6–12 additionally contain up to 5% of aluminosilicate fiber [41–42].

*Catalytic activity of the samples in ammonia oxidation was tested in a laboratory flow-type setup [17] upon variation of temperature (*T*) and linear velocity of the gas flow (*V*, m/s). The tested catalysts had the grain size of 2–5 mm. In some experiments, to estimate the catalytic properties of the oxide system itself, a platinum gauze was not loaded in the reactor.

**NO yield 1 is the NO yield after the two-step system; NO yield 2 is the NO yield after one platinoid gauze.

***Ammonia breakthrough was determined after the catalytic system using the Kjeldahl apparatus.

2. Lowering the alumina content (samples 8 and 11) increases selectivity at 800–850 °C without changes in selectivity at 950 °C. However, additional modification of the thermostabilizing additive with manganese (samples 10 and 14) produces a reverse effect.
3. Comparison of the results of testing for the iron oxide catalyst (sample 6), lanthanum ferrite (sample 24) and lanthanum manganite (samples 18–19) showed a higher selectivity of manganese-containing perovskites. However, when obtaining the perovskites, a possible substantial effect of the manganese oxide raw material should be taken into account (samples 18–28).

Table 9.2: Effect of the chemical composition and geometry of the catalyst monolith fragment on selectivity in ammonia oxidation.*.

No.	Catalyst	Catalyst weight (g)	Weight (mm)	NH_3 conversion on catalyst, %, at temperatures, (°C)			Selectivity to NO_x on catalyst, %, at temperatures (°C)		
				800	850	950	800	850	950
1. □	Fe_2O_3, 20% Al_2O_3, HNO_3	0.12	4	78.2	68.8	–	66	74.2	–
2. □	-«-	0.27	8	83.8	71.7	–	69	79.7	–
3. □	-«-	1.41	44	97.8	94.7	–	92.5	89	–
	V=5 л/ч	0.64	20	97.7	94.6	63.9	84.0	79	42
		0.64	20	81.9	79.8	45.7	71	52.7	26
4. +	Fe_2O_3, 20% Al_2O_3, HNO_3	2.15	39	96.1	94.6	67	79.3	30.3	18
5. Δ	Fe_2O_3, 20% Al_2O_3, HNO_3	0.68	20	97.8	94.8	66.4	90.2	84.7	40.7
6. ⊕	Fe_2O_3, 20% Al_2O_3, HNO_3	0.82	18	97.8	94.7	66.1	74	74	62.8
7. ⊕	Fe_2O_3, 20% Al_2O_3, HNO_3	0.97	20	97.8	94.6	66.6	91.4	88.4	51.9
8. Δ	Fe_2O_3, 5%Al_2O_3, Ln-NO_3	1.02	21	97.8	95	65.2	89.9	85.2	52.2
9. Δ	Fe_2O_3, 5%Al_2O_3, 5% Ln_2O_3, Ln-NO_3	1.19	22	97.8	95	65.1	86.5	78.4	50.8
10.Δ	Fe_2O_3, 5% Al_2O_3, LnMn-NO_3	1.11	22	97.9	94.9	67	83.7	70.5	34.8
11.Δ	Fe_2O_3, 20% Al_2O_3, Ln-NO_3, Al-Si	0.95	21	97.8	94.8	66	83	75.6	51.8
12.Δ	Fe_2O_3, 20% Al_2O_3, LnCo-NO_3, AlSi	0.93	23	97.8	94.4	65.6	81	69	28.4
13.Δ	Fe_2O_3, 20% Al_2O_3, LnNi-NO_3, AlSi	0.79	21	97.8	94.7	66.5	84.6	87	63.3
14.Δ	Fe_2O_3, 20% Al_2O_3, LnMn-NO_3, AlSi	1.02	25	97.8	94.7	66.4	84.9	81	47.3
15.⊕	Fe_2O_3, $Ce(NO_3)_3$	0.64	22	97.8	94.7	65.9	69	68.8	40.2
16.⊕	Fe_3O_4, HNO_3	0.85	19	97.8	94.7	65.4	65.2	69.2	64.8
17.⊕	Fe_2O_3, 20% $LnMnO_3$, AlSi, CH_3COOH	0.82	21	97.8	94.7	66.2	85.8	82	49.9
18.⊕	$LnMnO_3$-1, HNO_3	1.0	20	98	95	–	85.7	81.6	–
19.⊕	$LnMnO_3$-1, CH_3COOH, AlSi	0.54	17	97.8	94.7	65.4	85	81.7	62.7
20.⊖	$LnMnO_3$-1, LaCe-Co-NO_3	1.08	20	97.7	94.9	66.5	85.4	77.8	59

(continued)

Table 9.2 (continued)

No.	Catalyst	Catalyst weight (g)	Weight (mm)	NH_3 conversion on catalyst, %, at temperatures, (°C)			Selectivity to NO_x on catalyst, %, at temperatures (°C)		
				800	850	950	800	850	950
21.⊕	$LnMnO_3$-2, HNO_3	0.81	17	97.8	94.8	67.7	90.1	88,8	59.1
22.⊕	$LnMnO_3$-3, HNO_3	1.06	23	97.7	95	65.7	76.9	37.2	15.5
23.⊕	$LnMnO_3$-3, CH_3COOH, AlSi	0.54	24	97.6	94.8	66.6	68.2	24.5	5.9
24.⊕	$LnFeO_3$, HNO_3	1.05	22	98	95	–	83.7	71.7	–
25.Δ	$LnMnO_3$-4, CH_3COOH, AlSi	0.81	20	97.9	95	66	79	68.6	47
26.⊕	$LnMnO_3$-5, HNO_3	1.24	23	97.9	95	66	84.4	78.4	57.1
27.Δ	$LnMnO_3$-6, HNO_3	0.88	19	97.8	94.7	65.9	74.2	49.9	27.2
28.⊕	$LnMnO_3$-7, HNO_3	1.02	21	97.8	94.7	66.4	85.7	84.7	65-63
29.	Empty reactor, $V = 15$ L/h			2.3	5.4	35	0	0	0.14
30.	Empty reactor, $V = 5$ L/h			19	22	55	2.2	2.56	1.0

The "Catalyst" column shows the active component, content of a binder, electrolyte and the presence of aluminosilicate fiber. This is indicated in the catalyst composition perovskite, which contains $LnMnO_3$ as the active component, was prepared from different raw materials (1–7).

The geometry of monoliths is designated by the following symbols:

□ – Monoliths with square channels, the channel side length $a = 4.35$ mm, and the channel wall thickness $\delta = 0.8$ mm. One channel was tested.

+ – Monoliths with square channels, the channel side length $a = 5$ mm, and the channel wall thickness $\delta = 2$ mm. A "cross" was tested.

Δ – Monoliths with triangular channels, the channel side length $a = 4.5$ mm, and the channel wall thickness $\delta = 1.2$ mm.

⊕ – Spherical micromonoliths with the side 1.8 mm and wall thickness $\delta = 0.2$ mm.

The testing was performed in a flow reactor. The internal diameter of a quartz reactor was 8 mm, composition of the mixture: 1% NH_3 + 3% O_2 + He the balance, the mixture feed rate 15 and 5 L/h (specially indicated). Temperature was measured near the reactor wall close to the sample using a K-type thermocouple. The reaction products were analyzed chromatographically on a column with NaX molecular sieves and using a KM 9006 Quintox (Kane-May, UK) portable gas analyzer of tail gases.

* Products of ammonia oxidation at elevated temperatures (800–950 °C) are nitrogen oxides ($NO_x = NO + NO_2$) and elementary nitrogen, which can form also via the reduction of nitrogen oxides by ammonia. Individual and total concentrations of nitrogen oxides were measured with an accuracy of ±5% using a KM 9006 portable gas analyzer of tail gases. Nitrogen content was determined chromatographically. The conversion of ammonia to nitrogen in

an empty reactor was estimated in special experiments at three specified temperatures (800, 850 and 950 °C). Under the indicated conditions, NH_3 conversion to NO_x was 1–5 ppm.

The conversion of ammonia (α NH_3, %) on the catalyst was calculated by the formula:

$\alpha\ NH_3 = [(B - C)/A]*100$,

where A is the initial concentration of ammonia, ppm; C is the conversion of ammonia to nitrogen in an empty reactor (ppm NH_3); and B is the sum of concentrations $[NO] + [NO_2] + 2\,[N_2]$, ppm.

The selectivity of ammonia conversion to nitrogen oxides (S_{NOx}, %) was calculated by the formula:

$S_{NOx} = \{([NO] + [NO_2])/B - C\}*100$.

Formulas for calculating the activity and selectivity were chosen taking into account the testing conditions.

Investigation of the dependence of conversion on the length of catalyst fragment (samples 1, 2, 3) showed that elongation of the micromonolith fragment by more than 20 mm does not increase anymore the ammonia conversion. So, this length was used in further studies. Therefore, the total conversion of ammonia in the reactor with the catalyst was 100% at all the tested temperatures. The maximum conversion of ammonia was chosen to exclude the reduction of the produced nitrogen oxides by ammonia. In addition, it was found that decomposition of nitrogen oxides does not occur on sample 5 at these temperatures in the presence of oxygen and water vapor.

4. An attempt to improve the high-temperature selectivity of the iron oxide catalyst by introducing a small (10%) amount of $LnMnO_3$ (samples 7, 32, 33) has failed, probably due to a low quality of the manganese oxide raw material.

The $LnMnO_3$ catalysts possess the highest selectivity in a broader temperature range (the iron oxide catalyst has a lower selectivity in the low-temperature region); however, it is not so important for their application in the two-step systems. So, after successful pilot-production testing of the two-step system with the lanthanum manganite monolithic catalyst in Berezniki and Cherepovets, further operation was performed using the iron oxide monolithic catalyst. This is why regularities in the synthesis of bulk monolithic catalysts by extrusion molding are considered below for the iron oxide catalysts.

9.1 Development of the monolithic honeycomb iron oxide catalyst IK-42-1: effect of a raw material

The detailed studies of the effect of raw material on the properties of iron oxide catalyst demonstrated higher stability and strength of the catalysts synthesized from iron oxide that was obtained by the sulfate ($Fe_2O_3(SO_4)$) technology (precipitation from sulfates) as compared to those obtained by the chloride ($Fe_2O_3(Cl)$) technology (decomposition of iron chloride) [52]. Pseudoboehmite obtained by reprecipitation (HAP) and pseudoboehmite-containing hydroxide prepared by thermal activation (without reprecipitation, HATA) were used as binders. Table 9.3 and Fig. 9.5 present data for the catalysts synthesized as grains 4 mm in diameter.

At 100% conversion of ammonia observed for all the samples under consideration, the yield of nitrogen oxide (II) depended on the conditions of catalyst synthesis. Thus, raising the calcination temperature of the catalysts based on “chloride” α-Fe_2O_3 (900, 950, 1,000 °C) decreased the NO yield, whereas for the catalysts with “sulfate” α-Fe_2O_3 the yield of NO went over a maximum (for samples calcined at 950 °C). Irrespective of the binder type (HAP or HATA), the NO yield on the catalysts calcined at 900 °C was somewhat higher for the catalysts with “chloride” α-Fe_2O_3, while at a similar iron oxide but different binders – for the catalysts with the HATA binder. More stable at higher calcination temperatures (950 and 1,000 °C) were the catalysts containing “sulfate” iron oxide, and a higher NO yield was provided by the use of HATA as a binder.

A more pronounced effect was exerted by the raw material on the strength and pore structure of grains (Fig.9.5b). With the same binder, strength of the grain made of “sulfate” α-Fe_2O_3 strongly exceeded that of the grains based on “chloride” α-Fe_2O_3. Strength of the grains with HAP was higher as compared to those with HATA for each oxide, but in the case of “sulfate” iron oxide the difference was much smaller. An increase in calcination temperature virtually did not change

Table 9.3: Physicochemical characteristics of the catalyst samples in dependence on the raw material features and calcination temperature [52].

Iron oxide	Binder	NO yield*** (%)			Pore volume (cm^3/g)			Grain shrinkage** (%)		
	T_{calc} (°C)	900	950	1,000	900	950	1,000	900	950	1,000
Fe_2O_3 (Cl)	HAP	91.8	85.1	77.2	0.226	0.222	0.225	5	6	5
	HATA	92.1	84.3	55.5	0.261	0.256	0.243	0.3	2.5	3.8
Fe_2O_3 (SO_4)	HAP	86.0	89.9	82.1	0.160 0.157*	0.145 0.150*	0.112 0.135*	5.67	8.17	8.33
	HATA	91.1	93.3	86.6	0.256	0.161	0.153	0	1	3.17
	Bayerite	94.0	88.6	82.3	0.147	0.143	0.12	5	5.5	7
	Amorphous hydroxide	89.5	91.5	80.9	0.190*	0.175 0.165*	0.118	5	5.67	8

*Data were obtained after catalytic testing.
**Shrinkage is a decrease in linear dimensions and volume of grains due to a loss of moisture and phase transformations during thermal treatment.
***Catalytic data (NO yield) were obtained for 2–3 mm grains.

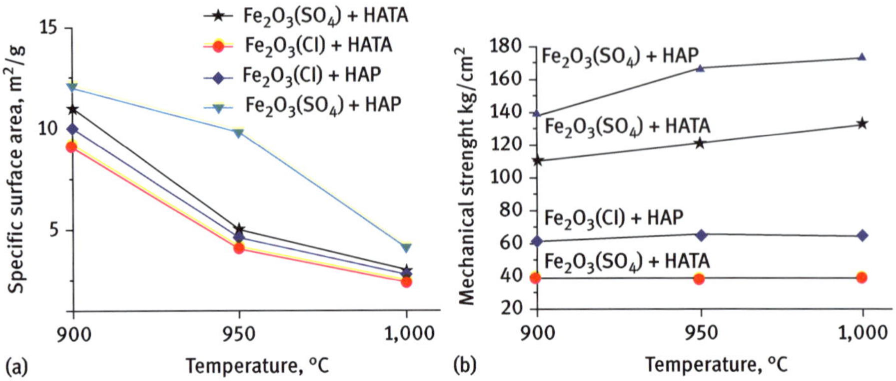

Fig. 9.5: Effect of the synthesis conditions (raw material and calcination temperature) on the structural-mechanical properties of grains: (a) specific surface area and (b) mechanical strength [52].

strength of the catalyst grains with "chloride" iron oxide, but increased strength of the "sulfate" iron oxide grains.

$Fe_2O_3(Cl)$ iron oxide grains have a greater total pore volume and a substantially smaller volume of large pores as compared to the grains of $Fe_2O_3(SO_4)$ iron oxide. As calcination temperature was raised, specific surface area of the catalysts decreased; therefore, pore volume patterns of the catalysts with "chloride" iron oxide changed only slightly (Fig. 9.6), whereas the catalysts with "sulfate" iron oxide showed a decrease in the volume of large (above 0.1 μm) pores (Fig. 9.7).

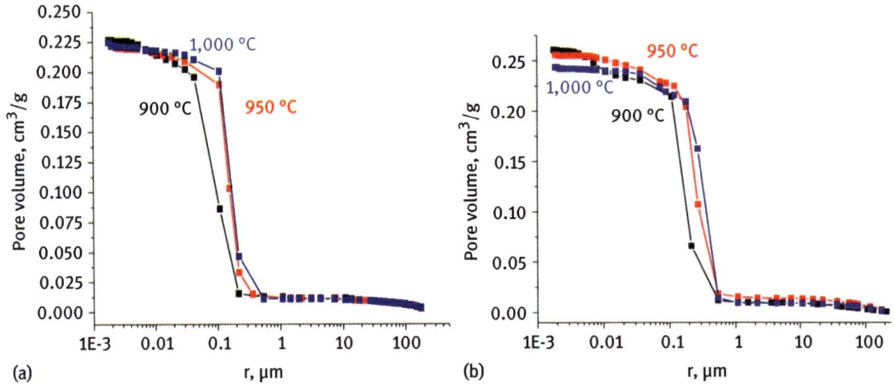

Fig. 9.6: Pore volume patterns for the $Fe_2O_3(Cl)$ catalysts synthesized with HAP (a) or HATA (b) and calcined at 900–1,000 °C [52].

Analysis of the data obtained revealed that changes in the yield of nitrogen oxide with raising the calcination temperature do not correlate with changes in the

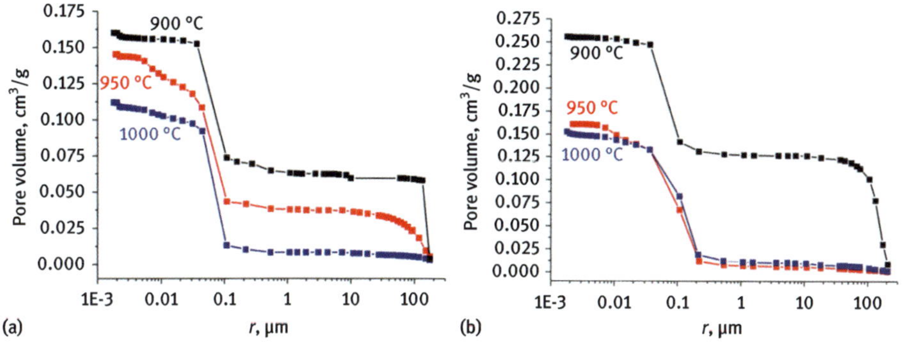

Fig. 9.7: Pore volume patterns for the $Fe_2O_3(SO_4)$ catalysts synthesized with HAP (a) or HATA (b) and calcined at 900–1,000 °C [52].

specific surface area (Fig. 9.5a) or pore structure of the samples (Fig. 9.6), because after calcination at 900 °C the catalysts synthesized with HAP and those with HATA have virtually equal specific surface areas of ca. 10 m^2/g (Fig. 9.5a), but differ in the total pore volume and the fraction of large pores (Figs. 9.6 and 9.7). The data obtained testify to differences in the chemical interaction between iron oxides and binders, which determine the quality of contact and the catalytic activity.

Thus, at a greater volume of large pores (and hence a smaller number of contacts in a grain), strength of the $Fe_2O_3(SO_4)$ iron oxide grains calcined at 900 °C exceeds that of the $Fe_2O_3(Cl)$ iron oxide grains. This testifies to a higher strength of a unit contact in the $Fe_2O_3(SO_4)$ grains. It can be noted also that strength of a unit contact in the grains with the HAP binder is higher as compared to those with HATA because at virtually similar pore volume patterns of the catalysts, the grain strength is higher in the case of HAP. This difference in the chemical interaction may be caused both by the anionic modification of oxides (residual chlorine or sulfate anions are present in the iron oxides obtained by chloride and sulfate technologies, respectively) and by the textural characteristics of iron oxide particles, for example, by the presence of micropores.

However, based on transmission electron microscopy (TEM) data (Fig. 9.8), samples of “chloride” ($Fe_2O_3(Cl)$) and “sulfate” ($Fe_2O_3(SO_4)$) hematites are identical: primary crystals of both samples have close sizes of ~100 nm.

X-ray diffraction data confirm the conclusion on a more strong interaction of “sulfate” iron oxide with the binder. All the catalysts comprise iron oxide and alumina phases, and an increase in calcination temperature produces the appropriate phase transformations of alumina (the formation of high-temperature phases). Nevertheless, shifts of the peaks corresponding to iron oxide are more pronounced on the diffraction patterns (Fig. 9.9) of catalysts based on “sulfate” iron oxide, which indicates the formation of aluminum solid solution in iron oxide [53]. The calculated lattice parameters of iron oxides in dependence on calcination temperature are listed in Table 9.4.

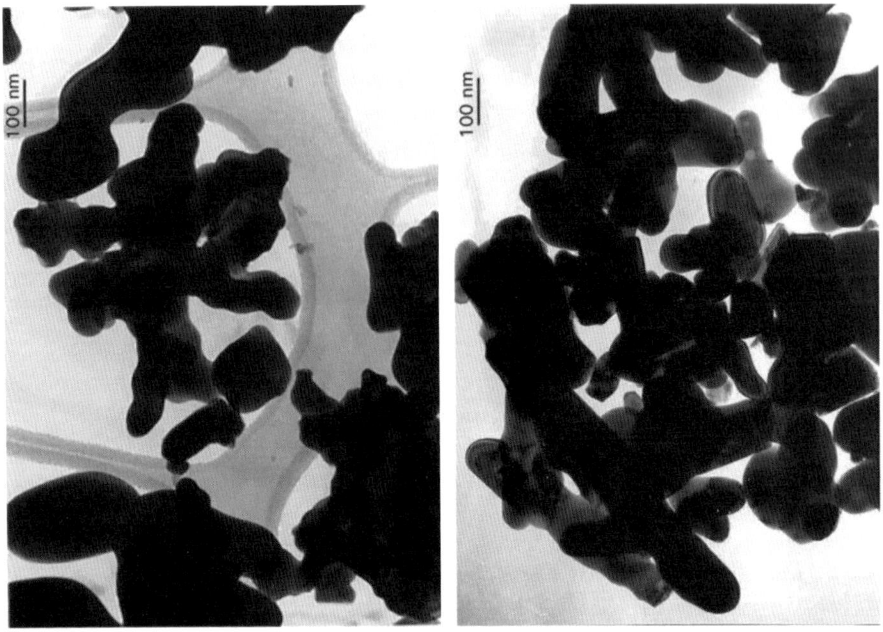

(a) (b)

Fig. 9.8: TEM data for different Fe_2O_3: (a) Fe_2O_3(Cl) and (b) $Fe_2O_3(SO_4)$.

The coherent scattering region of iron oxide in each sample is 100 nm and more. The data obtained indicate the anionic modification of iron oxides because their lattice parameters are somewhat increased. Thus, the neat hematite phase calcined at 800 °C has the lattice parameters 5.035(21) and 13.75292(1), which change only slightly with raising the calcination temperature to 1,100 °C [53, 54].

IR spectroscopy data are consistent with the data of X-ray diffraction analysis (Fig. 9.10). The initial samples can be identified as a metastable (anionic) modification of iron oxide (III) – protohematite, which is described in [54]. Spectra of the catalysts differ from the spectra of iron oxides by the high-frequency shift of all absorption bands corresponding to iron oxide. Positions of the main absorption bands at 330–345, 386, 483 and 561 cm^{-1} for the catalysts make it possible to identify iron oxide in the catalyst as hematite, which means that upon thermal treatment of the catalysts the initial protohematite transforms into hematite [52]. It seems interesting that in the spectrum of the catalyst made of α-Fe_2O_3(Cl) and HAP (Fig. 9.10, spectrum 5) iron oxide still retains the protohematite structure and remains anion modified.

The data obtained make it possible to relate stability of the catalysts to the formation of aluminum solid solution in iron oxide; however, since a relation with activity is not observed, it is necessary to examine surface properties of the samples.

When NO is adsorbed on iron oxides, the absorption bands at 1,885 and 1,832 cm^{-1} (a more intense band) as well as the band at 1,545 cm^{-1} are observed

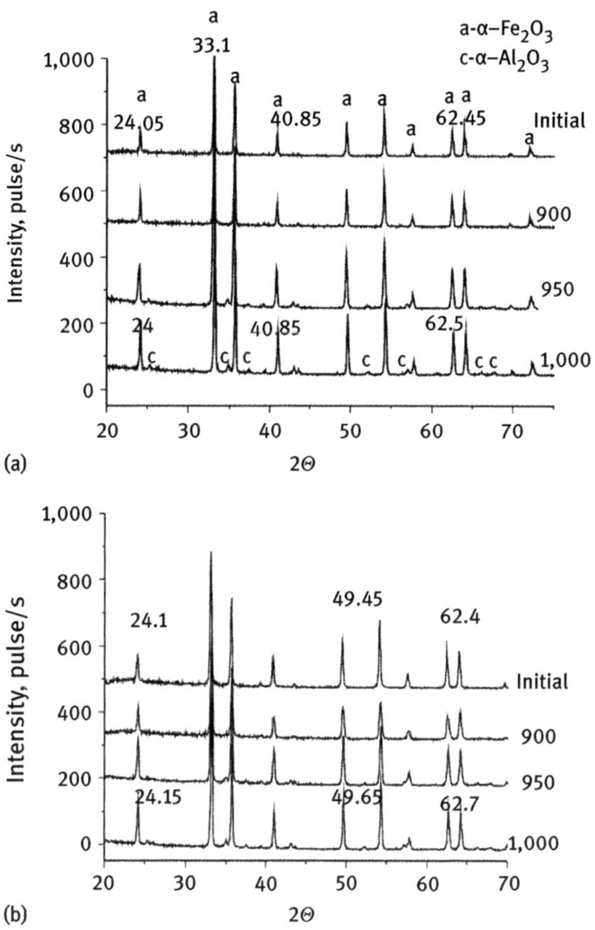

Fig. 9.9: Diffraction patterns of the initial oxides and catalysts based on $Fe_2O_3(Cl)$ iron oxide (a) and $Fe_2O_3(SO_4)$ (b), which were synthesized with HATA after calcination at 900–1,000 °C (indicated in the figure) [52].

(Fig. 9.11) [55]. The band at 1,885 cm^{-1} is assigned to NO adsorbed on coordinatively unsaturated cations $(O_5)Fe^{2+}$–NO, whereas the band at 1,832 cm^{-1} may be attributed to $(O_{5-x})Fe^{2+}$–NO complexes in the state close to tetracoordinated one. The band at 1,545 cm^{-1} corresponds to the $(O)_5Fe^{2+}$–NO_2 complex [56–58].

Low intensity of this band may indicate that the surface reduction of NO proceeds to a small extent and the reduced sites of NO adsorption, which are characterized by bands at 1,885 and 1,832 cm^{-1}, are formed in the course of treatment and may reflect the defectness of this oxide.

The bands of NO adsorbed on the "chloride" iron oxide are much (by an order of magnitude) less intensive than in the case of "sulfate" oxide at close specific

Table 9.4: Lattice parameters of iron oxide* obtained by chloride and sulfate technologies in dependence on calcination temperature [52].

Binder	T_{calc} (°C)	Lattice parameter (Å)			
		α-Fe_2O_3(Cl)		α-Fe_2O_3(SO_4)	
		a (Å)	*c* (Å)	*a* (Å)	*c* (Å)
No	–	5.040(1)	13.76(1)	5.042(1)	13.76(1)
HATA	900	5.040(1)	13.76(1)	5.033(1)	13.74(1)
HATA	950	5.036(1)	13.76(1)	5.025(1)	13.72(1)
HATA	1,000	5.022(1)	13.71(1)	5.020(1)	13.70(1)

*The neat hematite phase calcined at 800 °C has parameters $a = 5.035(21)$ and $c = 13.75292(1)$.

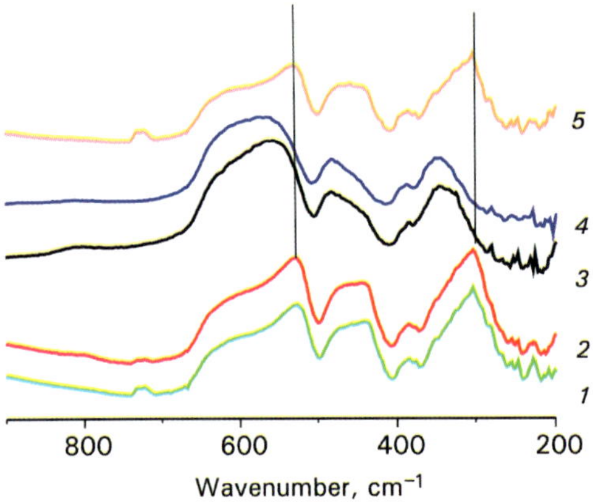

Fig. 9.10: IR spectra of the initial iron oxides (1, 2) and catalysts (3–5) after calcination at 900 °C: spectra 1, 3, 5 – based on α-Fe_2O_3(Cl) iron oxide (1), spectra 2, 4 – based on α-Fe_2O_3(SO_4); spectra 3, 4 – with HATA as a binder, spectrum 5 – with HAP as a binder [52].

surface areas of the samples. Absorption bands at 1,842 and 1,885 cm^{-1} and also at 1,545 cm^{-1} are observed. Low intensity of the bands at 1,842 and 1,885 cm^{-1} testifies to a much lower (as compared to the sulfate sample) content of reduced coordinatively unsaturated sites on the surface of this oxide and to a greater coordinative saturation of adsorption sites on the surface. This may be caused by a lower reducibility of the oxide upon treatment and a smaller reactivity in the reaction with the NO probe molecule, probably due to stabilization of active sites by chlorine anions that are present in the oxide. Indeed, washing of the sample increases the intensity of the bands of mononitrosyl complexes of adsorbed NO (bands at 1,842 and 1,885 cm^{-1}) nearly to the level of "sulfate" sample.

Catalysts based on "chloride" iron oxide and synthesized with different binders (HAP or HATA) and electrolytes (nitric or acetic acids) do not have significant

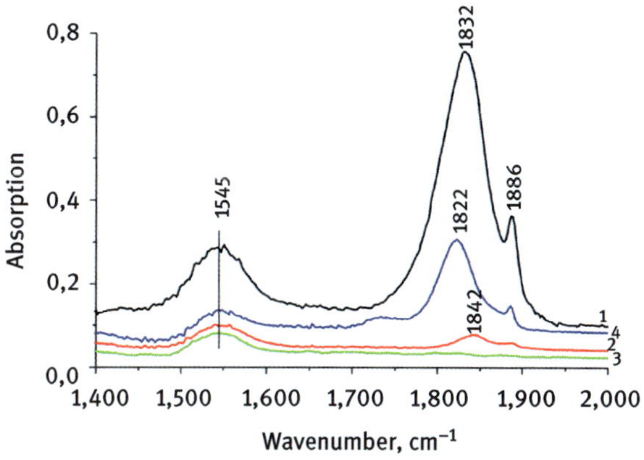

Fig. 9.11: Spectra of NO adsorbed on the samples: 1 – α-$Fe_2O_3(SO_4)$, 2 – α-Fe_2O_3(Cl), 3 – α-Fe_2O_3(Cl) after additional calcination at 900 °C, 4 – α-Fe_2O_3(Cl) after washing with water [55].

differences in the band positions in the spectra of adsorbed NO after calcination at 900 C. As in the case of the initial oxide, bands are observed in the regions of 1,540 and 1,885 cm^{-1}. However, instead of the 1,842 cm^{-1} band, there is a band with the maximum in the region of ~ 1,810 cm^{-1}. Its intensity is much higher as compared to the band at 1842 cm^{-1} observed for iron oxide (Figs. 9.11 and 9.12); therewith, no significant changes occur in intensity of the 1,540 cm^{-1} band.

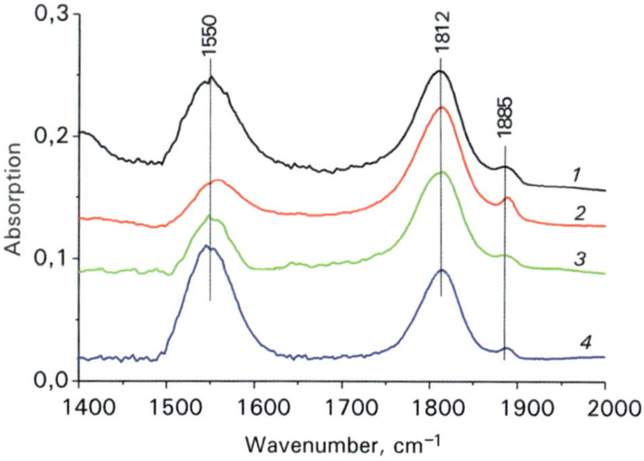

Fig. 9.12: Spectra of NO adsorbed on the samples of α-Fe_2O_3(Cl) catalyst synthesized using HATA (1 and 3) and HAP (2 and 4) as a binder and different electrolytes – acetic (1 and 2) and nitric (2 and 4) acids [55].

A shift of the band in the region of 1,840 cm^{-1} toward lower frequencies (1,810 cm^{-1}) may reflect the formation of even more coordinatively unsaturated Fe^{2+} sites (up to the tetracoordinated state or clusters of reduced Fe^{2+} cations) on the catalyst surface, in distinction to the iron oxide surface. The appearance of such sites may be caused either by treatment or by changes in defectness. Most likely, such low-coordinated iron cations are formed on iron cations in the composition of solid solution in alumina. Similar bands were observed during NO adsorption onto iron oxide deposited on the alumina support [59].

Assignment of the observed sites to the aluminum solid solution in iron oxide and, accordingly, its supposedly higher reactivity in comparison with iron oxide are less probable owing to the increased strength of its Fe–O bond [55], which may be indicated by a decrease in the lattice parameter.

Thus, new coordinatively unsaturated adsorption sites appear on the surface of iron-aluminum catalysts based on "chloride" iron oxide (as compared to the initial oxide) probably due to formation of a highly dispersed solid solution of iron cations in alumina. Either the band of coordinatively unsaturated iron cations on the surface of iron oxide phase (1,840 cm^{-1}) is not resolved against the higher intensity of the indicated band, or such sites, associated with termination of defects on the surface, are decorated by highly dispersed alumina particles, which is quite probable [60]. This may indicate that the oxidation of ammonia on the surface of iron-aluminum catalyst involves not only the coordinatively unsaturated sites of the hematite phase but also the Fe^{2+} sites in a nearly tetracoordinated state (the band at 1,810 cm^{-1}), which are formed on iron cations in the composition of the solid solution in alumina.

It seems interesting that positions and intensities of the bands of adsorbed NO for the catalysts synthesized using different iron oxides and binders are quite close, which makes it possible to correlate these data also with the poorly differentiated data on activity of the catalysts. However, it should be noted that the addition of binders to "sulfate" iron oxide, in distinction to "chloride" iron oxide, decreases the band intensity of nitrate and nitrosyl complexes of adsorbed NO. This corroborates the earlier assumption that alumina can block a part of active sites on the iron oxide surface, and additional coordinatively unsaturated sites can form on iron cations in the composition of the solid solution in alumina on the catalyst surface (the band at 1,810 cm^{-1}). Such a solution can form due to partial dissolution of iron oxide during the preparation of the catalyst paste. It seems interesting also that the band intensity of coordinatively unsaturated sites in the case of catalysts based on "chloride" iron (calcination at 900 °C) is somewhat higher as compared to the catalysts with "sulfate" iron, which may reflect a higher selectivity of this catalyst (catalysts 331 and 287, Table 9.5).

Figure 9.13 displays the spectra of NO adsorbed on iron oxide catalysts in dependence on calcination temperature. An increase in calcination temperature, which decreases the yield of nitrogen oxide (II), causes also a decrease in the band intensity of adsorbed NO and nitrite-nitrates, which, similar to the case of iron oxide, reflects a decrease in the number of catalytically active adsorption sites on the catalyst surface due to sintering and annealing of defects.

Table 9.5: Integrated intensity of the bands of nitrite–nitrate and nitrosyl complexes formed on the surface of tested catalysts upon adsorption of NO probe molecule in dependence on calcination temperature of the catalysts and features of the raw material (according to diffuse reflectance spectroscopy data) [55].

No.	Synthesis conditions		Characteristics of catalyst 80% Fe_2O_3 + 15% Al_2O_3 + 5% $(Al,Si)O_2$				
	Raw material	T_{calc} (°C)	S_{sp} (m^2/g)	NO* a.b. 1, 810–1,850 cm^{-1}	NO_x* a.b. 1,540 cm^{-1}	NO*/ NO_x*	NO yield, % from Table 9.3
1	α-Fe_2O_3(Cl) + HAP	900	11	0.14	0.89	0.16	91.8
		950	5	0.075	0.54	0.14	85.1
		1,000	3.4	0.070	0.40	0.26	77.2
2	α-Fe_2O_3(Cl) +HATA	900	9.1	2.50	0.88	2.80	92.1
		950	4.1	0.16	0.36	0.40	84.3
		1,000	2.0	0.024		0.24	55.5
3	α-$Fe_2O_3(SO_4)$ +HAP	900	10.0	0.74	0.54	1.4	86.0
		950	4.6	0.32	0.45	0.7	89.9
		1,000	2.8	0.12	0.30	0.4	82.1
4	α-$Fe_2O_3(SO_4)$ +HATA	900	12.0	1.70	1.00	1.7	91.1
		950	4.8	0.17	0.30	0.7	93.3
		1,000	4.1	0.08	0.40	0.2	86.6

*Integrated intensity of absorption bands (proportional concentrations of adsorbed complexes).

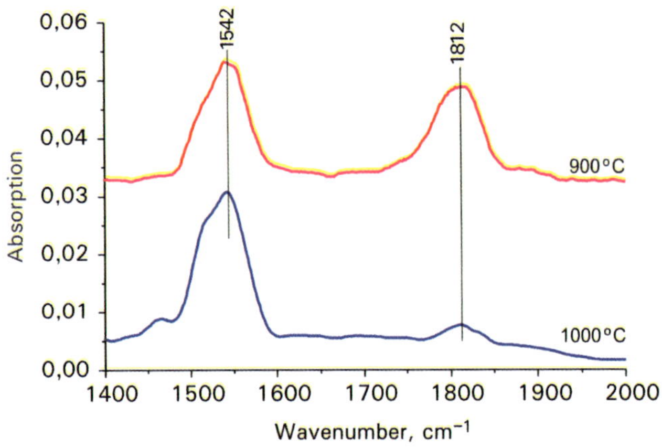

Fig. 9.13: The spectrum of NO adsorbed on the catalyst 80% α-$Fe_2O_3(SO_4)$ + 15% Al_2O_3 (from amorphous hydroxide) + 5%$(Al,Si)O_2$ calcined at different temperatures [55].

In addition, it cannot be ruled out that the formation of aluminum solid solution in iron oxide, which is observed when calcination temperature is raised, can also deteriorate the catalyst surface reactivity. Overall, it can be stated that the catalyst with a higher concentration of coordinatively unsaturated adsorption sites or a higher reactivity in the reaction with NO (i.e. a higher concentration of nitrate complexes) will provide a higher selectivity in ammonia oxidation (Table 9.5).

Figure 9.14 shows a correlation between selectivity (NO yield) and intensity of the band at 1,540 cm^{-1} assigned to surface nitrate fragments.

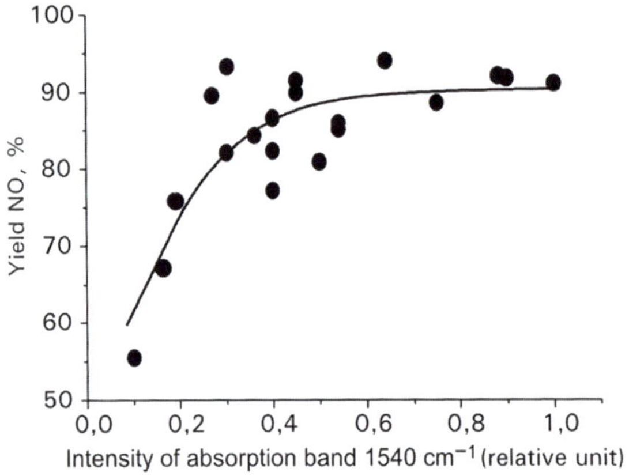

Fig. 9.14: NO yield during ammonia oxidation versus intensity of the 1,540 cm^{-1} band [55].

It is seen that the growth in intensity of the indicated band is accompanied by an increase in selectivity for all the samples.

The revealed correlation between the yield of nitrogen oxide (II) during ammonia oxidation and the amount of nitrate complexes formed upon NO adsorption looks as a curve with saturation, which may reflect the occurrence of the reaction in diffusion mode at high conversions for the majority of catalysts.

The data obtained demonstrate that the process selectivity is determined mostly by the presence of coordinatively unsaturated adsorption sites on the catalyst surface, which are characterized by the formation of surface nitrate complexes (with the band at 1,540 cm^{-1}). More detailed studies are required to reveal the quantitative dependences between selectivity and ratio of different coordinatively unsaturated sites of Fe^{2+} and Fe^{3+} cations that adsorb NO and NO_2, respectively; particularly, it is necessary to measure the activity in kinetic region and the number of surface complexes by quantitative IR spectroscopy (transmission spectroscopy). Overall, our data are consistent with the data reported by other researchers, who studied the surface properties of iron oxide by IR spectroscopy of adsorbed NO and the mechanism of ammonia oxidation on oxide catalysts by diffuse reflectance spectroscopy and showed that ammonia oxidation products in the gas phase are formed via the interaction with ammonia (in the presence or absence of oxygen) on the surface of oxide catalyst leading to the formation of ammonia oxidation products, particularly the surface complexes of NO, N_2O and NO_3 [7, 55, 59–62].

9.2 Structural-mechanical properties of plastic masses for preparation of monoliths

In extrusion molding, the pastes should have certain structural-mechanical properties (viscosity, plasticity, elasticity) providing the formation of defect-free grains. Rebinder and Tolstoy plastometers [63–65] were used in our studies on the structural-mechanical properties of the plastic pastes obtained by mixing of iron oxide with a binder. It was found that high-quality monoliths (with cross section (75 × 75 ± 0.5) mm, height (50 ± 0.5) mm, wall thickness (2 ± 0.1) mm and channel side length 4.5 mm) are formed from the pastes belonging to the first structural-mechanical type (SMT) according to classification reported in [66–69], although differences in the raw materials exert a noticeable effect on the synthesis conditions and rheological characteristics of the pastes.

Thus, the optimal moisture content (W) is lower and plastic strength (P_m) is higher for the pastes made of “sulfate” iron oxide (Table 9.6). It cannot be excluded that the observed higher plastic strength of the pastes based on α-$Fe_2O_3(SO_4)$ (as compared to α-Fe_2O_3(Cl) ones) may be caused by a lower moisture content of these pastes, which can increase the number of coagulation contacts. Elongation and elasticity moduli E_1 and E_2 (Table 9.7) are also higher for the pastes with “sulfate” iron oxide and a HAP binder. Storage of pastes in hermetically sealed containers additionally increases these characteristics [70].

Table 9.6: Plastic strength of pastes (fresh and stored for a day) measured at the optimal moisture content in dependence on the employed raw material [69].

Paste no.	Initial reagents for preparation of catalyst paste	Time until molding (h)	Moisture content, W (%)	Plastic strength, $P_m \cdot 10^{-4}$ (Pa)
1	α-Fe_2O_3(Cl) + HATA	0	15.72	10.79
2	α-Fe_2O_3(Cl)+ HATA	24	15.24	12.24
3	α-Fe_2O_3(Cl) + HAP	0	16.53	23.11
4	α-Fe_2O_3(Cl) + HAP	24	16.27	25.38
5	α-Fe_2O_3(SO_4) + HATA	0	12.35	14.72
6	α-Fe_2O_3(SO_4) + HATA	24	12.07	16.14
7	α-Fe_2O_3(SO_4) + HAP	0	13.25	27.62
8	α-Fe_2O_3(SO_4)+ HAP	24	13.06	46.43

On the contrary, the fraction of plastic deformations ($\acute{\varepsilon}_{pl}$) is higher in the pastes based on "chloride" iron oxide (Table 9.7). Although this fraction increases in storage for the pastes based on "sulfate" oxide, it remains smaller than the values obtained for "chloride" iron oxide. The analysis shows that high-quality monoliths are obtained from the pastes belonging to the first SMT (Fig. 9.15b). Changes in the SMT of the paste, for example, due to changes in the moisture content or storage of the paste in sealed containers, deteriorate the quality of the produced monoliths. For example, at insufficient moisture content, the molded mass corresponds to the zero SMT, which is characterized by a decrease in the fraction of plastic deformations ($\acute{\varepsilon}_{pl}$) (the paste is dry and sets quickly). This leads to the "dragon's teeth" molding defects (Fig. 9.15a). If the optimal moisture content is exceeded by 1% (Fig. 9.15c), the mass corresponds to the 5th SMT, which is characterized, on the opposite, by the development of plastic deformations ($\acute{\varepsilon}_{pl}$), and the resulting monolith is distorted under its own weight. It should be noted that the pastes used for molding of simple grains can belong to SMT 1-5.

Noteworthy is a good correlation between data on plastic strength of the pastes and grain strength (Table 9.7), namely, higher grain strength values were obtained for the pastes with a higher plastic strength probably because these characteristics depend on the number of contacts and the strength of unit contact in a grain. Thus, both hematites can be used to prepare a monolithic catalyst by extrusion molding, although the "sulfate" hematite is more preferable; only HATA can be used as a binder, and pastes should be preliminarily stored in hermetically sealed containers.

Table 9.7: Structural-mechanical properties of plastic pastes for IK-42-1 catalyst [70].

no.	Paste (time until molding, h)	P_k (kPa)	$\acute{\eta}_1$,·10^8 (Pa·s)	Пc·10^5 (s^{-1})	λ, E_1/E_1+E_2	E·10^4, Pa $E_1 * E_2/E_1+E_2$	Θ, s, $\acute{\eta}/E$	E_1,10^4 (N/m^2)	E_2,·10^4 (N/m^2)	Grain strength (kg/cm^2)	$\acute{\varepsilon}_0$, a.u. (%)	$\acute{\varepsilon}_2$, a.u. (%)	$\acute{\varepsilon}_{pl}$, a.u. (%)	Structural-mechanical type
1	α-Fe_2O_3(Cl) + HATA (0)	4.56	0.585	7.79	0.53	4.59	1,275	9.76	8.66	38.6	20.28 (26.17)	23.10 (29.81)	34.11 (44.02)	5
2	α-Fe_2O_3(Cl) + HATA (24)	4.33	1.18	3.67	0.47	3.60	3,278	6.76	7.73		29.58 (40.88)	25.87 (35.76)	16.90 (23.36)	1
3	α-Fe_2O_3(Cl) + HAP (0)	2.17	7.17	0.30	0.65	23.65	3,032	67.52	36.40	61.0	2.96 (26.33)	5.49 (48.84)	2.79 (24.83)	1
4	α-Fe_2O_3(Cl) + HAP (24)	4.80	12.41	0.39	0.65	31.84	3,898	91.66	48.79		2.18 (27.95)	4.01 (51.41)	1.61 (20.64)	1
5	α-Fe_2O_3(SO_4) + HATA (0)	4.80	3.47	1.38	0.57	4.48	7,746	10.39	7.89	110.0	19.24 (38.23)	25.35 (50.37)	5.74 (11.40)	1
6	α-Fe_2O_3(SO_4) + HATA (24)	1.80	7.90	0.23	0.67	9.74	8,111	29.80	14.47		6.71 (29.15)	13.82 (60.00)	2.50 (10.85)	1
7	α-Fe_2O_3(SO_4) + HAP (0)	4.33	4.49	0.96	0.71	38.60	1,163	133.33	54.33	138.8	1.50 (15.60)	3.68 (38.25)	4.44 (46.15)	5
8	α-Fe_2O_3(SO_4) + HAP (24)	5.20	9.05	0.57	0.72	37.80	2,394	135.70	52.43		1.47 (19.65)	3.81 (50.94)	2.20 (29.41)	2

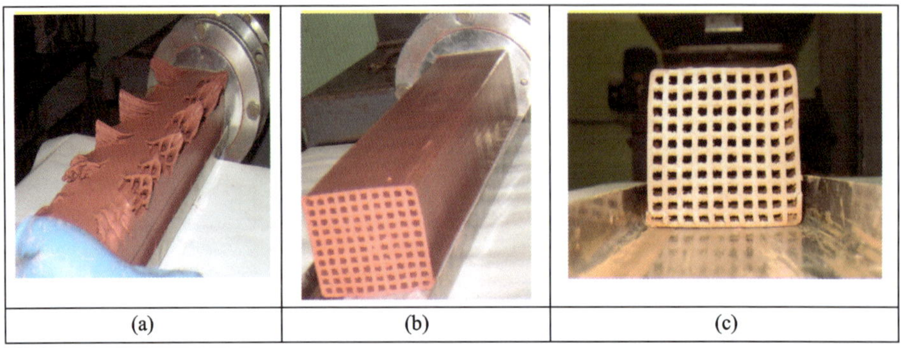

Fig. 9.15: Effect of moisture content on the quality of molded monoliths for the pastes based on α-$Fe_2O_3(SO_4)$ + HATA. (a) −1% from the optimal moisture content, (b) the optimal moisture content and (c) +1% to the optimal moisture content [70].

9.3 Thermal treatment of monolithic catalyst IK-42-1

Thermal treatment is the final step of the catalyst synthesis technology; in this step, main characteristics of catalysts are formed: strength, pore structure, phase composition, activity and others. Therefore, an essential effect on the catalyst strength is exerted not only by the final calcination temperature, but also by the chosen mode of temperature elevation, and sometimes by the mode of cooling [71–73]. Upon thermal treatment of honeycomb monoliths, which have more complicated shape and greater dimensions than granulated catalysts, the role of temperature mode becomes particularly important because the phase and polymorphous transformations or sintering during thermal treatment can disturb the integrity of the monoliths (formation of cracks) and, accordingly, lead to the formation of low-quality monoliths (rejects).

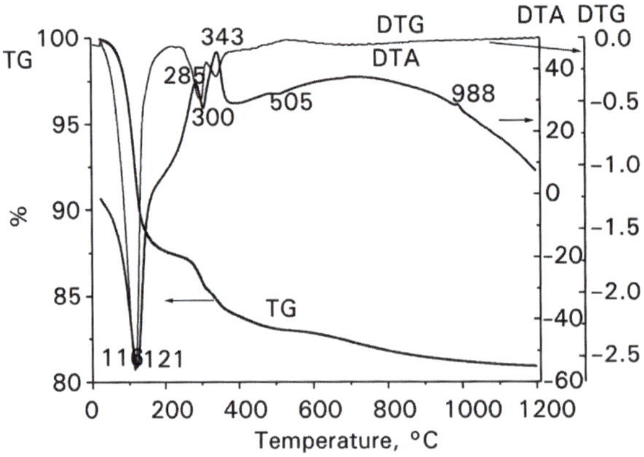

Fig. 9.16: Thermal analysis of the paste based on α-$Fe_2O_3(Cl)$ + HATA [73].

To find and optimize the conditions of thermal treatment of monoliths, thermal analysis was performed, which showed that heating of all the pastes produces virtually similar thermal effects (Figs. 9.16 and 9.17):

Endothermic effect in the range of 90–121 °C;
Exothermic peak in the range from 230 to 250 °C;
Endothermic effect at 290–315 °C;
Exothermic peak in the range of 330–340 °C;
Minor endothermic effect in the range of 500–510 °C; and
Exothermic peak at 986–988 °C.

The main weight loss (~17 wt.%) occurs upon heating to 600 °C, and the weight loss upon heating from 600 to 1,200 °C is ~ 2%.

The results obtained can be described taking into account the composition of pastes and the literature data and assuming the occurrence of the following processes:

- Removal of adsorbed water (90–121 °C);
- Thermal decomposition of organic additives (exothermic peaks in the range from 240 to 320 °C);
- Dehydration of trihydrates present in a binder: bayerite at ca. 290 °C, and gibbsite at 310–330 °C;
- Dehydration of pseudoboehmite and boehmite at 373–530 °C;
- Exothermic peak at 986–988 °C may correspond to processes in the mullite–silica reinforcing additive.

In compliance with quite close thermal analysis data obtained for different pastes (Figs. 9.16 and 9.17), thermal treatment of the monoliths was carried out in a stepwise manner with optimization of the heating rate in each temperature region.

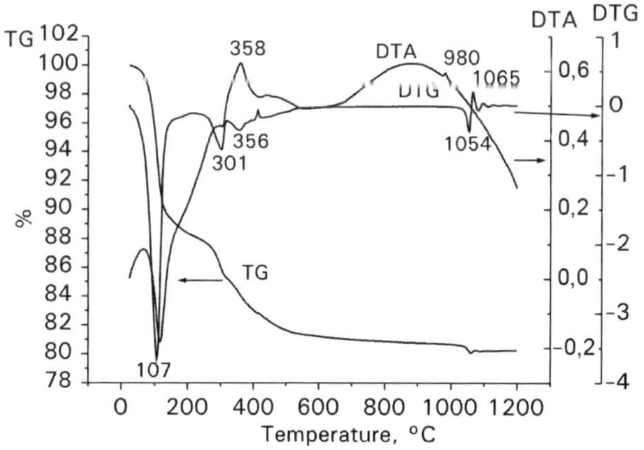

Fig. 9.17: Thermal analysis of the paste based on α-$Fe_2O_3(SO_4)$ + HATA [73].

Drying of monoliths at room temperature (25 °C);
Preliminary thermal treatment of monoliths at temperatures up to 400 °C; and
Calcination of monoliths at temperatures up to 900–950 C.

It was found that the prolonged drying and the stepwise thermal treatment in precalcination step are the key steps for the formation of high-quality monoliths with high stability in thermal cycles. All the monoliths calcined in the optimized stepwise mode, irrespective of the employed raw material, withstand at least 10 thermal cycles (rapid heating to 700 °C and cooling). Crushing strength essentially differs and is equal to 8,000 N/item (for the monoliths made of α-$Fe_2O_3(SO_4)$ iron oxide) and 4,000 N/item (for the monoliths made of α-Fe_2O_3(Cl) iron oxide) [73].

Results of the studies made it possible to propose a method for the synthesis of IK-42-1 bulk iron oxide catalyst for ammonia oxidation (Fig. 9.18) and prepare specifications TU 2175-018 035 33913-99 and temporary technological regulations.

Fig. 9.18: Outward appearance of IK-42-1 monoliths.

Full-scale endurance testing of the honeycomb catalyst was performed in 1996 at the AO AZOT in Berezniki. The catalytic reactor of the UKL-7 plant was loaded with 100 dm^3 of the oxide catalyst as the second step of the catalytic system (Fig. 9.19). The catalyst was placed in one layer over the entire cross-section of the reactor on a nichrome gauze covering the fire bars. A nichrome gauze was arranged on the catalyst, and a stack of catalyst gauzes (one old and nine new gauzes made of alloy no. 5) was placed over the nichrome gauze. The performance of the catalytic system is shown in Table 9.8. One can see that ammonia conversion during a normal service run (3,000 h) of the catalyst gauzes corresponds to the design value. The catalyst loaded in March 1996 has been operated for nearly two years [29].

Fig. 9.19: The UKL-7 reactor with the loaded IK-42-1 catalyst.

Table 9.8: Main characteristics of the catalytic system operation at a pressure of 7.3 atm [29].

Month	Conversion (av.) (%)	NH_3: air ratio (av.) (%)	Temperature (av.) (°C)	NH_3 loading (nm^3/h)	Nitric acid output (t µng HNO_3)	Productivity (t µng HNO_3/h)	Run (h)
March	94.10	9.36	866	6,000	8406	14.8	568
April	94.07	9.15	866	5,800	8114	13.5	601
May	93.64	9.15	868	5,700	9299	13.0	715
June	93.77	9.08	865	5,650	9819	14.4	684
July	92.95	9.13	862	5,550	9145	12.9	708

The pilot and industrial testing at OAO Acron, which was aimed to optimize the composition of the two-step system, demonstrated that one layer of the IK-42-1 catalyst can replace even three gauzes in the stack, probably due to the improved operation of the stack owing to equalization of the velocity field [74–75], and also due to the effects exerted by monoliths on mass transfer in the stack and by the stack on mass transfer in the monolith, according to the data reported in [76, 77].

The application of the honeycomb catalyst does not require the reconstruction of existing catalytic reactors or the use of special baskets to form a layer. Calculation of benefits from the use of the monolithic catalyst showed its high efficiency in the two-step ammonia oxidation system for the production of nitric acid in UKL-7 and AK-72 plants (Table 9.9). For more than 15 years, the catalyst has been produced in small batches meeting the industrial demand.

Table 9.9: Lowering the consumption and losses of platinoids in the two-step ammonia oxidation process [29].

Absolute pressure in reactor (kg s/cm^2 or bar)	A decrease in the initial mass of platinoid catalyst (%)	A decrease in irrecoverable losses of platinoids (%)
Atmospheric	50–66	30–35
Moderate (3–6)	35–45	20–30
Elevated (above 6)	20–25	10–20

References

[1] Pickwell P. Nitric acid plant optimization. Chem. Ind. 1981, 4, 114–117.
[2] Ostwald W. Berg- und Hüttenm. Rsch. 1906, 20, 12.
[3] Sperner F, Hohmann W. Rhodium-platinum gauzes for ammonia oxidation. A study by scanning electron microscopy. Platin. Met. Rev. 1976, 20, 12–20.
[4] Atroshchenko VI, Kargin SI. Nitric Acid Technology. Moscow, Khimiya, 1970 (In Russian).
[5] Taylor GB, Chilton TH, Handforth SL. Manufacture of nitric acid by the oxidation of ammonia. Ind. Eng. Chem. 1931, 23(8), 860–865.
[6] Farrauto RJ, Lee HC. Ammonia oxidation catalyst with enhanced activity. Ind. Eng. Chem. Res. 1990, 29(7), 1125–1129.
[7] Karavayev MM, Zasorin AP, Kleshchev NF, Catalytic oxidation of ammonia, Khimiya, Moscow, 1983.
[8] Sadykov VA, Isupova LA, Zolotarskii IA, Bobrova LN, Noskov AS, Parmon VN, Brushtein EA, Telyatnikova TV, Chernyshev VI, Lunin VV. Oxide Catalysts for Ammonia Oxidation in Nitric Acid Production: Properties and Perspectives. Appl. Catal. A: Gen. 2000, 204 (1), 59–87.
[9] Anderson DR. Catalytic etching of platinum alloy gauzes. J. Catal. 1988, 113 (2), 475–489.
[10] McCabe AR, Smith GDW, Pratt AS. The Mechanism of Reconstruction of Rhodium-Platinum Catalyst Gauzes. An explanation on the new surface morphology. Platin. Met. Rev. 1986, 30(2), 54–62.
[11] Schmidt LD, Luss D. Physical and chemical characterization of platinum-rhodium gauze catalysts. J. Catal. 1971, 22(2), 269–279.
[12] Zakharchenko NI. Catalytic properties of platinum oxide (II, III) in ammonia oxidation process. Zhurnal prikladnoi khimii 2001, 74 (10), 1636–1642 (In Russian).
[13] Salanov AN, Suprun EA, Serkova AN, Sidel'nikova ON, Sutormina EF, Isupova LA, Kalinkin AV, Parmon VN. Catalytic Etching of Platinoid Gauzes during the Oxidation of Ammonia by Air. Reconstruction of Surface of Platinoid Gauzes at 1133 K in Air, in Ammonia, and in an $NH_3 + O_2$ Reaction Medium Kinetics and Catalysis 2018, 59(1) 83–98.
[14] Salanov AN, Suprun EA, Serkova AN, Kochurova NM, Sidel'nikova ON, Sutormina EF, Isupova LA, Parmon VN. Catalytic etching of platinoid gauzes during the oxidation of ammonia by air. Reconstruction of the surface of a platinoid backside in the course of ammonia oxidation at 1133 K. Kinet. Catal. 2018, 6, 792–809.
[15] Morozov NM, Lukianova LI, Temkin MI. Ammonia oxidation on metal oxides. Kinetika I kataliz 1966, 7(1), 172–175.
[16] Kurin NM, Zakharov MS. Kataliz v vyshei shkole, In: Balandin AA ed. Moscow, MGU, 1962, 234–237.

[17] Olevskii VM. Production of nitric acid in high power single aggregates. Moscow, Khimiya, 1985.
[18] Handbook of nitric acid production. In: Mel'nikov EY. Ed. 2nd ed. Moscow, Khimiya, 1987.
[19] Kinetics of heterogeneous catalytic reactions under pressure. In: Atroshenko Vi. Ed. Har'kov, Visha shkola, 1974
[20] Catalysts in nitric industry. In: Atroshenko VI ed. Har'kov, Visha shkola, 1977.
[21] Podurovskaya ON. Catalysts for ammonia oxidation (bibliography). Moscow, NIITEKHIM, 1972.
[22] Petryk J, Kolakowska E. Cobalt oxide catalysts for ammonia oxidation activated with cerium and lanthanum. Appl. Catal. B: Environ. 2000, 24, 121–128.
[23] Zakharchenko NI. Catalytic properties of Fe2O3 – Y2O3 system in ammonia oxidation. Zhurnal fizicheskoy khimii 2002, 76(7), 1213–1220 (In russian).
[24] Zakharchenko NI. Catalytic properties of bismuth ferrite in ammonia oxidation. Zhurnal prikladnoy khimii 2000, 73(12), 1960–1964.
[25] Zakharchenko NI. Catalytic properties of Fe_2O_3 – Bi_2O_3 system in ammonia oxidation to NOx. Kinetika i kataliz 2002, 43(1), 104–104.
[26] Zakharchenko NI. Catalytic properties of Fe_2O_3 – Ga_2O_3 system in ammonia oxidation. Zhurnal prikladnoy khimii 2003, 76(3), 414–420.
[27] Zakharchenko NI. Phase composition and properties of Fe_2O_3 – Sc_2O_3 catalytic system in high temperature ammonia oxidation. Khimiya I khimicheskaya tehnologiya 2004, 47(7), 129–136.
[28] Zakharchenko NI. Catalytic properties of Fe_2O_3 – CuO in ammonia oxidation. Zhurnal fizicheskoy khimii 2001, 75(6), 985–990.
[29] Zakharchenko NI. Ammonia oxidation catalysts in Fe_2O_3 – PbO system. Khimiya I khimicheskaya tehnologiya 2001, 44(5), 70–75.
[30] Zakharchenko NI. Prognosis of properties of nolnplatinum oxide catalysts. Vestnik Khar'kovskogo national'nogo universiteta. Khimiya 1998, 2, 86–92.
[31] Zakharchenko NI. Catalysts for ammonia oxidation in Fe_2O_3 – Li_2O system. Zhurnal prikladnoy khimii 2001, 74(2), 226–231.
[32] Yue Wu, Tao Yu, Bo-sheng Dou, Chen-xian Wang, Xiao-rong Fan, Lian-chi Wang. A comparative Study on Perovskite-Type Mixed Oxide Catalysts $A'_xA_{1-x}BO_{3-x}$, A'=Ca, Sr; A=La, B=Mn, Fe, Co for NH_3 oxidation. J. Catal. 1989, 120, 88–107.
[33] Biausque G, Schuurman Y.The reaction mechanism of the high temperature ammonia oxidation to nitric oxide over $LaCoO_3$. J. Catal. 2010, 276(2), 306–313.
[34] Epshtein DA, Tkachenko NM, Miniovich MA, Dobrovol'skaya NV. Two bed ammonia oxidation catalyst. DAN SSSR 1958, 122(5), 874–879.
[35] Chernyshev VI, Brushtein EA. Decrease of platinum catalyst expense in nitric acid production. Kataliz v promyshlennosti 2001, 3, 30–41.
[36] Popik IV. Investigation of formation and physical and chemical transformation processes of iron-chromium catalysts for ammonia oxidation. Ph.D. Thesis. Har'kov, HPI, 1976.
[37] Karavaev MM, Katnor AY, Semenov GM. Ammonia oxidation on catalytic system. Khimicheskaya promyshlennost' 1990, 11, 669–672 (In Russian).
[38] Golovnya EV, Ammonia Oxidation on Pt Gauzes and Oxide Honeycomb Monolith Catalyst, Ph. D. Thesis, D. Mendeleev University of Chemical Technology of Russia, Moscow, 2009 (in Russian).
[39] Telyatnikova TV, Karavaev MM, Laricheva IV, Lotsman AA, Bondareva AA. Formation of structure of iron-aluminum catalyst. Khimicheskaya promyshlennost' 1991, 4, 226–228.
[40] Telyatnikova TV, Karavaev MM, Firsov OP, Lotsman AA. Specificity of preparation of ammonia oxidation pelleted catalyst. Khimicheskaya promyshlennost' 1991, 5, 283–284.

[41] Sadykov VA, Brushtein EA, Isupova LA, Telyatnikova TV, Kirchanov AA, Zolotarskii IA, Noskov AS, Kojevnikova NG, Kruglyakov VY, Snegurenko OI, Gibbadulin YN, Hazanov AA. Design and application of two bed catalytic ammonia oxidation in nitric acid production with honeycomb oxide catalysts using. Khimicheskaya promyshlennost' 1997, 12, 819–824 (33–38).
[42] Isupova LA, Sadykov VA. Physical and chemical basis of bulk honeycomb monolith catalysts preparation by mechanochemical method for deep oxidation. 2. Bulk honeycomb oxide catalysts for high temperature processes. Kataliz v promyshlennosti 2003, 5, 3–12.
[43] Golovnya EA, Brushtein EA. Optimization of two bed catalytic system with nonplatinum honeycomb secondary catalyst for ammonia oxidation. Kataliz v promyshlennosti 2004, 3, 9–14.
[44] Vanchurin VI, Golovnya EV, Brushtein EA, Yashenko AV. Investigation of catalytic systems for ammonia oxidation in experimental-industrial conditions. Kataliz v promyshlennosti 2007, 3, 38–42 (In Russian).
[45] Beskov VS, Vanchurin VI, Brushtein EA, Golovnya EA, Yashenko AV. Ammonia oxidation on catalytic system composing of platinum and gathering flying lead gauzes. Khimicheskaya promyshlennost' segodnya 2011, 4. 5–8 (In Russian).
[46] Chernyshev VI, Brushtein EA. Reduced spending of platinum catalyst and process explosibility during ammonia oxidation in nitric acid production. Kataliz v promyshlennosti 2001, 3, 30–42.
[47] Farrauto RJ, Heck RM. Honeycomb catalysts: present and future. Kinetika i kataliz 1998, 39(5), 646–652.
[48] Isupova LA, Sutormina EF, Zakharov VP, Rudina NA, Kulikovskaya NA, Plyasova LM. Cordierite-like mixed oxide monolith for ammonia oxidation process. Catal. Today 2009, 147S, S319–S323.
[49] Isupova LA, Sutormina EF, Kulikovskaya NA, Plyasova LM, Rudina NA, Ovsyannikova IA, Zolotarskii IA, Sadykov VA. Honeycomb supported perovskite catalysts for ammonia oxidation processes. Catal. Today 2005, 105, 429–435.
[50] Vanchurin VI, Beskov VS. Molding of monolith honeycomb catalyst from active in ammonia oxidation mixture. Khimicheskaya promyshlennost' 2000, 3, 21–26.
[51] Isupova LA, Sadykov VA, Tikhov SF, Kimkhai ON, Kovalenko ON, Kustova GN, Ovsyannikova IA, Dovbii ZA, Kryukova GN, Rozovskii AY, Tret'yakov VF, Lunin VV. Monoliths perovskite catalysts for environmentally benign fuels combustion and toxic wastes incineration. Catalysis Today 1996, 27(1–2), 249–256.
[52] Kruglyakov VY, Isupova LA, Kulikovskaya NA, Marchuk AA, Kharina IV, Tsybulya SV, Kryukova GN, Burgina EB, Sadykov VA. Properties of iron oxide catalyst for ammonia oxidation depending on used row materials. Kataliz v promyshlennosti 2007, 2, 46–53.
[53] Razdobarov VA. Influence of defect structure of oxides with spinel and corundum structures on their catalytic activity in CO oxidation. Ph.D. Thesis, Novosibirsk, Russia, Boreskov Institute of catalysis 1992.
[54] Burgina EB, Kustova GN, Tsybulya SV, Krukova GN, Litvak GS, Isupova LA, Sadykov VA. Structure specifics of metastable modification of iron oxide (III). Zurnal strukturnoi khimii 2000, 41(3), 489–497.
[55] Isupova LA, Budneva AA, Kruglyakov VY, Paukshtis EA. Adsorption Sites of an Iron–Aluminum Catalyst for Ammonia Oxidation as Studied by the IR Spectroscopy of the Adsorbed NO Probe Molecule Kinetics and Catalysis, 2009, 50(2), 264–269.
[56] Davydov AA. IR spectroscopy in chemistry of oxide's surface. Novosibirsk, Nauka, 1984.
[57] Valyon J, Hall WK. Studies of the surface species formed from NO on copper zeolites. J. Phys. Chem 1993, 97, 1204–1212

[58] Shen S.-T., Weng H.-S. Comparative Study of Catalytic Reduction of Nitric Oxide with Carbon Monoxide over the $La_{1-x}Sr_xBO_3$ (B = Mn, Fe, Co, Ni) Catalysts. Ind. Eng. Chem. Res. 1998, C. 2654–2661.
[59] Ramis G, Larrubia MA. An FT-IR study of the adsorption and oxidation of N-containing compounds over Fe_2O_3/Al_2O_3 SCR catalysts. Journal of Molecular Catalysis A: Chemical 2004, 215, 161–167.
[60] Krykova GN, Tsybulya SV, Solovyeva LP, Sadykov VA, Litvak GS, Andrianova MP. Effect of heat treatment on microstructure evolution of hematite derived from synthetic goethite. Materials Science and Engineering 1991, A 149, 121–127.
[61] Sil'chenkova ON, Korchak Vn, Matyshak Va. The Mechanism of Low-Temperature Ammonia Oxidation on Metal Oxides According to the Data of Spectrokinetic Measurements. Kinetics and Catalysis 2002, 43(3), 363–371.
[62] Avilova IM. Kinetics and mechanism of low temperature oxidation of ammonia on oxides catalysts. Ph.D. Thesis, Kiev, USSR, Pisarzhevskii Institute of physical chemistry, 1981.
[63] Il'in AP, Prokof'ev VY, Gordina NE. Optimization of molding pastes properties for supports, catalysts and sorbents extrusion. Khimiya i khimicheskaya tehnologiya 2003, 46 (6), 152–155 (In Russian)
[64] Il'in AP, Prokof'ev VY. Physico-chemical mechanics in catalysts and sorbents technology. Ivanovo, Russia, Ivanovskii khimiko-tehnologicheskii universitet, 2004 (in Russian).
[65] Nechiporenko SP. Physical-chemical mechanics in technology of structural ceramics. Kiev, USSR, Naukova dumka, 1968.
[66] Yurchenko EN, Prokof'ev VY, Il'in AP, Shirokov YG. Regulation of structural, mechanic and rheological properties of based on dioxide titanium molding pastes. Zhurnal prikladnoy khimii 1995, 68(4), 607–612(In Russian)
[67] Prokof'ev VY, VY, Il'in AP, Shirokov Yg, Yurchenko En. Choice of optimal properties of molding pastes for extrusion of honeycomb monoliths supports and catalysts. Zhurnal prikladnoy khimii 1995, 68(4), 613–618 (In Russian).
[68] Dzis'ko VA. Bases of methods for catalysts preparation. Novosibirsk, USSR, Nauka, 1983 (In Russian).
[69] Malkin AI, Malkin, AY, Isayev AI. Rheology: Concepts, Methods & Applications, ChemTec Publishing, 2006.
[70] Kruglyakov VY, Kulikovskaya NA, Isupova LA. Rheological properties of based on iron oxide plastic pastes depending on used row materials. Kataliz v promyshlennosti 2008, 5, 41–49 (In Russian).
[71] Vanchurin VI, Bespalov LV, Beskov VS. Thermal treatment of honeycomb monolith catalysts of ammonia oxidation. Khimicheskaya promyshlennost' 2001, 10, 17–20 (In Russian).
[72] Ioffe YV, Ketov AA, Fazleev MP, Dobrynin GF, Barannik GB. Specificity of thermal treatment of based on high-aluminous ceramic honeycomb monolith supports. In collected papers: Honeycomb supports and catalysts. Novosibirsk, Institute of Catalysis SB RAS, 1990, 22–26 (In Russian).
[73] Kruglyakov VY, Isupova LA, Litvak GS, Kulikovskaya NA, Marchuk AA. Influence of thermal treatment conditions on the properties of honeycomb monolith BIC 42-1 catalyst. Kataliz v promyshlennosti 2010, 1, 48–52.
[74] Vanchurin VI, Golovnya EV, Yashenko AV. Ammonia oxidation on woven and knitted platinum gauzes. Kataliz v promyshlennosti 2011, 6, 46–50 (In Russian).
[75] Isupova LA. Monolith honeycomb catalysts in two bed ammonia oxidation process. Comparative investigations. Kataliz v promyshlennosti 2012, 6, 52–59 (In Russian).

[76] Brushtein EA, Vanchurin VI, Yashenko AV. Perspectives for progress of two bed catalytic systems for ammonia oxidation in nitric acid production. Kataliz v promyshlennosti 2012, 6, 47–52.
[77] Vernikovskaya NV, Pinaeva LG, Isupova LA. Oxidation of ammonia to NO_x in a two bed reactor (Pt gauzes + oxide monolithic layer): Experimental studies and mathematical modelling. Chem. Eng. 2014, 238, 140–147.
[78] Zakharov VP, Zolotarskii IA, Kuzmin VA. CFD simulation of "gauze pad-honeycomb" catalytic system. Chem. Eng. J. 2003, 91, 249–255.

10 N_2O decomposition in nitric acid production

Nitric acid production is one of the major sources of N_2O emissions to atmosphere in the chemical industry – only one UKL-7 plant produces about 450 tons N_2O per year. The high environmental impact of N_2O as a greenhouse gas and the international agreements and regulations require the development of efficient and economical systems for N_2O abatement in nitric acid production [1–5].

N_2O as a byproduct is formed during ammonia oxidation on platinum gauzes and its concentration (about 1,000–2,000 ppm) in tail gases depends on the platinum gauze selectivity [1–3].

Two ways can be used to reduce N_2O emission without big plant reconstruction – organization of secondary de-N_2O catalyst bed directly after platinum bed in ammonia oxidation burner (high-temperature de-N_2O process) or after de-NO_x bed in selective catalytic reduction (SCR) NO with ammonia reactor (middle- or low-temperature de-N_2O process) that needs very different de-N_2O catalysts due to their different working conditions [1, 6, 7]. In Fig. 10.1, one-reactor schemes for high-temperature and low-temperature N_2O decomposition are presented.

There are data in the literature for both processes. For high-temperature application, granulated CuO/Al_2O_3 (BASF), Co_2AlO_4/CeO_2 (YARA) and $La_{0.8}Ce_{0.2}CoO_3$ (Johnson Matthey) catalysts were developed and tested under industrial conditions. Figure 10.2 displays the $La_{0.8}Ce_{0.2}CoO_3$-based catalyst for high-temperature decomposition developed by Jonson Mattey [8]. Based on the industrial test (South Korea) in ammonia oxidation plant of 90,000 tons HNO_3/year capacity with N_2O emission 300,814 tons/year, the N_2O decomposition degree was about 80% with the catalyst.

Honeycomb oxide catalysts (bulk and supported) are very promising for high-temperature application due to low pressure drop; oxides with the perovskite, spinel and corundum-type structure ($La,Me^1Me^2O_3$ (Me = Co, Fe. Mn), Fe_2O_3, Co_3O_4) are of interest as catalysts due to their attractive physical and chemical properties (activity, chemical and thermal stability) [1, 6, 9–15]. Both types of honeycomb catalysts were developed at BIC. For preparation of the $La,MeFeO_3$-based bulk honeycomb monolith, oxide powders were mixed with the alumina-based binder to obtain a plastic paste, which is extruded through special spinnerets. The prepared monoliths with ~200 channels/inch^2 and 50 mm height were dried and calcined at 900 °C. In the pilot tests under conditions close to those of AK-72 plant, the bulk monolithic catalyst arranged after a platinoid slab provided nitrous oxide decomposition at a level not lower than 80% [6].

Figure 10.3 displays the bulk perovskite based honeycomb monolith catalyst developed at BIC for high-temperature N_2O decomposition. The catalyst was loaded in UKL-7 plant for living test and its good stability was retained even after 12 months of running.

https://doi.org/10.1515/9783110587777-010

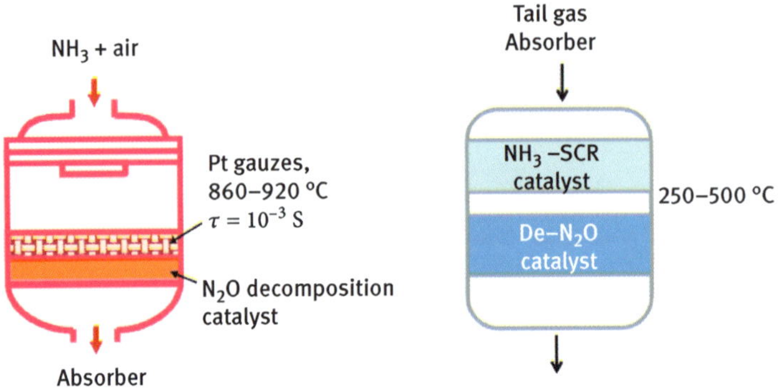

Fig. 10.1: High-temperature and low-temperature schemes for N_2O abatement in nitric acid plants.

Fig. 10.2: $La_{0.8}Ce_{0.2}CoO_3$-based catalyst for high-temperature application (Johnson Matthey).

Fig. 10.3: Perovskite-based bulk honeycomb monolith catalyst for high-temperature N_2O decomposition.

A corundum support was used to synthesize the supported monolithic catalyst for high-temperature decomposition of nitrous oxide. Supported $La,Me^1Me^2O_3$ and Fe_2O_3-based catalysts were synthesized by impregnation of the dried or calcined monolithic support with subsequent drying and calcination of the slabs. Fe_2O_3 oxide was supported on a honeycomb corundum support with ~200 inch and 50 mm height calcined at 1,200 °C by Pechini route using nitrate or oxalate salts as raw materials [10]. After impregnation, the catalysts were calcined at 900 °C, and then the impregnation and calcination procedures were repeated. It was shown that high-temperature activity (measured under testing conditions of the middle pressure ammonia oxidation plant; honeycomb catalysts were placed after platinum gauzes) of the catalysts prepared using nitrate salt is higher as compared to the catalysts prepared using oxalate salt due to a probably higher Fe_2O_3 content [10]. But if the first impregnation of active oxide was done using the low-temperature (380 °C) calcined support and after that the catalysts were calcined at 1,000 °C (second impregnation and calcination procedures were the same), activity of the "oxalate" sample became the same as for the sample prepared with nitrate salt even at a lower Fe_2O_3 content in the catalyst due to formation of highly dispersed Fe_2O_3 particles in this case, which is in good agreement with the one in[16]. As a result, catalytic activity and stability of as-prepared "oxalate" catalyst was nearly the same as for the most active Cu-containing perovskite like the La-Cu-Fe-O/Al_2O_3 catalyst that unfortunately lost Cu during the living test, which is a problem because of possible increase in the products explosibility [6, 9]. The as-prepared honeycomb high-temperature supported Fe_2O_3/Al_2O_3 catalyst placed after platinum gauzes was shown to reduce the N_2O yield by ~80% under the conditions of ammonia oxidation mid-pressure plant (Table 10.1). The catalyst was stable during the 6 months living test in UKL-7 plant.

The presented own data and literature data on the high-temperature decomposition of nitrous oxide demonstrate the possibility to reach the decomposition degree of 80%, which is sufficient for providing the required nitrous oxide content in tail gases, 40 ppm.

For middle-temperature (350–450 °C) NO + N_2O abatement with a zeolite catalyst, one-reactor schemes were developed by Uhde [7, 17–19]. Because in Russian plants the de-NO with NH_3 process is implemented at 200–250 °C with V_2O_5-based catalysts, to organize one-reactor scheme for NO + N_2O abatement, the N_2O decomposition catalysts working at 250–300 °C are desirable [3, 20–23]. Low-temperature $(Co,Ni)_3O_4$-based catalysts modified with alkaline cations may be very attractive for application not only for tail gas purification [24–28], but also for the low-temperature N_2O decomposition in one-reactor scheme with the low-temperature (250–300 °C) SCR NO with ammonia on V_2O_5 catalyst.

$(Co,Ni)_3O_4$ oxides with a spinel-like structure were prepared and modified with Cs by impregnation from nitrate salt or by Pechini route, which provides a uniform

Table 10.1: Influence of salt nature and support impregnation procedure on NO and N_2O yields in ammonia oxidation process with two-bed catalytic systems as compared to one-bed system. [10].

Catalytic system	FeO_x (%)	Air + NH_3	Ar + O_2 + NH_3	
		X_{N2O} (%)	Y_{NO+NO2} (%)	X_{N2O} (%)
Gauzes bed 1 of 8 gauzes				
Gauzes bed	–	–	80.5	–
Gauzes bed + N-Fe_2O_3/Al_2O_3 (1,200)	5.8	71	88.0	75
Gauzes bed + O-Fe_2O_3/Al_2O_3 (1,200)	2.8	53	85.0	57
Gauzes bed 2 of 8 gauzes				
Gauzes bed	–	–	82.4	–
Gauzes bed + O-Fe_2O_3/Al_2O_3 (380)–1,000	–	80	89.7	77

distribution of the active component or a modifying additive over the surface [29–31]. It was shown that the catalytic activity of $Cs/(Co,Ni)_3O_4$ spinel-like oxide catalysts in N_2O decomposition depends on the Ni and Cs contents as well as on the modification route [29–31]. Without Cs no strong influence of the Ni content on catalytic activity was revealed (Fig. 10.4). The highest activity and stability in the presence of O_2 and H_2O in the reaction mixture were detected for the samples with 2% Cs supported by Pechini route on $Ni_{0.1}Co_{2.9}O_4$ and $Ni_{0.9}Co_{2.1}O_4$ oxides as compared to the samples impregnated from nitrate salt [22]. It was shown that modification with Cs cations increases the content of surface low-bonded oxygen species in oxides, which are also desorbed at a lower temperature, thereby affecting the limiting step of the process – oxygen desorption.

The kinetic constant of N_2O decomposition was estimated from experimental data obtained in an integral laboratory reactor on the 0.3–0.5 mm catalyst particles. Modeling of N_2O decomposition process was carried out under the outlet conditions that are typical of the SCR NO + NH_3 process with the V_2O_5/Al_2O_3 de-NO_x catalyst: [N_2O] 1,500 ppm, [O_2] 2.5%, [H_2O] ≈ 3%, [NO_X] 50 ppm, [NH_3]/[NOX] ≈ 1.1; temperature 250 °C, pressure 0.101 MPa, GHSV 7200 h^{-1}. The loading of the commercial reactor (3.8 m ID) with the catalyst consisting of 3–5 mm granules was predicted by simulation. At the bed height 0.4–0.5 m and temperature 250 °C, the conversion of N_2O was not less than 95%, thus reducing N_2O emission in tail gas to 30 ppm [32]. The development of a monolithic catalyst for low-temperature (250 °C) nitrous oxide decomposition and a one-reactor scheme is the immediate task corresponding to the world trends. Oxide and metal catalysts supported on cordierite monolith with the channel density 400/inch^2, which provide nitrous oxide decomposition in kinetic regime at 350–400 °C, were studied in [33–37].

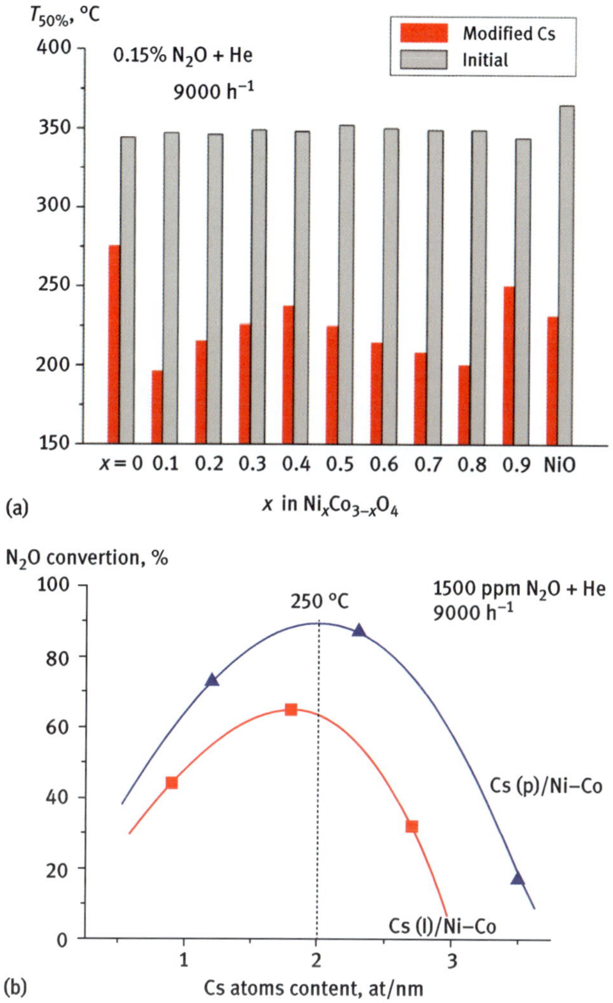

Fig. 10.4: Catalytic activity of $(Co_{1-x}Ni_x)_3O_4$ and 2%Cs/$(Co_{1-x}Ni_x)_3O_4$ oxides (a) and dependence of activity of the sample with $x = 0.75$ on Cs content and method of its introduction (b). Cs(P) – introduction by Pechini route; Cs(I) – introduction by deposition.

Hence, both types of secondary de-N_2O oxide catalysts – (1) for application in ammonia oxidation plant and (2) for application in de-NO_x plant – were developed at the Boreskov Institute of Catalysis. High-temperature (850–900 °C) honeycomb catalysts (bulk and supported) placed after platinum gauzes in the pilot reactor for NH_3 oxidation under AK-72 middle-pressure plant conditions were shown to reduce N_2O emission to 80%. Catalysts were stable in living test in UKL-7 plant. Low-temperature (250–300 °C) de-N_2O oxide granulated catalyst placed after

V_2O_5/Al_2O_3 SCR NO with ammonia catalyst (as the second bed) works in the pore-diffusion regime and, based on modeling at contact time of 0.35 s, can reduce N_2O concentration by 95% under SCR plant conditions. Honeycomb catalyst can provide a kinetic regime of catalyst operation in addition to a low pressure drop.

References

[1] Pérez-Ramırez J, Kapteijn F, Schöffel K, Moulijn J. Formation and control of N_2O in nitric acid production. Where do we stand today? Appl. Catal. B Environ. 2003, 44, 117–151.

[2] Kamphus M. Emission monitoring in nitric acid plants. Nitrogen+Syngas 2014, 328, 48–53.

[3] Brushtein EA, Vanchurin VI, Yashenko AV. Change of N2O concentration along the flue gas path in UKL-7 plant in nitric acid production. Catal Ind 2012, 4, 7–12 (In Russian).

[4] Pegov SA. Anthropogenic influence upon biosphere. Transaction ISA RAN 2009, 42, 5–32.

[5] Lashof DH, Ahuja DR. Relative contributions of greenhouse gas emission to global warming. Nature 1990, 344, 529–531.

[6] Isupova LA, Kulikovskaya NA, Marchuk AN, Pinaeva LG. Catalyst, catalyst preparation and method of N_2O decomposition. Patent RU 2430781, 2011 (In Russian)

[7]. Groves MCE, Sasonow A. Uhde EnviNOx® technology for NO_X and N_2O abatement: a contribution to reducing emissions from nitric acid plants. J. Integr. Environ. Sci. 2010, 7(S1), 211–222.

[8] Axon SA, Coupland DR, Ridland JR, Wishart IC. International patent WO 04096703A2, 2004

[9] Isupova LA. Honeycomb catalysts in ammonia oxidation two bed technology. Comparative investigations.Catal. Ind. 2012, 6, 52–59. (In Russian)

[10] Pinaeva LG, Dovlitova LS, Isupova LA. Monolithic FeO_x/Al_2O_3 catalysts for ammonia oxidation and nitrous oxide decomposition. Kinet. Catal. 2017, 58(2), 167–178.

[11] Wu Y, Ni X, Beaurain A, Dujardin C, Granger P. Stoichiometric and non-stoichiometric perovskite-based catalysts: Consequences on surface properties and on catalytic performances in the decomposition of N2O from nitric acid plants. Appl. Catal. B Environ. 2012, 125, 149–157.

[12] Kruk J, Stołecki K, Michalska K, Konkol M, Kowalik P. The influence of modifiers on the activity of Fe_2O_3 catalyst for high temperature N_2O decomposition (HT-deN2O). Catal. Today 2012, 191, 125–128.

[13] Giecko G, Borowiecki T, Gac W, Kruk J. Fe_2O_3/Al_2O_3 catalysts for the N_2O decomposition in the nitric acid industry. Catal. Today 2008, 137, 403–409.

[14] Jiratová K, Balabanová J, Kovanda F, Klegová A, Obalová L, Fajgar R. Cobalt oxides supported over ceria-zirconia coated cordierite monoliths as catalysts for deep oxidation of ethanol and N_2O decomposition. Catal. Lett. 2017, 147, 1379–1391.

[15] Pietrogiacomia D, Campab MC, Carbonea LR, Tutic S, Occhiuzzia M. N_2O decomposition on CoO_x, CuO_x, FeO_x or MnO_x supported on ZrO_2: The effect of zirconia doping with sulfates or K^+ on catalytic activity. Appl. Catal. B Environ. 2016, 187, 218–227.

[16] Pinaeva LG, Prosvirin IP, Dovlitova LS, Danilova IG, Sadovskaya EM, Isupova LA. MeO_x/Al_2O_3 and MeO_x/CeO_2 (Me = Fe, Co, Ni) catalysts for high temperature N_2O decomposition and NH_3 oxidation. Catal. Sci. Technol. 2016, 6, 2150–2161.

[17] Groves M, Schwefer M, Siefert R. Nitric acid – without the emissions. tce April 2006, 30–31. (Accessed at https://www.thyssenkrupp-industrial-solutions.com/media/download_1/nitrates/uhde_publications_pdf_en_10000009.pdf).

[18] EnviNox Uhde Tehnology (Accessed at https://www.tkisrus.com/assets/pdf/brochures/ru/TKIS-Nitric-Acid-ru.pdf).

[19]. Roy PK, Prins R, Pirngruber GD. The effect of pretreatment on the reactivity of Fe-ZSM-5 catalysts for N_2O decomposition: Dehydroxylation vs. steaming. Appl. Catal. 2008, 80, 226–236.

[20] Sankar S, Putluru R, Schill L, Godiksen A. Poreddy R. Mossin S, Jensen AD, Fehrmann R. Promoted V_2O_5/TiO_2 catalysts for selective catalytic reduction of NO with NH_3 at low temperatures. Appl. Catal. 2016, 183, 282–290.

[21] Huang X, Zhang S, Chen H, Zhong Q. Selective catalytic reduction of NO with NH_3 over V_2O_5 supported on TiO_2 and Al_2O_3: A comparative study. J. Mol. Struct. 2015, 1098, 289–297.

[22] Носков, AC Zenkovetz, G Shutilov, AA Piryutko, LV Mokrinsky, VV Kharitonov, A Chumachenko, V. Low-temperature abatement of nitrogen oxides (N_2O, NO_x) from the effluent gases in nitric acid production. In: Abstracts XIX International Conference on Chemical Reactors Chemreactor-19, Vienna, September, 5–9, 2010, 462–463.

[23] Kharitonov AS, Starokon' EV, Chernyavskii VS, Panov GI, Piryutko LV, Parfenov AN, Noskov AS. Method for N_2O decomposition catalyst preparation and process for tail gas containing N_2O purification. Patent RU 2477177 (2013) in Russian.

[24] Yan L, Ren T, Wang X, Ji D, Suo J. Catalytic decomposition of N_2O over M_xCo_{1-x} Co_2O_4 (M = Ni, Mg) spinel oxides. Appl. Catal. 2003, 45, 85–90.

[25] Grzybek G, Stelmachowski P, Gudyka S, Duch J, Cmil K, Kotarba A, Sojka Z. Insights into the twofold role of Cs doping on deN_2O activity of cobalt spinel catalyst – towards rational optimization of the precursor and loading. Appl. Catal. 2015, 168–169, 509–514.

[26] Lykaki M, Papista E, Carabineiro SAC, Tavares PB, Konsolakis M. Optimization of N2O decomposition activity of CuO-CeO2 mixed oxides by means of synthesis procedure and alkali (Cs) promotion. Catal. Sci. Technol. 2018, 9, 2312–2322.

[27] Gudyka S, Grzybek G, Gryboś J, Indyka P, Leszczyński B, Kotarba A, Sojka Z. Enhancing the deN_2O activity of the supported $Co_3O_4|\alpha$-Al_2O_3 catalyst by glycerol-assisted shape engineering of the active phase at the nanoscale. Appl. Catal. B Environ. 2017, 201, 339–347.

[28] Klegová A, Pacultová K, Fridrichová D, Volodarskaja A, Kovanda F, Jiratová K. Cobalt oxide catalysts on commercial supports for N_2O decomposition. Chem. Eng. Technol. 2017, 40, 981–990.

[29] Ivanova YA, Ivanov DV, Chumachenko VA, Isupova LA, Noskov AS. One-Reactor Scheme for NO and N_2O Low Temperature Abatement from Tail Gas in Nitric Acid Production. In: Thesis of XXII International Conference on Chemical Reactors "CHEMREACTOR-22" (CR-22) – London, September, 19–23, 2016, 242.

[30] Ivanova YA, Sutormina EF, Isupova LA, Vovk EI. Catalytic Activity of the Oxide Catalysts Based on $Ni_{0.75}Co_{2.25}O_4$ Modified with Cesium Cations in a Reaction of N_2O Decomposition. Kinet. Catal. 2017, 58(6),793–799.

[31] Ivanova YA, Sutormina EF, Isupova LA, Rogov VA. Effect of the composition of $Ni_xCo_{3-x}O_4$ (x = 0–0.9) oxides on their catalytic activity in the low-temperature reaction of N_2O decomposition. Kinet. Catal. 2018, 59(3),357–362.

[32] Isupova LA, Sutormina EF, Ivanova YA. N_2O abatement in nitric acid production. Evropacat-13, 2N 1.2. August, 27–31, 2017, Florence, Italy (ISBN978-5-906376-10-7).

[33] Boissel V, Tahir S, Koh CA. Catalytic decomposition of N_2O over monolithic supported noble metal-transition metal oxides. Appl. Catal. B Environ. 2006, 64, 234–242.

[34] Centi G, Perathoner S, Vazzana F, Marella M, Tomaselli M, Mantegazza M. Novel catalysts and catalytic technologies for N_2O removal from industrial emissions containing O_2, H_2O and SO_2. Adv. Environ. Res. 2000, 4, 325–338.

[35] Suarez S, Saiz C, Yates M, Martin JA, Avila P, Blanco J. Rh/gamma-Al_2O_3-sepiolite monolithic catalysts for decomposition of N_2O traces. Appl. Catal. B: Environmental. 2005, 55(1), 57-64.
[36] Wojcik S, Ercolino G, Gajewska M, Quintero C, Specchia S, Kotarba A. Robust $Co_3O_4|\alpha\text{-}Al_2O_3|$cordierite structured catalyst for N_2O abatement – Validation of the SCS method for active phase synthesis and deposition. Chem. Eng. J. 2018, https://doi.org/10.1016/j.cej.2018.10.025.
[37] Wójcik S, Grzybek G, Gryboś J, Kotarba A, Sojka Z. Designing, optimization and performance evaluation of the $KZn_{0.4}Co_{2.6}O_4|\alpha\text{-}Al_2O_3|$cordierite catalyst for low-temperature N_2O decomposition. Catal. Commun. 2018, 110, 64–67.

11 Structured catalysts

Transformation of fuels (fossil fuel, biofuels) into syngas or hydrogen is one of the most important tasks of catalysis in the energy-related fields [1–4]. Catalysts comprised of precious metals and/or Ni supported on fluorite-like, spinel or perovskite-like complex oxides with a high lattice oxygen mobility are known to be very efficient and stable to coking in reforming of a variety of fuels by using different oxidants (oxygen, water, CO_2 and their combinations) [5–16]. Monolithic substrates with a good thermal conductivity are promising for providing an efficient heat transfer within the reactor to prevent emergence of hot spots/cool zones deteriorating performance [17–22].

Nanocomposite-active components selected by screening tests were supported on structured ceramic, metallic or cermet substrates and tested in reactions of CH_4 and liquid fuels transformation into syngas at short contact times [8, 23–31]. A high yield of syngas approaching equilibrium values was obtained and a stable performance due to a high sintering resistance and coking suppression was demonstrated. For liquid fuels, especially real complex fuels (gasoline, diesel, etc.) and oxygenates (bio-oil, etc.), design of a unique monolithic evaporator/mixer unit comprised of thick foil Fecralloy substrate protected by a corundum layer supported by blast dusting and heated by passing the electric current [24] allowed to solve the problem of preparation of feeds with a high concentration of steam and liquid fuels with sufficient feed rates to test monolithic catalysts at millisecond contact times. This provided a possibility to compare performance of a series of monolithic/structured catalysts in realistic feeds containing different fuels at short contact times and elucidate specificity of their action as related to the nature of fuel, substrate and active component which is certainly required for design of compact and reliable syngas or hydrogen generators from a variety of fuels.

In this chapter, results of such a comparative research aimed at elucidating specificity of such structured catalysts performance at lab-scale and pilot-scale levels using specially designed reactors and installations allowing to broadly tune the operational parameters (inlet temperatures and feed compositions) are considered. A special attention was paid to new types of structured heat-conducting substrates comprised of compressed Ni–Al foam, Fecralloy gauzes and microchannel washers protected by refractory corundum layers (Figs. 11.1 and 11.2, and Tables 11.1 and 11.2).

Steam reforming of methane. To check the effect of the foam Ni–Al substrate characteristics on performance of these structured catalysts in steam reforming of methane, substrates with broadly varying density were prepared and loaded with the same active component (Table 11.1, samples 7.1–7.9). There is a certain trend in decreasing the loading of active component with increasing substrate density when using the same suspension and number of supporting cycles (3 in this case).

https://doi.org/10.1515/9783110587777-011

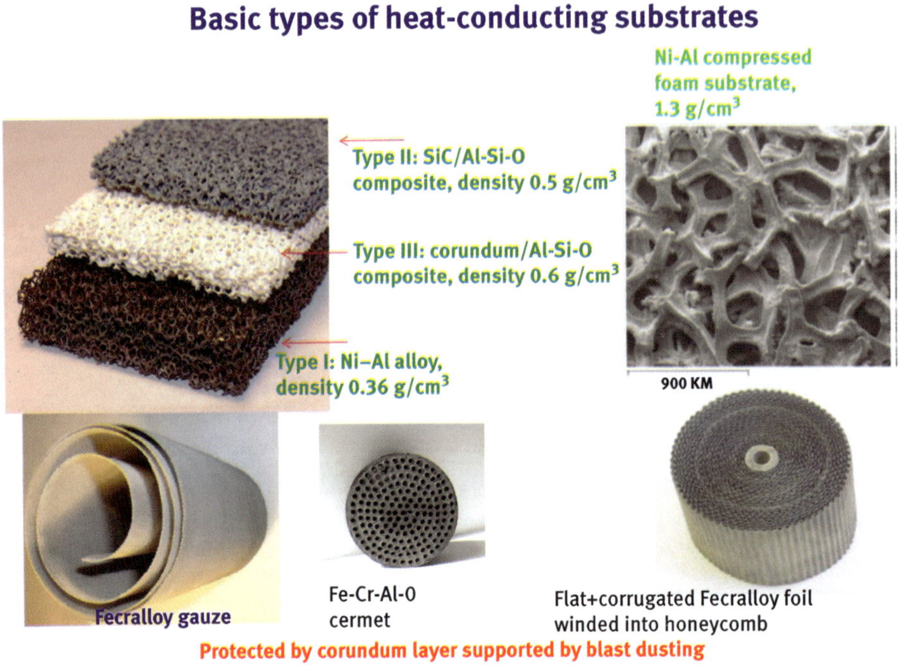

Fig. 11.1: Structured substrates.

As follows from Fig. 11.3, the temperature dependence of CH_4 conversion is rather similar for samples based on different density substrates. Rather steep temperature dependence of conversion suggests a high apparent activation energy, and hence, a small (if any) effect of the heat and mass transfer on these catalysts performance.

To check the effect of upscaling the size of structured catalysts on their performance in the reaction of methane steam reforming, a package comprised of stacked foam platelets and gauzes was tested in feeds with or without addition of a small amount of air. In this case, the temperature difference between the front and rear ends of the package was in the range of 50–70 °C which apparently indicates existence of the temperature gradient within the package. Hence, detailed analysis of these data required a proper modeling with a due regard for heat transfer within this package. At least, as judged by the values of CH_4 conversion (50–60% within studied range of exit temperatures) and hydrogen content in the effluent (>45%, Fig. 11.4), performance of this package is rather good. Tests for 100 h with startup and shutdown of pilot installation each day (8 h working time per day) confirmed stability of this level of H_2 content in effluent.

For liquid fossil fuels such as decane or gasoline, the most clear advantage of structured catalysts on heat-conducting substrates is in the case of partial oxidation

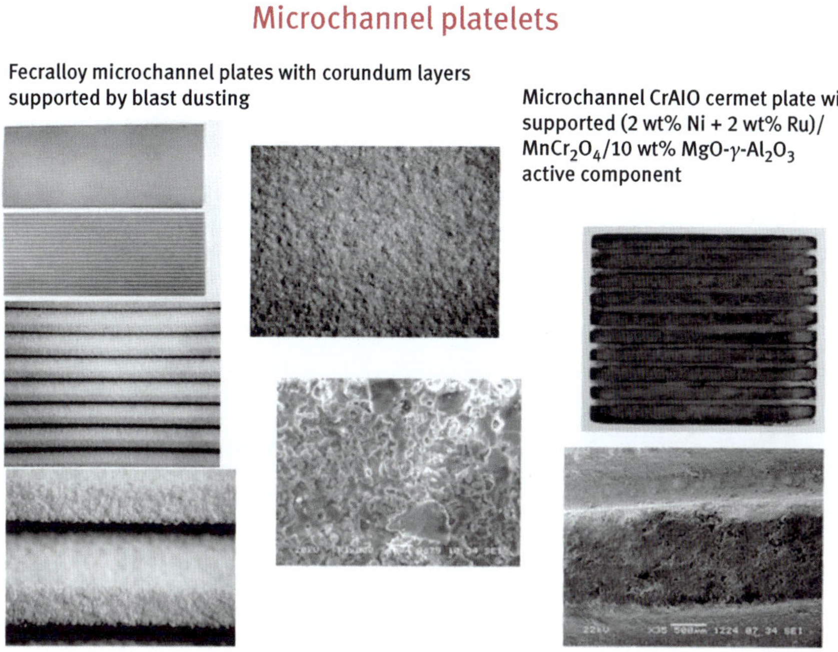

Fig. 11.2: Microchannel plates.

or autothermal reforming reactions [1–4, 20, 21]. This is determined by a very fast oxygen consumption in the inlet part of the reactor, so the heat generated by combustion reaction is to be transferred along the catalytic bed to be consumed by endothermic reactions of steam and dry reforming. Heat-conducting substrates allowing to minimize the temperature difference between the inlet part of the layer and its main part help to prevent thermal shocks and stresses leading to cracking of ceramic monolithic substrates.

Figure. 11.5 presents results obtained in the axial reactor for the catalyst N 9 based on a thick Fecralloy foil substrate. In all experiments, the exit temperature was ~1,000 °C. Even at very short contact time, syngas yield is rather high. The main byproduct is methane formed via cracking reactions. At the longest contact time (12 ms), concentrations of H_2 (21%) and CO (24%) in the effluent coincide with the equilibrium values (20 and 23%, respectively), with CH_4 admixture ~0.3%. At the shortest contact times, H_2/CO ratio is somewhat lower, perhaps, due to a higher content of methane (~4%) and olefins (~2) in the effluent.

In the radial reactor equipped with the internal heat exchanger [21], in the steady-state mode, the temperature of feed before the reactor was kept at a nearly constant level of 110–120 °C. The temperature measured by thermocouple situated in the inlet of the central part of the stack of microchannel washers decreased from 180 to 130 °C with the increase in feed rate from 1.3 to 4.0 m^3/h (STP) due to cooling by

Table 11.1: Basic types of monolithic catalysts used in transformation of methane and liquid fuels into syngas in steam and oxi-steam reforming [21, 31].

No	Sample description	
	Type of substrate	**Active component composition and loading (wt.%)**
1	FeCrAl gauze (cylinder *D* 15 mm, *L* 26 mm)	5.9 $Pr_{0.3}Ce_{0.35}Zr_{0.35}O_2$ + 0.92 Pt
2	The same as 1	5.9 $Pr_{0.3}Ce_{0.35}Zr_{0.35}O_2$ + 7 LaNi(Pt) O_3(1.2 Pt)*
3	The same as 1	6.6 $Pr_{0.3}Ce_{0.35}Zr_{0.35}O_2$ + 1.3 Pt + 0.3 Ru
4	FeCrAl foil 20 μm (cylinder *D* 15 mm, *L* 26 mm)	5.5 $Pr_{0.3}Ce_{0.35}Zr_{0.35}O_2$ + 1.2 Pt
5	The same as 4	6.3 $Pr_{0.3}Ce_{0.35}Zr_{0.35}O_2$ + 1.0 Pt
6	FeCrAl gauze (the same as 1)	5.4 $La_{0.1}Ce_{0.45}Zr_{0.45}O_2$ + 10 LaNi(Pt) O_3(1Pt)*
7.0	Ni–Al compressed foam (plate 10 × 20 × 1 mm), density 2.5 g/cm^3	4% (50 $La_{0.8}Pr_{0.2}Mn_{0.2}Cr_{0.8}O_3$ + 30 NiO + 20 YSZ) + 0.7Ru
7.1.	Ni–Al compressed foam (plate 10 × 20 × 1 mm), density 0.4 g/cm^3	26 (50 $La_{0.8}Pr_{0.2}Mn_{0.2}Cr_{0.8}O_3$ + 30 NiO + 20 YSZ) + 6.51 Ru
7.2.	Ni–Al compressed foam (plate 10 × 20 × 1 mm), density 1.17 g/cm^3	13.84 (50 $La_{0.8}Pr_{0.2}Mn_{0.2}Cr_{0.8}O_3$ + 30 NiO + 20 YSZ) + 2.99 Ru
7.3.	Ni–Al compressed foam (plate 10 × 20 × 1 mm), density 1.28 g/cm^3	10.2 (50 $La_{0.8}Pr_{0.2}Mn_{0.2}Cr_{0.8}O_3$ + 30 NiO + 20 YSZ) + 1.44 Ru
7.4.	Ni–Al compressed foam (plate 10 × 20 × 1 mm), density 1.37 g/cm^3	11.64 (50 $La_{0.8}Pr_{0.2}Mn_{0.2}Cr_{0.8}O_3$ + 30 NiO + 20 YSZ) + 2.38 Ru
7.5.	Ni–Al compressed foam (plate 10 × 20 × 1 mm), density 1.39 g/cm^3	9.56 (50 $La_{0.8}Pr_{0.2}Mn_{0.2}Cr_{0.8}O_3$ + 30 NiO + 20 YSZ) + 2.59 Ru
7.6.	Ni–Al compressed foam (plate 10 × 20 × 1 mm), density 1.79 g/cm^3	10.73 (50 $La_{0.8}Pr_{0.2}Mn_{0.2}Cr_{0.8}O_3$ + 30 NiO + 20 YSZ) + 2.83 Ru
7.7.	Ni–Al compressed foam (plate 10 × 20 × 1 mm), density 2.23 g/cm^3	7.3 (50 $La_{0.8}Pr_{0.2}Mn_{0.2}Cr_{0.8}O_3$ + 30 NiO + 20 YSZ) + 2 Ru
7.8.	Ni–Al compressed foam (plate 10 × 20 × 1 mm), density 2.26 g/cm^3	7.43(50 $La_{0.8}Pr_{0.2}Mn_{0.2}Cr_{0.8}O_3$ + 30 NiO + 20 YSZ) + 1.59 Ru
7.9.	Ni–Al compressed foam (plate 10 × 20 × 1 mm), density 2.49 g/cm^3	6,64 (50 $La_{0.8}Pr_{0.2}Mn_{0.2}Cr_{0.8}O_3$ + 30 NiO + 20 YSZ) + 1.3 Ru
8	$CrAlO_y/CrAl_x$ cermet (cylinder *D* 19 mm, *L* 21 mm, 20 channels).	3.3 ($Ce_{0.35}Zr_{0.35}Sm_{0.15}Pr_{0.15}O_2$ + 1.5 Ru)

Table 11.1 (continued)

No	Sample description	
	Type of substrate	**Active component composition and loading (wt.%)**
9	FeCrAl foil 200 µm (*D* 53 mm, *L* 12.5 mm)	2 $La_{0.1}Ce_{0.45}Zr_{0.45}O_2$ + 1 LaNi(Pt) O_3(0.1Pt)*
10	FeCrAl foil 20 µm (*D* 50 mm, *L* 50 mm).	10 $La_{0.1}Ce_{0.45}Zr_{0.45}O_2$ + 7 LaNi(Pt) O_3(1Pt)*
11	Corundum honeycomb monolith, (*D* 50 mm, *L* 50 mm)	10 $La_{0.1}Ce_{0.45}Zr_{0.45}O_2$ + 7 LaNi(Pt) O_3(1Pt)*
12	Ti platelet (0.5 × 5 × 10 mm) with Ce-Zr-Ti-O protective layer	0.44 Pt + 5 $Ce_{0.35}Zr_{0.35}Sm_{0.15}Pr_{0.15}O_2$
13	Ti platelet (0.5 × 5 × 10 mm) with Ce-Zr-Ti-O protective layer	5 (10NiO + 10YSZ + 80 $Ce_{0.35}Zr_{0.35}Sm_{0.15}Pr_{0.15}O_2$)
14	Ni–Al compressed foam substrate, density 2.5 g/cm^3	5.3 (50% LaMnCrPr + 30% NiO+20% YSZ) + 0.85 Ru

*Total Pt content in monolithic catalyst.

Table 11.2: Characteristics of substrates used for preparation of catalysts for methane dry reforming.

No.	Chemical composition (wt.%)	Density (g/cm^3)	Pore size/ porosity	Heat conductivity (W/m K)	Shape and size, mm
1	Fecralloy microchannel plate/5 µm corundum sublayer	7.3	n/p		20/50/1.5, one side flat, one side with 15 channels of 0.4 mm depth
2	Fecralloy gauze with 5 µm corundum sublayer	n/a	n/a		Squares 35 × 35, woven of wires of 0.2 mm diameter with 0.2 mm spacing,
3	Ni–Al alloy, Al 10%	2.5	60 ppi/ 65.6%	9.8	35 × 35 × 1
4	SiC 40%; Al_2O_3–SiO_2 rest	0.5	30 ppi/ 75.5%	3.5	38 × 38 × 5
5	α-Al_2O_3 45%, Al_2O_3–SiO_2 rest	0.6	30 ppi/ 78.6%	1.7	38 × 38 × 5
6	Ni–Al alloy, Al 10%	0.36	30 ppi/ 95.0%	5.5	38 × 38 × 10

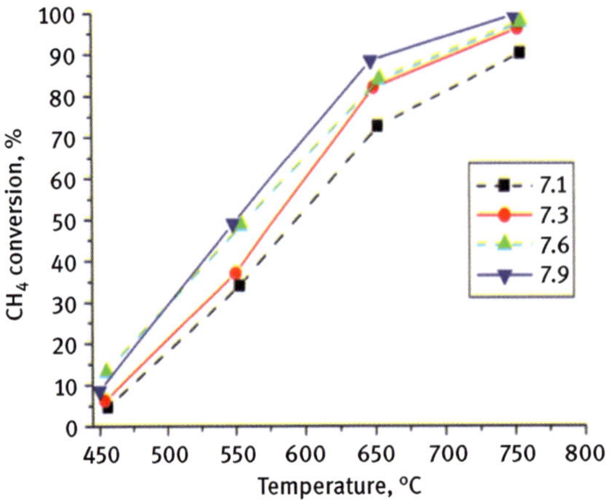

Fig. 11.3: Temperature dependence of CH_4 conversion in CH_4 SR for catalysts based on nanocomposite active component $La_{0.8}Pr_{0.2}Mn_{0.2}Cr_{0.8}O_3$ + NiO + YSZ + Ru supported on Ni–Al foam substrates with a different density (Table 11.1). Contact time 50 ms, feed composition 20% CH_4 + 40% H_2O in Ar.

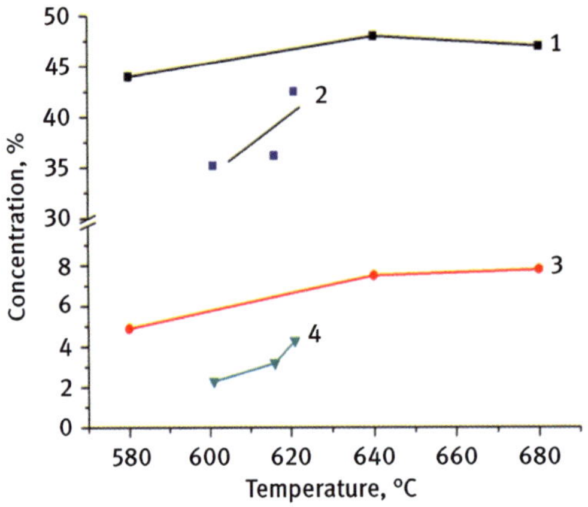

Fig. 11.4: Temperature dependence of H_2 (1,2) and CO (3,4) content in effluent for the reaction of steam (1,3) or oxysteam (2,4) reforming of CH_4 on a stack comprised of 12 Ni–Al-foam plates and 11 sheets of Fecralloy gauzes loaded with $La_{0.8}Pr_{0.2}Mn_{0.2}Cr_{0.8}O_3$ + NiO + YSZ + Ru (volume 34 × 34 × 34 mm^3). Feed 31% CH_4 + H_2O (H_2O/CH_4 = 1.9) in Ar (1, 3) or 31% CH_4 + H_2O (H_2O/CH_4 = 1.9) + 1.5 % O_2 in Ar (2,4), contact time 0.15 s.

the inlet stream (Fig. 11.6). The temperature measured by thermocouple situated after the layer of gauzes wound around the stack of washers goes through the maximum at ~800 °C with increasing the feed rate. The maximum temperature within the microchannels in the stack of washers is expected to be ~1,000 °C as judged by results

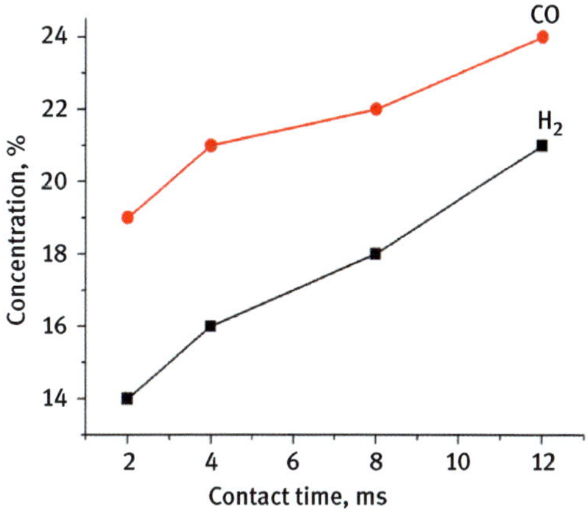

Fig. 11.5: CO and H_2 content in effluent versus contact time for partial oxidation of decane on the catalyst 9 (Table 11.1) in the axial-type pilot reactor. Feed composition 3% of decane in air ($O_2/C = 0.6$), inlet feed temperature 180–200 °C.

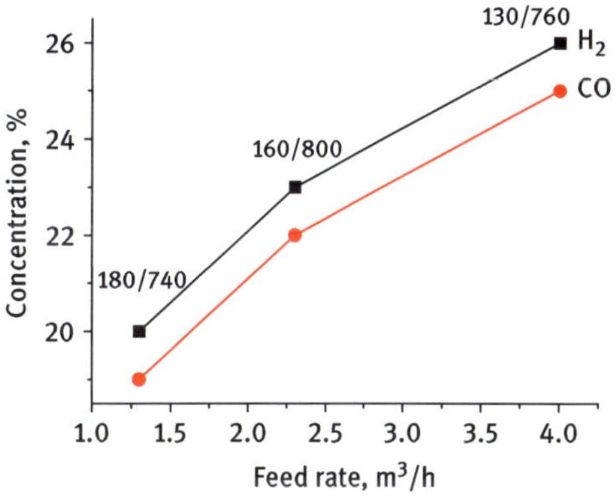

Fig. 11.6: CO and H_2 content in effluent versus feed rate for partial oxidation of decane in the radial-type reactor. Feed composition 3% of decane in air ($O_2/C = 0.6$), the feed temperature after evaporator/mixer 110–120 °C. The ratio between temperatures in the inlet part of reactor and after a layer of gauzes is indicated in the figure for each feed rate (see text).

with a short piece of monolithic catalyst tested in the axial reactor (vide supra). Hence, the temperature profile within the radial reactor is controlled by the balance between the rate of heat generation due to oxygen consumption within the stack of microchannel washers and the rates of its transfer to gauzes (increases with the flow rate) and consumption by endothermic reactions of steam and dry reforming on gauzes and within the layer of microspherical catalyst. This seems to determine both the increase of syngas yield with the feed rate and a higher content of hydrogen and

CO (as well as higher H_2/CO ratio in the effluent) as compared with the case of the axial-type reactor (cf. Figs. 11.5 and 11.6). In all studied range of feed rates, CH_4 content in the effluent was ~0.2–0.5%. Similar features – the increase of syngas content in the effluent with the feed rate – were earlier observed for the partial oxidation of methane into syngas in this axial-type reactor [30].

Gasoline. Reformulated gasoline transformation was studied for the catalyst based on thick Fecralloy foil substrate (N9 in Table 11.1) and on the package including this catalyst in the inlet part and a honeycomb monolithic catalyst on corundum substrate (N11, Table 11.1) as the main part of the catalytic layer. As follows from results presented in Fig. 11.7, for the layer comprised of only foil-supported catalyst, there is certainly a trend of decreasing CO and H_2 content in the effluent with increasing the contact time. This suggests that at very short contact times CO and H_2 are primary products of fast transformation of hydrocarbons in the presence of gas –phase oxygen. The increase of contact time for combined catalytic layer increases syngas yield approaching it to the limit corresponding to equilibrium at the exit temperature (~ 1,000 °C) (24% H_2 and 26% CO). A similar close to equilibrium syngas yield was obtained for monolithic catalysts on thin-foil Fecralloy or microchannel cermet substrates [29].

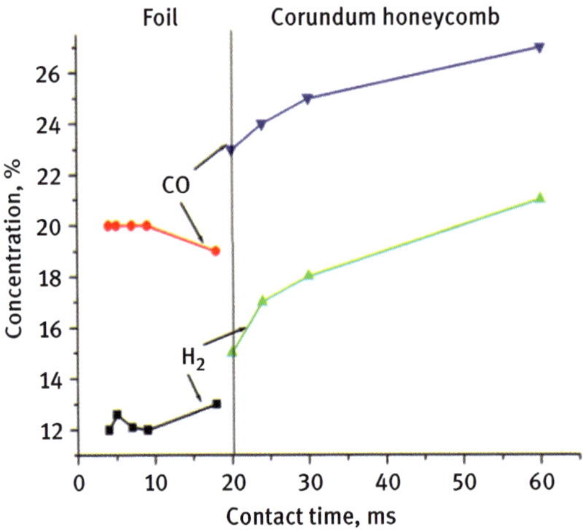

Fig. 11.7: CO and H_2 content in effluent versus contact time for the partial oxidation of gasoline on the catalyst 9 alone (contact time range marked by "foil") and its stack with catalyst 11 (contact time range marked by "corundum honeycomb" (Table 11.1). Feed composition 4.7% of gasoline in air (O_2/C = 0.48), inlet feed temperature 180–200 °C, exit temperature ~1,000 °C.

In all these reactions of liquid hydrocarbon fuels selective oxidation, performance was stable with the time-on-stream for at least 10–200 h including pilot-scale testing in real syngas generators [37]. New types of nanocomposite active components, application of electric current heated evaporation/mixing unit designed in this work and placing in the inlet part of catalytic layer structured catalysts with a high thermal conductivity allowed to achieve a high yield of syngas at short contact time and provide a stable performance even without adding steam to the feed which was earlier considered to be inevitable for providing a stable performance without degradation for such complex fuels as gasoline [1–4].

For oxysteam reforming of ethanol which is attractive from the heat management point of view, catalysts on monolithic heat-conducting substrates with nanocomposite active components (sample 10, Table 11.1) also provide a high yield of syngas at short contact time (Fig. 11.8). In this case, nearly constant H_2/CO ratio ~ 1 can be achieved at a broad variation of H_2O/ethanol ratio. Concentration of ethylene in effluent is below 0.2% agreeing with conclusion about importance of redox properties of a catalyst for transformation of C_2H_4 into syngas.

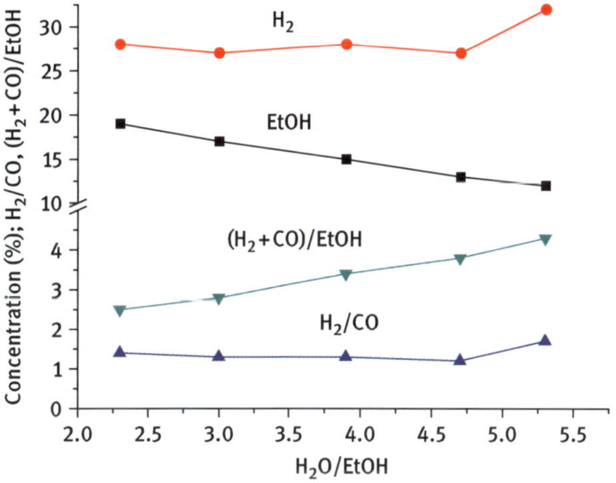

Fig. 11.8: Effect of H_2O/C_2H_5OH ratio on product concentrations in oxysteam reforming on catalyst 10 based on thin-walled Fecralloy monolithic substrate (sample 10, Table 11.1). Inlet feed: 7% O_2 + H_2O + EtOH+N_2 balance, *T* inlet 700 °C, contact time 0.3 s.

Acetone. The same types of nanocomposite active components containing Ni, Pt and doped fluorite-like oxides are efficient and stable in acetone transformation into syngas even when supported on monolithic corundum substrate (Fig. 11.9). Similar to the case of ethanol steam reforming, the main byproduct here is also CH_4. At longer (0.5 s) contact times, only small amounts of ethylene byproduct are observed, apparently due to its transformation into syngas without forming carbonaceous deposits. Oxygen addition into the feed only slightly increases hydrogen yield but helps to improve the heat balance and further decreases the ethylene

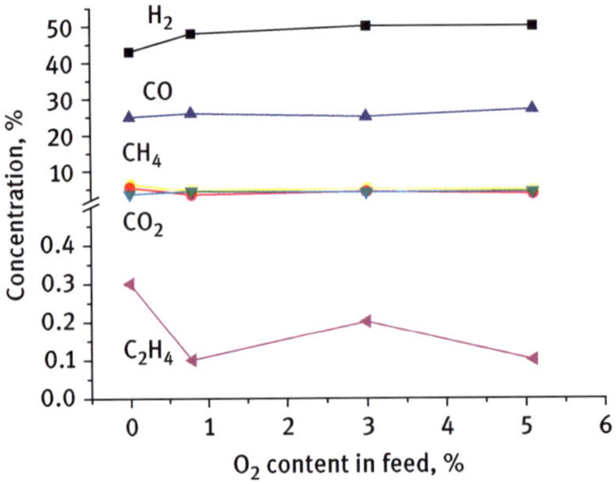

Fig. 11.9: Effect of oxygen content in the feed on products concentration in acetone oxysteam reforming on catalyst 11 based on corundum monolithic substrate (Table 11.1). Contact time: 0.5 s; inlet feed: 24% acetone + 48% H_2O + O_2, N_2 balance, *T* inlet 600 °C, *T* outlet 700 °C.

content, thus ensuring performance stability even at a moderate excess of steam that is attractive from the process economy point of view.

Anisole, sunflower oil. For these feeds a high and stable performance of developed catalysts was demonstrated as well. Some typical results are shown in Figs. 11.10 and 11.11. Due to well-known high coking ability of these fuels, their stable performance was obtained only in the case of oxygen addition to the feed (oxysteam reforming). Moreover, only unique design of evaporation and mixing unit used in this work is

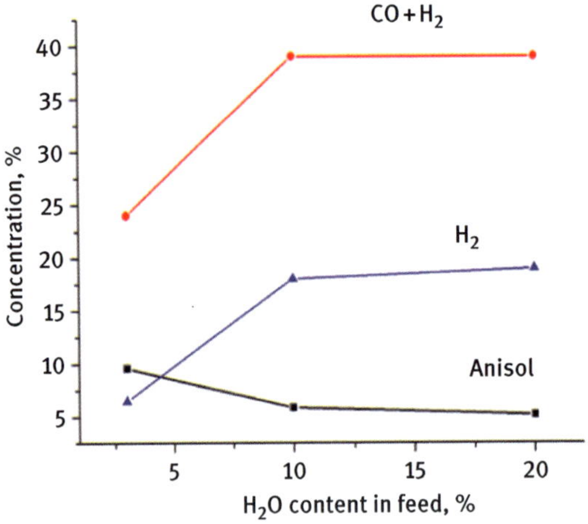

Fig. 11.10: Effect of H_2O content in the feed on product concentrations in the oxysteam reforming of anisole on catalyst 10 based upon Fecralloy thin-walled foil substrate (Table 11.1). Contact time: 0.06 s, *T* exit: 850 °C, feed composition 18% O_2 + anisole (5–10%) + H_2O (3–20%) + N_2 balance.

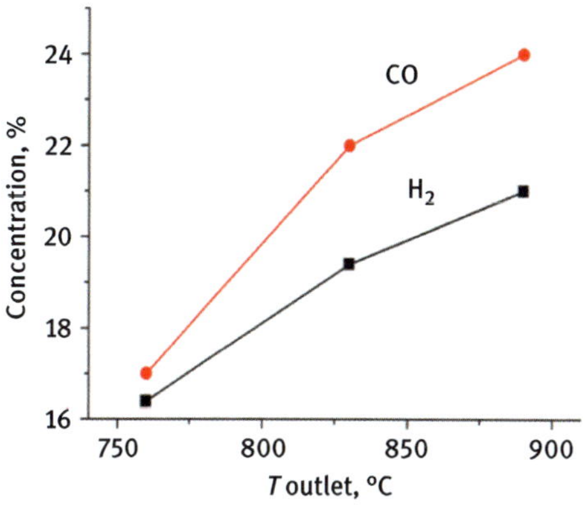

Fig. 11.11: Effect of catalyst temperature on the product concentration in the oxysteam reforming of sunflower oil on stacked layer of catalysts 10 (front part) and 11 (rear part). Feed composition 0.7% of sunflower oil + 15% H_2O + 20% O_2, N_2 balance, contact time 0.2 s.

allowed to obtain stable and reproducible results preventing cracking of fuels in supplying lines. The most efficient performance in transformation of these heavy fuels was provided by combination of catalysts based on heat-conducting metallic substrates (placed in the front part of the layer, efficiently transferring heat generated due to fast oxygen consumption) and those based on corundum monolithic substrates (placed in the rear part, which provide a high conversion level due to developed surface area). As far as we know, this is the first example of successful transformation of these fuels into syngas in the reactors with stationary layers of catalysts.

Natural gas dry reforming. Since for any practical application under investigation, structured catalysts are well known to present several advantages, nanocomposite active components were successfully deposited on ceramic honeycombs, microchannel cermets or metallic substrates (Table 11.2) and tested in realistic feeds at lab-scale and pilot-scale levels using specifically designed reactors and installations allowing to broadly tune the operational parameters.

At some excess of CO_2 in dry reforming of natural gas (NG) containing up to 5% C_2-C_4 alkanes and some admixture of sulfur compounds, stable and comparable performance was achieved for structured catalyst 1 (Table 11.3), that is, active component based on Ni + Ru/LaPrMnCr-YSZ nanocomposite, and catalyst 7, that is, active component based on Ni + Ru/fluorite-like oxide support on microchannel substrates (Figs. 11.12 and 11.13).

The values of main process parameters for the structured catalysts tested in the natural gas reforming are shown in Table 11.4. For a stack of gauzes and high-density Ni–Al foam plates (catalyst 2, Table 5.2), the NG dry reforming performance is quite close to that of catalyst 1. Apparently, these types of structured substrates also provide quite efficient heat and mass transfer, which agrees with results on steam

Table 11.3: Basic types of structured catalysts tested in methane dry reforming.

No.	Sample description	
	Type of substrate	**Active component composition and loading (wt.%)**
1	5 stacked plates no. 1	4% ($La_{0.8}Pr_{0.2}Mn_{0.2}Cr_{0.8}O_3$ + 10% NiO + 10% YSZ) +2% Ru
2	6 plates no. 3 + 6 gauzes no. 2	4% ($La_{0.8}Pr_{0.2}Mn_{0.2}Cr_{0.8}O_3$ + 10% NiO + 10% YSZ) +2% Ru
3	1 plate N6	10% (10wt.% $LaNi_{0.95}Ru_{0.05}O_3$/Mg-doped alumina)
4	2 plates N4	10% (10wt.% $LaNi_{0.95}Ru_{0.05}O_3$/Mg-doped alumina)
5	2 plates N5	10% (10wt.% $LaNi_{0.95}Ru_{0.05}O_3$/Mg-doped alumina)
6	2 plates N4	4% ($La_{0.8}Pr_{0.2}Mn_{0.2}Cr_{0.8}O_3$ + 10% NiO + 10% YSZ) + 2% Ru
7	5 stacked plates no. 1	4% SmPrCeZrO + 1%NiO + 1% Ru

reforming of natural gas for catalysts on these substrates (vide supra). Addition of oxygen to the feed improves syngas yield at shorter contact times (Fig. 11.13 and Table 11.4), while addition of water increases hydrogen yield (Table 11.4).

For both types of active components supported on the low-density foams, a reasonable performance is achieved only at much longer contact time (Table 11.4, catalysts 3–6), which apparently can be explained by lower values of the active surface per unit volume. For the same active component 2, the lowest syngas yield is observed for a pure ceramic foam as substrate (structured catalyst 5). Hence, in the case of foam substrates with low density, the heat transfer from the reactor wall into the catalyst package could affect the catalyst performance in strongly endothermic reaction of natural gas dry reforming.

To assess the long-term stability of the catalysts' performance with respect to coking, the catalysts were contacted with reaction feed at selected experimental conditions after a standard pretreatment in O_2 at 700 °C and kept for 6–8 h a day followed by reactor purging with Ar stream and cooling to the room temperature. The testing was resumed the next day by heating the catalyst in an Ar stream to the operating temperature followed by switching to the reaction feed stream. The long-term test of the structured catalysts has confirmed that these nanocomposite active components can also retain their high activity and coking/sintering stability when supported on much bigger substrates (microchannel/foam plates, gauzes). Apparently, the development of a given catalytic process has always been closely related to optimization of both specific active component and structured substrates to provide reliable and stable performance.

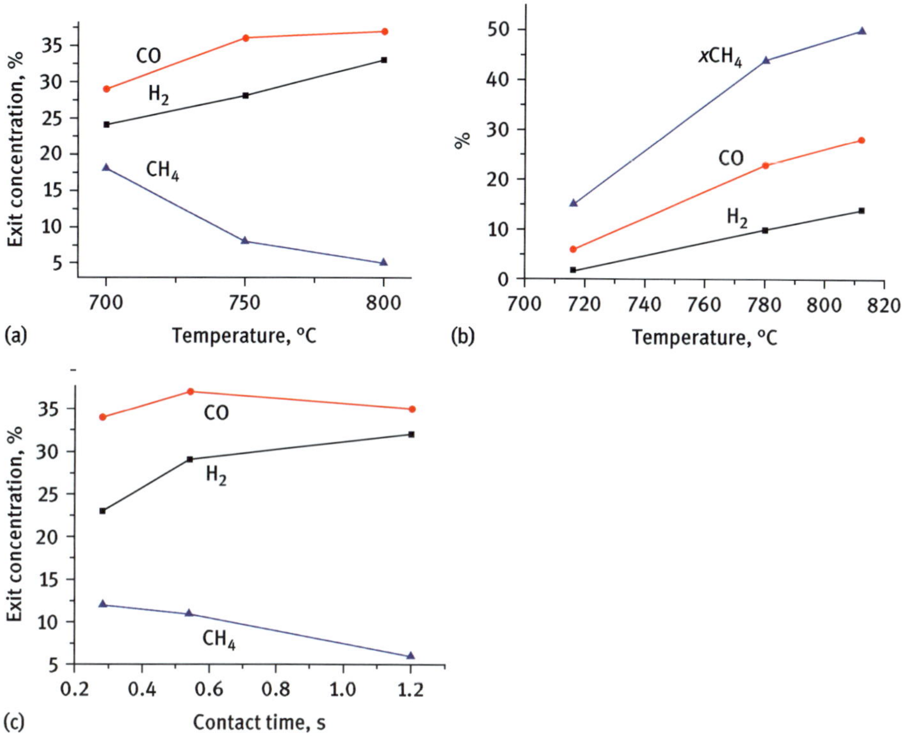

Fig. 11.12: Characteristics of natural gas dry reforming process for the package of microchannel plates (catalyst 7, Table 11.3) in feed 50% CO_2 +40% NG +N_2. Contact time 1.2 s (a) and 0.1 s (b), temperature 800 °C (c).

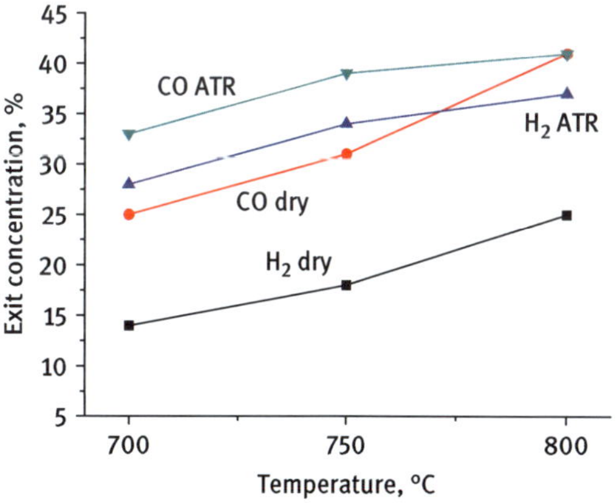

Fig. 11.13: Characteristics of natural gas dry or ATR reforming process for the package of microchannel plates (catalyst 7, Table 11.3) in feed 47% CO_2 + 44% NG + N_2. (dry) or 41% CO_2 + 44% NG + 9% O_2 (ATR). Contact time 0.5 s (dry) or 0.29 s (ATR).

Table 11.4: Main parameters of natural gas reforming on structured catalysts.

Type of structured catalyst	*T* (°C)	Contact time (s)	Feed (mol. %)	Concentration of products, vol.% (exit/equilibrium)			
				H_2	CO	CO_2	CH_4
1.	800	1.2	50% CO_2 + 40% NG + N_2	33/43	35/47	12/3	5/3
2	800	0.9	50% CO_2 + 40% NG + N_2	18	33	16	10
2	800	0.5s	45% NG + 45% CO_2 + 8% O_2 + N_2	18	25		
2	800	0.3	$30CO_2 + 31NG + 3O_2 + 37H_2O$	21	7	26	20
2	900	0.3	The same	40	27	17	8
3	850	1.9	56% CO_2 + 44% NG + N_2	14	30		
4	850	1.9	56% CO_2 + 44% NG + N_2	22	40		
5	850	1.9	56% CO_2 + 44% NG + N_2	12	26		
6	800	3.8	50% CO_2 + 40% NG + N_2	16	31	25	14

Scanning electron microscopic data for catalysts on foam substrates discharged after reaction demonstrated the absence of coke deposits/fibers and strong adherence of supported layers to substrates without cracks and spallation [21, 31].

References

[1] Ahmed S., M. Krumpelt, Hydrogen from hydrocarbon fuels for fuel cells, Int. J. Hydr. Energy. 2001; 26: 291–301.

[2] C. Song, Recent Advances in Catalysis for Hydrogen Production and Fuel Processing for Fuel Cells, Topics in Catal. 2008; 49: 1–3.

[3] Villegasa L., N. Guilhaume, H.Provendier, C. Daniel, F. Masset, C. Mirodatos, A combined thermodynamic/experimental study for the optimisation of hydrogen production by catalytic reforming of isooctane, Appl. Catal. A: General. 2005; 281: 75–83.

[4] Lindermeir A., S.Kah, S.Kavurucu, M.Mühlner, On-board diesel fuel processing for an SOFC–APU – Technical challenges for catalysis and reactor design, Appl. Catal. B: Envir. 2007; 70: 488–497.

[5] Biswas P., D.Kunzru, Oxidative steam reforming of ethanol over Ni/CeO_2-ZrO_2 catalyst, Chem. Eng. J. 2008; 136: 41–49.

[6] Laosiripojana N., D.Chadwick, S.Assabumrungrat, Effect of high surface area CeO_2 and Ce-ZrO_2 supports over Ni catalyst on CH_4 reforming with H_2O in the presence of O_2, H_2, and CO_2, Chem. Eng. J. 2008; 138: 264–273.

[7] Valderrama G., A.Kiennemann, M.R. Goldwasser, Dry reforming of CH_4 over solid solutions of $LaNi_{1-x}Co_xO_3$, Catal. Today. 2008; 133–135:142–148.
[8] Sauvet, A.L., J.T.S.Irvine, Catalytic activity for steam methane reforming and physical characterisation of $La_{1-x}Sr_xCr_{1-y}Ni_yO_{3-\delta}$, Solid State Ionics. 2004; 167: 1–8.
[9] Sadykov V., Bobrova L., Pavlova S., Simagina V., Makarshin L., Parmon V., Ross J. R. H., Van Veen A. C., Mirodatos C., Syngas generation from hydrocarbons and oxygenates with structured catalysts, *Series Energy Science, Engineering and Technology*, Nova Science Publishers, Inc, New York, 140p, 2012.
[10] Yaseneva P., S.Pavlova, V.Sadykov, G. Alikina, A. Lukashevich, S. Belochapkine, J. Ross, Combinatorial approach to the preparation and characterization of catalysts for biomass steam reforming into syngas, Catal. Today. 2008; 137: 23–28.
[11] Sadykov, V., N.Mezentseva, G. Alikina, R. Bunina, V. Rogov, T. Krieger, S. Belochapkine, J. Ross, Composite catalytic materials for steam reforming of methane and oxygenates: Combinatorial synthesis, characterization and performance, Catal. Today. 2009; 145: 127–137.
[12] Sadykov, V., N. Mezentseva, A. Smirnova, J. Irvine, O. Vasylyev, Effect of complex oxide promoters and Pd on activity and stability of Ni/YSZ (ScSZ) cermets as anode materials for IT SOFC, Catal. Today. 2008; 131: 226–237.
[13] Sadykov V., N. Mezentseva, A. Smirnova, J. Irvine, Doped Nanocrystalline Pt-Promoted Ceria-Zirconia as Anode Catalysts for IT SOFC: Synthesis and Properties, Mater. Res. Soc. Symp. Proc., 2007; JJ02-07. 1023: 1–6.
[14] Sadykov V., N. Mezentseva, J. Ross, Pt-supported nanocrystalline ceria-zirconia doped with La, Pr or Gd: factors controlling syngas generation in partial oxidation/autothermal reforming of methane or oxygenates, Solid State Phenomena. 2007; 128: 239–248.
[15] Sadykov V., N. Mezentseva, G. Alikina, R. Bunina, V. Pelipenko, A. Lukashevich, S.Tikhov, V. Usoltsev, Z. Vostrikov, O. Bobrenok, A. Smirnova, J. Ross, O. Smorygo, B. Rietveld, Nanocomposite catalysts for internal steam reforming of methane and biofuels in solid oxide fuel cells: Design and performance, Catal. Today 2009; 146: 132–140.
[16] Resini C., M. Concepcion, H. Delgado, S. Presto, L. Alemany, P. Riani, R. Marazza, G. Ramis, G. Busca, Yttria-stabilized zirconia (YSZ) supported Ni–Co alloys (precursor of SOFC anodes) as catalysts for the steam reforming of ethanol, Int. J. Hydrogen Energy 2008; 33: 3728–3735.
[17] Smorygo O., V. Mikutski, A. Marukovich, Y. Vialiuha, A. Ilyushchanka, N. Mezentseva, G. Alikina, Z. Vostrikov, Y. Fedorova, V. Pelipenko, R. Bunina, V. Sadykov, Structured catalyst supports and catalysts for the methane indirect internal steam reforming in the intermediate temperature SOFC, Int. J. Hydr. Energy 2009; 34: 9505–9514.
[18] Sadykov V., S. Pavlova, S. Tikhov, V. Usol'tsev, V. Parmon, Selective Oxidation of Hydrocarbons into Synthesis Gas at Short Contact Times: Design of Monolith Catalysts and Main Process Parameters, Kinet. Catal.. 2005; 46: 227–250.
[19] Sadykov, V., Mezentseva, N., Alikina, G., Bunina, R., Pelipenko, V., Lukashevich, A., Vostrikov, Z., Rogov, V., Krieger, T., Ishchenko, A., Zaikovsky, V., Bobrova, L., Ross, J., Smorygo, O., Smirnova, A., Rietveld, B. and van Berkel, F., Nanocomposite catalysts for steam reforming of methane and biofuels: design and performance, in *Nanocomposite materials, theory and applications*, INTECH, Vienna, 2011, 909–946.
[20] Groppi G., E.Tronconi, Design of novel monolith catalyst supports for gas/solid reactions with heat exchange, Chem. Eng. Sci. 2000; 55: 2161–2171.
[21] Sadykov, V., Sobyanin, V., Mezentseva, N., Alikina, G., Vostrikov, Z., Fedorova, Y., Pelipenko, V., Usoltsev, V., Tikhov, S., Salanov, A., Bobrova, L., Beloshapkin, S., Ross, J.R.H., Smorygo, O., Ulyanitskii, V. and Rudnev, V., Transformation of CH_4 and liquid fuels into syngas on monolithic catalysts, *Fuel*, 89, 1230–1240, 2010.

[22] Tikhov S., V. Usoltsev, V.Sadykov, CrAl alloy –based cermet monolith with polymodal pore structure for partial oxidation of methane to synthesis gas, Stud. Surf. Sci. Catal. 2006; 162: 641–648.

[23] Sadykov V., S.Pavlova, S.Tikhov, Design of structured catalysts based on metallic monoliths for syngas production via partial oxidation of natural gas, Stud. Surf. Sci. Catal. 2007; 172: 241–244.

[24] Ulyanitskii V., A.Shterzer, S.Zlobin, V.Matrenin, I.Schipanov, S.Serykh, A.Stikhin, L. Tretyakova, V.Sadykov, S.Pavlova, S.Tikhov, V.Kuzmin, Blast dusting of refractory protective layers for solving the problems of hydrogen energetics, Alternative Energetics and Ecology 2006; 9: 137–144.

[25] Pavlova S., N.Sazonova, V. Sadykov, G.Alikina, A.Lukashevich, E.Gubanova, and R.Bunina, Study of synthesis gas production over structured catalysts based on LaNi(Pt)O_x- and Pt (LaPt)-CeO_2-ZrO_2 supported on corundum, Stud. Surf. Sci. Catal. 2007; 167: 343–348.

[26] Sadykov V., S.Pavlova, Z. Vostrikov, N.Sazonova, E.Gubanova, R.Bunina, G.Alikina, A. Lukashevich, L.Pinaeva, L.Gogin, S.Pokrovskaya, V.Skomorokhov, A.Shigarov, C.Mirodatos, A. van Veen, A.Khristolyubov, and V.Ulyanitskii, Performance of monolithic catalysts with complex active component in partial oxidation of methane into syngas: experimental studies and modeling, Stud. Surf. Sci. Catal. 2007; 167: 361–366.

[27] Bobrova L., I.Zolotarsky, V.Sadykov, V.Sobyanin, Hydrogen-rich gas production from gasoline in a short contact time catalytic reactor, Int. J. Hydr. Energy 2007; 32: 3698–3704.

[28] Bobrova L., I.Zolotarskii, V.Sadykov, S.Pavlova, O.Snegurenko, S.Tikhov, V.Korotkich, T. Kuznetsova, V.Sobyanin, V.Parmon, Syngas formation by selective catalytic oxidation of liquid hydrocarbons in a short contact time adiabatic reactor, Chem. Eng. J. 2005, 107: 171–179.

[29] Bobrova L., V.Korotkich, V.Sadykov, V.Parmon, Syngas formation from gasoline in adiabatic reactor: Thermodynamic approach and experimental observations, Chem. Eng. J. 2007; 134: 145–152.

[30] Bobrova L., N.Vernikovskaya, V.Sadykov, Conversion of hydrocarbon fuels to syngas in a short contact time catalytic reactor, Catal. Today. 2009: 144: 185–200.

[31] Sadykov V., N. Mezentseva, M. Simonov, E. Smal, M. Arapova, S. Pavlova, Y. Fedorova, O. Chub, L. Bobrova, V. Kuzmin, A. Ishchenko, T. Krieger, A.-C. Roger, K. Parkhomenko, C. Mirodatos, O. Smorygo, J. Ross, Structured nanocomposite catalysts of biofuels transformation into syngas and hydrogen: Design and performance, Int. J. Hydrog. Energy. 2015; 40: 7511–7522.

Index

https://doi.org/10.1515/9783110587777-012